Design and Analysis of Secure Quantum Communication Schemes

Ri-Gui Zhou • Xiao-Xue Zhang • Lin-Tao Du

Design and Analysis of Secure Quantum Communication Schemes

Springer

Ri-Gui Zhou
College of Information Engineering
Shanghai Maritime University
Shanghai, China

Xiao-Xue Zhang
College of Information Engineering
Shanghai Maritime University
Shanghai, China

Lin-Tao Du
College of Information Engineering
Shanghai Maritime University
Shanghai, China

ISBN 978-3-031-78427-9 ISBN 978-3-031-78428-6 (eBook)
https://doi.org/10.1007/978-3-031-78428-6

This Springer imprint is published by the registered company Springer Nature Switzerland AG
The registered company address is: Gewerbestrasse 11, 6330 Cham, Switzerland

Preface

With the development of modern cryptography and quantum computer, two methods have been developed to deal with the threat of some cryptographic algorithms: one is anti-quantum cryptography, which has developed from classical cryptography, and the other is Secure Quantum Communication. The essence of Secure Quantum Communication is to realize the transmission of unknown quantum states, and to provide an unconditional and absolutely secure encryption method in theory, secret messages are protected by the properties of quantum indivisibility, quantum state non-cloning, and quantum entanglement to ensure the secure transmission of information. Secure Quantum Communication, as a new modern communication method, is expected to provide security for the next generation communication network and data and information security for the related communication network system, thus promoting the development of the entire communication industry, and will be widely used in various industries, such as national defense, banking, securities, commerce, and health care. This book presents Secure Quantum Communication techniques, which exploit quantum mechanical phenomena such as quantum superpositions and quantum entanglement.

This book provides a variety of schemes related to the various branches of Secure Quantum Communication. The book includes eight chapters. Chapter 1 is an introductory part, which describes the research background, significance of topics, literature review, and research methods. Chapter 2 is the basic of knowledge, which provides the techniques of the field of quantum information, such as quantum entanglement, density operator, and semi-quantum theory.

Chapters 3, 4, 5, 6, 7, and 8 are on various Secure Quantum Communication schemes. We present new schemes in branch quantum key distribution, quantum secret sharing, quantum direct communication, quantum key agreement, quantum private query, and quantum network coding. These chapters describe the design of communication schemes and their corresponding performance analysis. In addition, the simulated experimental results and analysis demonstrate that the superiority of the proposed schemes in this book.

Shanghai, China
January 2024

Ri-Gui Zhou
Xiao-Xue Zhang
Lin-Tao Du

Acknowledgments

This work is supported by the National Natural Science Foundation of China under Grant No. 6217070290 and Shanghai Science and Technology Project under Grant No. 21JC1402800 and 20040501500.

We thank Wenwen Hu, Haipeng Wang, Yun Hua, and many others for research collaborations and support.

Ri-Gui Zhou would like to thank his wife, Airong Nie, and his son, Hanxuan Zhou, for their understanding and help.

Xiao-Xue Zhang and Lin-Tao Du would like to express their sincere gratitude to their advisor, Prof. Ri-Gui Zhou, for his guidance, nurturing encouragement, and support in every stage of their graduate study. His knowledge, kindness, patience, open-mindedness, and vision have provided us with lifetime benefit.

Contents

About the Authors

Lin-Tao Du received a BS degree in information and computing science from Shenyang Agricultural University, Liaoning, China, in 2020. She is currently pursuing a PhD degree in information management and information system at Shanghai Maritime University, Shanghai, China. Her research interests include quantum cryptography and secure communication.

Xiao-Xue Zhang received a BS degree in computer science and technology from Huaibei Normal University, Anhui, China, in 2019. She is currently pursuing a PhD degree in information management and information system at Shanghai Maritime University, Shanghai, China. Her research interests include quantum communication and quantum computation.

Ri-Gui Zhou received a BS degree from Shandong University, China, in 1997; an MS degree from the Department of Computer Science and Technology of Nanchang Hangkong University, China, in 2003; and a PhD degree from the Department of Computer Science and Technology of Nanjing University of Aeronautics and Astronauts, China, in 2007.

From 2008 to 2010, he was a Postdoctoral Fellow in Tsinghua University, China. From 2010 to 2011, he was a Postdoctoral Fellow in Carleton University, Ottawa, Canada. From 2014 to 2015, he was a Visiting Scholar at North Carolina State University, Raleigh, NC, USA. He is currently a Professor with the College of Information Engineering, Shanghai Maritime University, China. His main research interests include quantum image processing, quantum reversible logic, and the quantum genetic algorithm, among others.

Prof. Zhou is a senior member of the China Computer Federation (CCF) and the recipient of the New Century Excellent Talents program, Ministry of Education of China, in 2013.

Chapter 1
Introduction

Abstract As time goes on, unconditionally secure means of communication are more and more needed. As one of the key points of quantum informatics, quantum communication generally refers to those quantum technologies that transmit and transfer signals, information, or quantum states by moving quantum states, it has solved the problems of small information capacity and weak confidentiality, which cannot be broken through in classical communication, and has become a hot topic of research. Quantum communication is a new cryptographic communication technology that guarantees unconditional security in principle by quantum cryptography mechanisms that do not rely on computational complexity. It is the first practical quantum information technology that lays a theoretical foundation for building an "absolutely safe" communication system.

1.1 The Concept of Secure Quantum Communication

With the advent of the information age, people pay more and more attention to the security of communication systems. In the actual communication environment, there are often many uncertain factors, and the security of communication systems has attracted wide attention from society. Secure communication technology can provide security guarantee for information systems, that is, using encryption and decryption technology to ensure reliable transmission of information and to achieve the purpose of protecting data security.

Generally, the encryption system in modern communication technology includes private key cryptosystem and public key cryptosystem. In the private key cryptosystem, the key used for encryption and decryption is the same, and the communication parties need to share the key in advance. Since the same key is used for both encryption and decryption, the private key cryptosystem is also called symmetric cryptosystem.

Unlike symmetric cryptosystems, in public key cryptosystems, there is no need for both parties to share a key in advance, and the keys used for encryption and decryption are different. The communication parties get a key pair (a public key and a private key) through some mathematical computational complexity. The public

R.-G. Zhou et al., *Design and Analysis of Secure Quantum Communication Schemes*,
https://doi.org/10.1007/978-3-031-78428-6_1

key is the public part of the key pair, and the private key is the private part. One key in the key pair is used for encryption and the other for decryption. An eavesdropper can break the password only if he obtains both the private key and the public key.

But with the development of quantum computing theory, the encryption of classical communication based on mathematical computational complexity is not secure in principle. Therefore, it is particularly important to develop a cryptographic method to resist quantum attacks. So people use quantum mechanical properties, combined with classical cryptography and communication technology, to design the corresponding cryptography scheme, that is, quantum communication.

The first quantum communication scheme was the quantum key distribution (QKD) scheme proposed by Bennett and Brassard in 1984 [1]. This is the most famous BB84 scheme, which will be introduced in Chap. 3. This scheme is guaranteed by quantum properties such as Heisenberg's uncertainty principle, noncloning theorem, and the correlation of entangled particles and nonlocality in quantum mechanics. Both parties can find eavesdroppers in time during the communication process and realize the absolute security of communication. It has attracted the attention and interest of researchers. Then the concept of Secure Quantum Communication is proposed.

Secure Quantum Communication (SQC) uses quantum states for key agreement and information transmission and uses the basic principles of quantum mechanics to discover eavesdropping behavior, so that eavesdroppers cannot obtain information, so as to ensure the security of information. It can overcome the security risks of classical encryption technology and securely distribute keys in real time on the public channel. It is a communication technology that has been strictly proven so far and can guarantee unconditional security in principle [2–4]. Therefore, the problem of information security can be solved physically by using Secure Quantum Communication.

After years of development, Secure Quantum Communication has made great progress, developing many branches, including quantum key distribution (QKD), quantum secret sharing (QSS), quantum direct communication (QDC), quantum key agreement (QKA), quantum private query (QPQ), and quantum network coding (QNC).

1.1.1 Research Status of Secure Quantum Communication

The security of classical cryptography is based on the difficulty and computational complexity of the decomposition of large prime numbers in number theory. These mathematical problems are very easy to solve for quantum computers, which poses a threat to the security of classical cryptography. While quantum computing poses a great threat to traditional cryptosystems, the development of quantum information has also brought a corresponding solution to cryptography: quantum cryptography, which is the basis of Secure Quantum Communication.

Secure Quantum Communication originated in 1984. Bennett and Brassard designed the first quantum key distribution (QKD) based on single-photon polarization state, which is the core of Secure Quantum Communication. With the deepening of research, Secure Quantum Communication has developed quantum secret sharing (QSS), quantum direct communication (QDC), quantum key agreement (QKA), quantum private query (QPQ), quantum network coding (QNC), and other branches.

This section briefly reviews the establishment process of Secure Quantum Communication, the main research progress of QKD, and other schemes.

1. **QKD.**

   QKD is the first and developing part of the research in the field of Secure Quantum Communication. BB84 proved it secure against certain attacks. But it was not rigorously proven secure against an adversary able to perform any operation permitted by Secure Quantum Communication. Shor proved the unconditional security of quantum key distribution, which established a solid foundation for the security of quantum communication [5]. Before that, Ekert proposed a quantum entangled key distribution scheme based on the EPR (Einstein-Podolsky-Rosen, EPR) paradox [6], that is, the E91 scheme. Bennett improved the E91 scheme, taking the quantum correlation of EPR entangled pairs as the security basis, and achieved quantum state coding of EPR quantum entangled states, so it was called the EPR scheme [7]. Based on the BB84 scheme, Bennett proposed a quantum key distribution scheme based on the complementarity and nonorthogonality of linearly polarized photons and circularly polarized photons [8], namely, the B92 scheme. BB84 scheme, E91 scheme, and B92 scheme form the three main QKD schemes in Secure Quantum Communication. Subsequently, researchers proposed various QKD derivative schemes based on mainstream schemes [9, 10] and began experimental verification studies of the schemes.

   At present, there are many quantum key distribution schemes, which are based on different quantum characteristics and improved for each link of quantum communication. For example, the sources used include single photons, EPR pairs, and coherence with continuous variables; the coding methods include polarization coding, phase coding, and differential phase coding. Decoy state and measurement device independent are used to resist the security vulnerabilities caused by equipment defects.

   Generally, all communicators are quantum parties with complete quantum abilities in quantum communication. But owing to the expensive cost of quantum resources, it is luxury to ensure that all communicators in quantum communication have complete quantum capabilities in reality. Boyer [11] proposed the semi-quantum key distribution (SQKD), which realized key distribution between quantum communicator and classical communicator. Boyer [12] presented two SQKD schemes and proved their complete robustness against attacks.

2. **QSS.**

   As mentioned above, classical secret sharing schemes are designed based on mathematical computational complexity, and their security decreases with the development of human computing power. Based on the three-particle (GHZ) entangled state, Hillery proposed the first three-user QSS scheme [13]. The research on secret sharing officially entered the category of quantum cryptography. Because of the correlation of local operations, it allows sharing the secret information (single bit value or an unknown quantum state) among three (four) parties. Subsequently, Karlsson proposed the concept of QSS, based quantum entanglement swapping of Bell states, and a possible extension for (k, n) threshold QSS schemes [14]. Guo proposed the first QSS that utilizes product states instead of entangled states and improves the scheme's efficiency up to asymptotic 100% [15]. At the same year, Lance first named controlled multiparty quantum state teleportation as quantum state sharing (QSTS) and proposed a three-party QSTS scheme [16]. Xiao [17] extended Hillery' s scheme into multiparty version via asymmetrically choosing measuring bases, which makes their scheme can be asymptotically devised to be 100% efficiency. Then Yan [18] first proposed a multiparty and multiparty QSS scheme using a single particle sequence. Deng proposed the multiparty QSTS scheme based on the entanglement swapping of maximal entangled Bell states to share an unknown two-qubit state via performing three-qubit GHZ-state measurement (GM) and two-qubit Bell-state measurement (BM) [19].

   In the process of secret sharing, sometimes the number of managers needs to be dynamically increased or decreased according to the demand, which is the dynamic quantum secret sharing scheme (DQSS). The first novel DQSS scheme with an additional function of dynamically adding or deleting participants was proposed [20]. Then Hsu [21] proposed a new DQSS scheme based on the entanglement swapping of EPR states. Without changing the original secret shared by current agents, an agent can join or quit existing quantum secret sharing, and two QSS schemes also can be merged into one QSS scheme. However, two major security holes in Hsu et al.'s scheme were pointed out, which are dishonest revoked agent attack and collusion attack. Following that, Liao [22] designed a dynamic quantum secret sharing scheme using GHZ states and Controlled-NOT (CNOT) operation, which can avoid the aforementioned collusion attack and verify the honesty of the cancelled agents in the meantime. Liu [23] implemented DQSS between a manager and multiple dynamic agent groups, which is more flexible and easier to apply. Song [24] proposed a new DQSS scheme using the mutually unbiased measurement basis based on generalized GHZ states.

3. **QDC.**

   As a branch of Secure Quantum Communication, quantum direct communication (QDC) has been extensively studied in recent years. It includes quantum secure direct communication (QSDC) and deterministic secure quantum communication (DSQC).

The first QSDC scheme is based on a two-step communication strategy [25]. Deng extended the first QSDC scheme and designed a two-step QSDC scheme with the idea of block transmission [26]. Also, she proposed the first QSDC scheme with single photons [27]. Then Jin designed a three-party QSDC scheme by using GHZ states [28]. Liu presented a high-capacity QSDC scheme with single photons in both polarization and spatial-mode degrees of freedom [29]. In addition, controlled QSDC is an innovation of QSDC in terms of security, which indicates that the receiver can only obtain messages with the permission of the controller. There were many studies on the controlled QSDC scheme based on GHZ states [30].

To save quantum resources and equipment, Zou [31] proposed the first semi-quantum secure direct communication (SQSDC) scheme, which was a three-step scheme based on single-photon source. Luo [32] presented two authenticated SQSDC schemes using Bell states, which did not require any classical channel. Zhang [33] designed a SQSDC scheme based on EPR pairs. Subsequently, a controlled direct secure semi-quantum communication scheme based on GHZ-like states was proposed [34]. In this scheme, the sender Alice was classical, the receiver Bob and the controller Charlie were quantum. Gu [35] discovered that the scheme proposed by Zou had hidden security risks and made improvements to it. Successively, Yan [36] proposed a SQSDC scheme based on Bell states. Sun [37] proposed two SQSDC schemes based on Bell states, and no extra decoy states were employed in the communication process. Later, Tao [38] proposed two SQSDC schemes, in which both the quantum and the classical parties could verify the authenticity of the secret message.

Compared with QSDC, DSQC demands that the receiver can read out the secret message only after each qubit exchanges at least one bit of classical information [41], which means that DSQC is more secure because it is unnecessary to transmit the qubits carried by the message through an external channel [43]. Man proposed several DSQC schemes based on quantum teleportation and entanglement swapping [39]. Next, Li designed two DSQC schemes, one of which is based on pure entangled states and the other is based on d-dimensional single-photon states [40]. Long presented a DSQC scheme in which the receiver can read out the secret message only after each qubit exchanges at least one bit of classical information [41]. Shukla extended a DSQC scheme based on orthogonal states so that it does not require the actual transmission of message qubits [42]. Joy proposed two DSQC schemes that use three-qubit GHZ-like states and five-qubit Brown states as quantum channels [43], respectively.

However, in these schemes, the sender and receiver cannot exchange useful information at the same time, that is, the transmission of secret message is one-way, which limits the application of QSDC. In this regard, Nguyen [44] first proposed a bidirectional QSDC scheme based on the EPR states. In this scheme, two communicants could exchange secret message with each other at the same time. Therefore, it is also known as the quantum dialogue scheme (QD). Since then, more and more scholars have focused their research on the design and analysis of quantum dialogue. Ji [45] proposed a QD scheme based on N groups of single-photon sequences. Yang [46] also presented a quasi-safe QD scheme using single photons. With many QD schemes [47–49] based on GHZ states, W states and four-particle entangled states had been proposed one after another. Gao [50] found that most of these schemes have information leakage problem. In order to solve this problem, researchers have successively proposed a series of QD schemes with higher security [51–53].

However, although the security of these QD schemes proposed at present can meet certain requirements, their requirements for equipment are too high to be practical during the implementation process. In fact, many of the existing schemes require each participant to be able to perform quantum operations, which makes them more expensive and difficult to realize in practice [54]. In order to solve these problems, researchers [55, 56] proposed a semi-quantum key distribution (SQKD) scheme.

4. **QKA.**
Compared with QKD, QKA requires all communicators to participate equally in the agreement of final shared keys, and the final shared secret key cannot be determined by any communicator alone. The first QKA based on maximum entanglement state and quantum teleportation technology [57]. In the same year, Hsueh and Chen proposed a QKA scheme based on maximally entangled states [58]. But Tsai pointed out that in the previous two QKA schemes, the final key can be completely determined by one party [59]. And Tsai and Hwang considered that their improvement still is not a QKA. So Chong and Hwang proposed a QKA scheme using delay measurement technique, in which communicants cannot independently determine the shared key [60]. However, these QKA schemes only involve two parties and are not suitable for multi-party communication. In order to adapt to the actual situation, it is necessary to extend the quantum key agreement scheme from the case of two parties to the case of multiple parties. Shi used the entanglement swapping between Bell states to realize the two-party QKA scheme and extended it to a multiparty QKA scheme [61]. After that, many scholars began to pay attention to the multiparty QKA scheme. The MQKA scheme is based on single photons to resist internal attacks [62]. And the circle-type MQKA scheme was proposed, which can effectively resist collusion attacks [63]. However, each node of the scheme has only one participant. In response to this, Min made improvements and presented a new MQKA scheme based on entanglement swapping of GHZ-like states and Bell states in which each node contains two participants [64]. In order to decrease the quantum capacity necessary for participants, Liu was the first to propose a semi-quantum key agreement (SQKA) scheme between classical and quantum parties [65]. Then Yan proposed a novel SQKA scheme in which two users negotiate shared keys equally [66]. Zhou proposed an SQKA scheme that allows one quantum participant and two classical participants to negotiate shared keys equally [67].

5. **QPQ.**

Bennett [68] and Brassard [69] proposed quantum schemes to solve problems similar to SPIR but found that it was very difficult to protect both parties. The first QPQ scheme (GLM scheme) [70] was proposed by V. Govannetti. In this scheme, they use oracle operations to represent the records of the database and at the same time act on the quantum state to be queried. The scheme requires two types of quantum states, the query state, which is used to obtain the required records from the database, and the temptation state, which is used to check the possible attacks from Bob. Compared to existing schemes, the GLM scheme reduces the complexity of communication and computation. Although the existing QPQ schemes have many advantages in theory, it is difficult to implement high-dimensional oracle operations, so these schemes are difficult to apply to large databases with massive data.

In order to solve the above problems, Jakobi proposed a quantum privacy query scheme based on the SARG-04 secret key distribution scheme [71]. By using this approach, oblivious keys can be distributed between Alice and Bob. For this key, Bob knows all about this key information, but Alice only knows a part of it. This scheme is a pioneering and practical QPQ scheme based on QKD. After that, QKD-based QPQ has become a research hotspot, attracting a large number of scholars to study. Then Gao proposed a flexible and practical QPQ scheme (G scheme) based on the J scheme [71]. Compared to J scheme, G scheme shows better performance in terms of flexibility, computational complexity, and security.

In practice, quantum channel noise may cause Alice to get incorrect query results, which makes Alice misunderstand the database holder. Considering the actual noisy channel, the error tolerant QKD-based QPQ scheme was used in the oblivious key of the error correction method [72]. In addition, Gao puts forward $gkN \rightarrow N$ postprocessing algorithm on the basis of $kN \rightarrow N$ postprocessing with linear error correction code, which has significant effect on error correction [73]. However, they all correct the classical information after the oblivious key distribution, oblivious key distribution has few robust and anti-noise ability. Fortunately, not all quantum states are equally fragile when they interact with the environment. Decoherence-free states (DFS) are the quantum states that remain unchanged no matter how strong the collective noise is as they are transmitted through the quantum channel. This is because the collective errors that occur on different qubits can cancel each other out when a particularly relevant symmetry arises. Wang proposed a flexible QPQ scheme based on four-particle DFS, in which the specific security requirements are achieved by adjusting parameter θ [74]. At the same time, Li presented a QPQ scheme that can resist collective noise by using four-particle DFS [75].

The above QPQ scheme is all for a single user to query the database in order to ensure the security of a multiuser query database. Recently, a multiuser QPQ scheme was proposed [76]. However, Ye only implemented the QPQ scheme repeatedly according to the number of users, which greatly increased the communication complexity. Then Xiao presented a multiuser cooperative QPQ scheme [77], which extends the participants of the scheme to multiple parties.

6. **QNC.**

Quantum network coding (QNC) is proposed on the basis of network coding because quantum communication has greater potential in information transmission than classical communication. With the potential advantages of network coding, QNC has attracted the attention of researchers and has been developed rapidly and soon will become one of the research hotspots in quantum field. In a butterfly network, the first quantum network coding designed XQQ schemes and QNC based on prior entanglement scheme [78, 79]. They have successfully realized the cross-transmission of quantum information on the butterfly network theoretically, which provides an operational platform for QNC. Kobayashi analyzed the effective transmission of quantum state, introduced the classic communication into quantum network coding, and proposed a QNC scheme for lossless quantum state transmission [80–82]. Satoh [83] tried to introduce QNC into repeater and proved that QNC can improve the transmission rate on quantum relay grid. Shang [84] improved the butterfly network proposed by Hayashi and proposed QNC based on controlled teleportation. By introducing a third-party controller, the decoding process of unicast stream is controlled, which helps extend these schemes to general networks with unicast or multicast. Chen [85] optimized the classical assisted QNC and designed a new structure that separated the classical communication from the classical network solution, improved the communication efficiency, and reduced the communication cost. In the same year, according to the classical network coding technology, Xu [86] also established the QNC of quantum cooperative multicast from the perspective of multicast and unicast. Shang [87] introduced quantum signature into QNC. In order to design the network model reasonably, they creatively proposed a quantum homomorphic signature scheme, which combined two signatures into one by means of entanglement swapping. Shang [88] continued to improve the homomorphic signature scheme and applied homomorphic signature to the butterfly network. They also proposed QNC against pollution attacks.

Subsequently, a large number of QNC schemes have been proposed: The opportunistic quantum network coding scheme, to ensure the safe transmission of quantum state, made full use of the channel characteristics of quantum teleportation, and the eavesdropper can be detected by quantum channel verification [89]. Li used the global entangled 2D and 3D cluster states to solve the problem of quantum multicast in directed acyclic networks [90]. In order to achieve efficient quantum communication based on quantum repeater, Shang [91] proposed a secure QNC scheme for controlled repeater networks. Then, Shang proposed a QNC scheme based on continuous variables, which improved the enforceability of QNC [92]. The new perfect network coding scheme solved the quantum k-pair problem [93, 94]. Pan [95] added a single particle to the butterfly network to realize the perfect transmission of quantum state without losing information.

1.1.2 Structure of the Book

The structure of the book is as follows:

1. Chapter 1 is the introduction, which summarizes the concept and research progress of quantum secret communication.
2. Chapter 2 introduces the theoretical basis of quantum information theory, including the basic assumptions of quantum mechanics, density operator, quantum entanglement, the properties of quantum information, semi-quantum correlation theory, and quantum information metric.
3. Chapter 3 analyses the research status, classification, and development of quantum key distribution and introduces the classic quantum key distribution schemes.
4. Chapter 4 introduces the research status of quantum secret sharing, and the classical schemes of quantum secret sharing are described.
5. Chapter 5 studies quantum direct communication, and the characteristics of its three sub-directions are analyzed, including quantum secure direct communication, deterministic secure quantum communication, and quantum dialogue; it also describes the classical scheme.
6. Chapter 6 introduces the characteristics of quantum key agreement, and the key schemes are reviewed.
7. Chapter 7 introduces the concept of quantum privacy query, reviews the classical scheme, and expands it.
8. Chapter 8 introduces the theory of quantum network coding, discusses its classification, and proposes a new scheme.

References

1. C.H. Bennett, G. Brassard, Quantum cryptography: Public key distribution and coin tossing. Theor. Comput. Sci. **560**, 7–11 (2014)
2. H.-K. Lo, H.F. Chau, Unconditional security of quantum key distribution over arbitrarily long distances[J]. Science **283**(5410), 2050–2056 (1999)
3. E. Biham, M. Boyer, P. Boykin, et al., A proof of the security of quantum key distribution. J. Cryptol. **19**, 381–439 (2006)
4. D. Mayers, Unconditional security in quantum cryptography[J]. J. ACM **48**(3), 351–406 (2001)
5. P.W. Shor, J. Preskill, Simple proof of security of the BB84 quantum key distribution scheme. Phys. Rev.Lett. **85**, 441–444 (2000)
6. A.K. Ekert, Quantum cryptography based on Bell's theorem[J]. Phys. Rev. Lett. **67**(6), 661–663 (1991)
7. C.H. Bennett, G. Brassard, Teleporting an unknown quantum state via dual classic and Einstein-Podolsky-Rosen channels[J]. Phys. Rev. Lett. **70**, 1895–1899 (1993)
8. C.H. Bennett, G. Brassard, N.D. Mermin, Quantum cryptography without Bell's theorem [J]. Phys. Rev. Lett. **68**(5), 557–559 (1992)
9. D. Bruss, Optimal eavesdropping in quantum cryptography with six states[J]. Phys. Rev. Lett. **81**(14), 3018–3021 (1998)
10. H. Bechmann-Pasquinucci, W. Tittel, Quantum cryptography using larger alphabets[J]. Phys. Rev. A **61**, 062308–062314 (2000)

11. M. Boyer, D. Kenigsberg, T. Mor, Quantum key distribution with classical bob. Phys. Rev. Lett. **99**, 1–5 (2007)
12. M. Boyer, R. Gelles, D. Kenigsberg, T. Mor, Semiquantum key distribution. Phys. Rev. A At. Mol. Opt. Phys. **79**, 1–12 (2009)
13. M. Hillery, V. Buzek, A. Berthiaunie, Quantum secret sharing. Phys. Rev. A **59**, 1829 (1999)
14. A. Karlsson, M. Koashi, N. Imoto, Quantum entanglement for secret sharing and secret splitting. Phys. Rev. A **59**(1), 162 (1999)
15. G.P. Guo, G.C. Guo, Quantum secret sharing without entanglement. Phys. Lett. A **310**, 247–251 (2003)
16. T. Symul, W.P. Bowen, et al., Tripartite quantum state sharing. Phys. Rev. Lett. **92**, 177903 (2004)
17. L. Xiao, G.L. Long, F.-G. Deng, J.-W. Pan, Efficient multiparty quantum-secret-sharing schemes. Phys. Rev. A **69**(5), 052307 (2004)
18. F.-L. Yan, T. Gao, Quantum secret sharing between multiparty and multiparty without entanglement. Phys. Rev. A **72**(1), 012304 (2005)
19. F.G. Deng, X.H. Li, C.Y. Li, et al., Multiparty quantum-state sharing of an arbitrary two-particle state with Einstein-Podolsky-Rosen pairs. Phys. Rev. A **72**(4), 044301 (2005)
20. H.-Y. Jia, Q.-Y. Wen, F. Gao, S.-J. Qin, F.-Z. Guo, Dynamic quantum secret sharing. Phys. Lett. A **376**(10–11), 1035–1041 (2012)
21. J.L. Hsu, S.K. Chong, T. Hwang, C.W. Tsai, Dynamic quantum secret sharing. Quantum Inf. Process **12**(1) (2013)
22. C.H. Liao, C.W. Yang, T. Hwang, Dynamic quantum secret sharing scheme based on GHZ state. Quantum Inf. Process **13**(8), 1907–1916 (2014)
23. H. Liu, H. Ma, K. Wei, X. Yang, W. Qu, T. Dou, Y. Chen, R. Li, W. Zhu, Multi-group dynamic quantum secret sharing with single photons. Phys. Lett. A, 2349–2353 (2016)
24. S. Yun, L. Zhihui, L. Yongming, A dynamic multiparty quantum direct secret sharing based on generalized ghz states. Quantum Inf. Process **17**(9), 244 (2018)
25. G. Long, X. Liu, Theoretically efficient high-capacity quantum-key-distribution scheme. Phys. Rev. A **65**(3), 032302 (2002)
26. F.G. Deng, G.L. Long, X.S. Liu, Two-step quantum direct communication scheme using the Einstein-Podolsky-Rosen pair block. Phys. Rev. A **68**(4), 042317 (2003)
27. F.G. Deng, G.L. Long, Secure direct communication with a quantum one-time pad. Phys. Rev. A **69**, 521–524 (2004)
28. X.R. Jin, X. Ji, Y.Q. Zhang, et al., Three-party quantum secure direct communication based on GHZ states. Phys. Lett. A **354**(1), 67–70 (2006)
29. D. Liu, J.L. Chen, W. Jiang, High-capacity quantum secure direct communication with single photons in both polarization and spatial-mode degrees of freedom. Int. J. Theor. Phys. **51**(9), 2923–2929 (2012)
30. Z.X. Man, Y.J. Xia, Controlled bidirectional quantum direct communication by using a GHZ state. Chin. Phys. Lett. **23**(7), 1680–1682 (2006)
31. X.F. Zou, D.W. Qiu, Three-step semi-quantum secure direct communication scheme. Sci. China Phys. Mech. Astron. **57**(9), 1696–1702 (2014)
32. Y.-P. Luo, T. Hwang, Authenticated semi-quantum direct communication schemes using Bell states. Quantum Inf. Process **15**(2), 947–958 (2016)
33. M.-H. Zhang, H.-F. Li, Z.-Q. Xia, X.-Y. Feng, J.-Y. Peng, Semi-quantum secure direct communication using EPR pairs. Quantum Inf. Process **16**(5), 117 (2017)
34. C. Shukla, K. Thapliyal, A. Pathak, Semi-quantum communication: Schemes for key agreement, controlled secure direct communication and dialogue. Quantum Inf. Process **16**, 295 (2017)
35. J. Gu, P.-H. Lin, T. Hwang, Double C-NOT attack and counterattack on 'three-step semi-quantum secure direct communication scheme'. Quantum Inf. Process **17**(7), 182 (2018)
36. L. Yan, Y. Sun, Y. Chang, S. Zhang, G. Wan, Z. Sheng, Semi-quantum scheme for deterministic secure quantum communication using Bell states. Quantum Inf. Process **17**(11), 315 (2018)

37. Y. Sun, L. Yan, Y. Chang, S. Zhang, T. Shao, Y. Zhang, Two semi-quantum secure direct communication schemes based on Bell states. Mod. Phys. Lett. A **34**(01), 1950004 (2019)
38. Z. Tao, Y. Chang, S. Zhang, et al., Two semi-quantum direct communication schemes with mutual authentication based on Bell states. Int. J. Theor. Phys. **58**, 2986–2993 (2019)
39. Z.X. Man, Z.J. Zhang, L. Yong, Deterministic secure direct communication by using swapping quantum entanglement and local unitary operations. Chin. Phys. Lett. **22**, 18 (2005)
40. X.H. Li, F.G. Deng, C.Y. Li, Y.J. Liang, P. Zhou, H.Y. Zhou, Deterministic secure quantum communication without maximally entangled states. J. Korean Phys. Soc. **49**, 1354 (2006)
41. G.L. Long, F.G. Deng, C. Wang, X.H. Li, K. Wen, W.Y. Wang, Quantum secure direct communication and deterministic secure quantum communication. Front. Phys. China **2**(3), 251–272 (2007)
42. C. Shukla, A. Pathak, Orthogonal-state-based deterministic secure quantum communication without actual transmission of the message qubits. Quantum Inf. Process **13**, 2099–2113 (2014)
43. D. Joy, S.P. Surendran, M. Sabir, Efficient deterministic secure quantum communication schemes using multipartite entangled states. Quantum Inf. Process **16**, 157 (2017)
44. B.A. Nguyen, Quantum dialogue. Phys. Lett. A **328**(1), 6–10 (2004)
45. X. Ji, S. Zhang, Secure quantum dialogue based on single-photon. Chin. Phys. **15**, 1418–1420 (2006)
46. Y.G. Yang, Q.Y. Wen, Quasi-secure quantum dialogue using single photons. Sci. China Ser. G-Phys. Mech. Astron. **50**(5), 558–562 (2007)
47. Y. Xia, C.B. Fu, S. Zhang, et al., Quantum dialogue by using the GHZ state [J]. J. Korean Phys. Soc. **48**(1), 24–27 (2006)
48. L. Dong, X.M. Xiu, Y.J. Gao, et al., Quantum dialogue scheme using a class of three-photon W states[J]. Commun. Theor. Phys. **52**(5), 853–856 (2009)
49. W. Li, X. Zha, Y. Yu, Secure quantum dialogue scheme based on four-qubit cluster state. Int. J. Theor. Phys. **57**, 371–380 (2018)
50. F. Gao, F. Guo, Q. Wen, et al., Revisiting the security of quantum dialogue and bidirectional quantum secure direct communication. Sci. China Ser. G-Phys. Mech. As **51**, 559–566 (2008)
51. G.F. Shi, X.Q. Xi, M.L. Hu, et al., Quantum secure dialogue by using single photons. Opt. Commun. **283**(9), 1984–1986 (2010)
52. Y. Luo, C. Lin, T. Hwang, Efficient quantum dialogue using single photons. Quantum Inf. Process **13**, 2451–2461 (2014)
53. N. Zhou, T. Hua, G. Wu, et al., Single-photon secure quantum dialogue scheme without information leakage. Int. J. Theor. Phys. **53**, 3829–3837 (2014)
54. K.F. Yu, C.W. Yang, C.H. Liao, T. Hwang, Authenticated semi-quantum key distribution scheme using Bell states. Quantum Inf. Process **13**, 1457 (2014)
55. M. Boyer, R. Gelles, D. Kenigsberg, T. Mor, Semi-quantum key distribution. Phys. Rev. A **79**, 032341 (2009)
56. Y. Nie, Y. Li, Z. Wang, Semi-quantum information splitting using GHZ-type states. Quantum Inf. Process **12**, 437–448 (2013)
57. N. Zhou, G. Zeng, J. Xiong, Quantum key agreement scheme. Electron. Lett. **40**(18), 1149–1150 (2004)
58. C.C. Hsueh, C.Y. Chen, *Quantum key agreement scheme with maximally entangled states* [C], in Proceedings of the 14th information security conference, Taipei (2004), pp. 236–242
59. C.W. Tsai, T. Hwang, On quantum key agreement scheme. Technical Report C-S-I-E, NCKU, Taiwan (2009)
60. S.K. Chong, T. Hwang, Opt. Commun. **283**, 1192–1195 (2010)
61. R.H. Shi, H. Zhong, Multi-party quantum key agreement with bell states and bell measurements. Quantum Inf. Process **12**, 921–932 (2013)
62. B. Liu, F. Gao, W. Huang, Q.-Y. Wen, Multiparty quantum key agreement with single particles. Quantum Inf. Process **12**, 1797–1805 (2013)
63. P. Wang, Z.W. Sun, X.Q. Sun, Multi-party quantum key agreement scheme secure against collusion attacks. Quantum Inf. Process **16**, 170 (2017)

64. S.Q. Min, H.Y. Chen, L.H. Gong, Novel multi-party quantum key agreement scheme with g-like states and bell states. Int. J. Theor. Phys. **57**(6), 1811–1822 (2018)
65. W.J. Liu, Z.Y. Chen, S. Ji, et al., Multi-party semi-quantum key agreement with delegating quantum computation. Int. J. Theor. Phys. **56**, 3164–3174 (2017)
66. L.L. Yan, S.B. Zhang, Y. Chang, et al., Mutual semi-quantum key agreement scheme using Bell states. Mod. Phys. Lett. A **34**(35), 1950294 (2019)
67. N.R. Zhou, K.N. Zhu, Y.Q. Wang, Three-party semi-quantum key agreement scheme. Int. J. Theor. Phys. **59**, 663–676 (2020)
68. C.H. Bennett, G. Brassard, *Practical quantum oblivious transfer*, in Advances in Cryptology [C]. Proceedings of Crypto'91, Lecture Notes in Computer Science, 576, 351–366 (1992)
69. G. Brassard, C. Crepeau, *A quantum bit commitment scheme provably unbreakable by both parties*[C], in Proceedings of the 34th Annual Symposium on Foundations of Computer Science (IEEE, Washington, 1993), p. 362
70. V. Giovannetti, S. Lloyd, L. Maccone, Quantum private queries[J]. Phys. Rev. Lett. **100**, 230502 (2008)
71. M. Jakobi, C. Simon, N. Gisin, J.-D. Bancal, C. Branciard, N. Walenta, H. Zbinden, Practical private database queries based on a quantum-key-distribution scheme[J]. Phys. Rev. A **83**(2), 022301 (2011)
72. P. Chan, I. Lucio-Martinez, X. Mo, et al., Performing private database queries in a real-world environment using a quantum scheme[J]. Report **4**(23), 5233 (2014)
73. G. Fei, B. Liu, H. Wei, et al., Post-processing of the oblivious key in quantum private queries [J]. IEEE J. Sel. Topics Quantum Electron. **21**(3), 98–108 (2014)
74. T.Y. Wang, S.-Y. Wang, J.-F. Ma, et al., Robust quantum private queries[J]. Int. J. Theor. Phys. **55**(7), 3309–3317 (2016)
75. N. Li, J. Li, X. Chen, et al., Quantum private query with perfect performance universally applicable against collective-noise[J]. IEEE Access, 1–1 (2019)
76. T.Y. Ye, H.K. Li, J.L. Hu, Multi-user quantum private query scheme[J]. Int. J. Theor. Phys. **16** (2020)
77. H. Yang, M. Xiao, Multi-user quantum private query[J]. Quantum Inf. Process **19**(8) (2020)
78. M. Hayashi, Prior entanglement between senders enables perfect quantum network coding with modification. Phys. Rev. A **76** (2007) https://doi.org/ARTN04030110.1103/PhysRevA.76.040301
79. M. Hayashi, K. Iwama, H. Nishimura, R. Raymond, S. Yamashita, W. Thomas, P. Weil, Presented at the quantum network coding (2007)
80. H. Kobayasffl, F. Le Gall, H. Nishimura, M. Roetteler, General scheme for perfect quantum network coding with free classical communication. Int. Colloq. Autom. Lang. Program. (2009)
81. H. Kobayashi, F.L. Gall, H. Nishimura, M. Roetteler, Perfect quantum network communication scheme based on classical network coding. IEEE. (2009)
82. H. Kobayashi, F.L.Gall, H. Nishimura, M. Roetteler, Constructing Quantum Network Coding Schemes from Classical Nonlinear Schemes. Ieice Tech. Rep. (2010)
83. T. Satoh, F.L. Gall, H. Imai, Quantum network coding for quantum repeaters. Phys. Rev. A **86** (2012)
84. T. Shang, X.J. Zhao, J.W. Liu, Quantum network coding based on controlled teleportation. IEEE Commun. Lett. **18**, 865–868 (2014)
85. J. Li, X.B. Chen, G. Xu, Y.X. Yang, Z.P. Li, Perfect quantum network coding independent of classical network solutions. IEEE Commun. Lett. **19**, 115–118 (2015)
86. G. Xu, X.B. Chen, J. Li, C. Wang, Y.X. Yang, Z. Li, Network coding for quantum cooperative multicast. Quantum Inf. Process **14**, 4297–4322 (2015)
87. T. Shang, X.J. Zhao, C. Wang, J.W. Liu, Quantum homomorphic signature. Quantum Inf. Process **14**, 393–410 (2015)
88. T. Shang, Z. Pei, X.J. Zhao, J.W. Liu, Quantum network coding against pollution attacks. IEEE Commun. Lett. **20**, 1369–1372 (2016)

89. T. Shang, G. Du, J.W. Liu, Opportunistic quantum network coding based on quantum teleportation. Quantum Inf. Process **15**, 1–21 (2016)
90. J. Li, X. Chen, X. Sun, Z. Li, Y. Yang, Quantum network coding for multi-unicast problem based on 2D and 3D cluster states. Sci. China (Information Sci.) **59** (2016)
91. T. Shang, J. Li, J.W. Liu, Secure quantum network coding for controlled repeater networks. Quantum Inf. Process **15**, 2937–2953 (2016)
92. T. Shang, K. Li, J.W. Liu, Continuous-variable quantum network coding for coherent states. Quantum Inf. Process **16** (2017)
93. Z.Z. Li, G. Xu, X.B. Chen, Y.X. Yang, Secure quantum network coding based on quantum homomorphic message authentication. Quantum Inf. Process (2019)
94. Z.Z. Li, G. Xu, X.B. Chen, et al., Efficient quantum state transmission via perfect quantum network coding. Sci. China Inf. Sci. **62**, 12501 (2019)
95. X.B. Pan, G. Xu, Z.P. Li, X.B. Chen, Y.X. Yang, Quantum network coding without loss of information. Quantum Inf. Process **20**, 1–31 (2021)

Chapter 2
Fundamentals of Quantum Information Theory

Abstract Quantum information theory is the foundation of secure quantum communication, which mainly refers to the theoretical foundation of quantum information theory; it includes quantum mechanics, density operator, quantum entanglement, properties of quantum information, semi-quantum correlation theory, and basic assumptions of quantum information measurement.

2.1 The Basic Assumptions of Quantum Mechanics

2.1.1 State Space

Let $\mathcal{H}$ be a complex vector space, and $|\psi\rangle$ and $|\varphi\rangle$ be arbitrary vectors in the space $\mathcal{H}$. The complex function $\langle\cdot|\cdot\rangle$: $\mathcal{H}\times\mathcal{H}\rightarrow\mathbb{C}$ is defined as an inner product, where

(a) Positivity: $\langle\phi|\psi\rangle > 0$ for $|\psi\rangle \neq 0$
(b) Linearity: $\langle\varphi|(a|\psi_1\rangle + b|\psi_2\rangle) = a\langle\phi|\psi_1\rangle + b\langle\phi|\psi_2\rangle$
(c) Skew symmetry: $\langle\phi|\psi\rangle = \langle\psi|\phi\rangle^*$

The $*$ denotes complex conjugation.

Then ($\mathcal{H}$, $\langle\cdot|\cdot\rangle$) is an inner product space, and the modulus of any vector $|\psi\rangle$ in the space is defined as $\||\psi\rangle\| = \sqrt{\langle\psi|\psi\rangle}$. When $\mathcal{H}$ is finite dimensional, $\mathcal{H}$ is also called a Hilbert space.

2.1.2 Mechanical Quantity Operator

The first postulate of quantum mechanics establishes the application of quantum mechanics in Hilbert space.

Postulate 1: Any isolated physical system has a complex inner product vector space of the system state space (i.e., Hilbert space) associated with it, and the system can be described by a state vector, which is a unit vector of the system state space.

R.-G. Zhou et al., *Design and Analysis of Secure Quantum Communication Schemes*,
https://doi.org/10.1007/978-3-031-78428-6_2

Postulate 2: Any observable mechanical quantity A can be represented by the corresponding linear HE-TE notation $\widehat{A}$. These operators operate on the wave function of the state.

Postulate 3: Quantum measurements are characterized by a set of measurement operators $\{\{M_m\}\}$ that operate on the state space of the system under test, with subscripts m representing possible measurement outcomes. If the state of the quantum system before measurement is $|\psi\rangle$, then the probability of the result m occurring is

$$p(m) = \langle\psi|M_m^\dagger M_m|\psi\rangle \tag{2.1}$$

and the measured system is in

$$\frac{M_m|\psi\rangle}{\sqrt{\langle\psi|M_m^\dagger M_m|\psi\rangle}} \tag{2.2}$$

The measurement operator satisfies the completeness equation:

$$\sum_m M_m^\dagger M_m = I \tag{2.3}$$

This expresses the fact that the sum of probabilities is 1.

Postulate 3 shows that once any quantum state is measured, its initial quantum state may change, collapsing to the new state with a certain probability, that is, the observed state may be different from the initial state of the system itself. In addition, the measurement result depends not only on the initial state but also on the measurement operator chosen, which can be represented by a set of orthonormal bases of the state space, that is, the measurement basis.

Postulate 4: The evolution of a closed quantum system can be described by a unitary transformation, that is, the state of the system at time t_1 and t_2, $|\psi_1\rangle$ and $|\psi_2\rangle$, can be associated with a unitary operator U that depends only on time t_1 and t_2:

$$|\psi_2\rangle = U|\psi_1\rangle \tag{2.4}$$

Postulate 5: Two microscopic particles are identical particles, if all their intrinsic properties (mass, charge, spin, isospin, internal structure, and other intrinsic properties) are the same.

2.1.3 *Quantum State Evolution*

In quantum information systems, information is encoded in qubits (quantum states). The evolution of quantum information follows Schrodinger equation in quantum physics:

$$i\hbar \frac{d|\varphi\rangle}{dt} = H|\varphi\rangle \tag{2.5}$$

In Eq. 2.5, where ħ is Planck's constant, H represents a Hermitian operator called the Hamiltonian of a closed system.

As can be seen from the above Eq. 2.5, the state of the evolved system can be obtained by solving the differential equation, while the Hamiltonian of a specific physical system can be determined by a large number of experiments. In quantum information theory, it is generally assumed that the Hamiltonian is known. The possible states of a quantum system can be determined using the Hamiltonian, which is a Hermitian operator and can be decomposed into

$$H = \sum_i \lambda_i |i\rangle\langle i| \tag{2.6}$$

where λ_i is the eigenvalue and $|i\rangle$ is the eigenvector. Therefore, given the Hamiltonian of a quantum system and the state $|\varphi(t_0)\rangle$ at time t_0, the state of the system at time t can be described by the operator $U(t, t_0)$ as

$$|\varphi(t)\rangle = U(t, t_0)|\varphi(t_0)\rangle \tag{2.7}$$

The operator U is said to be unitary or unitary when the operator satisfies the condition $U^\dagger U = UU^\dagger = I$, where $U^\dagger = (U^*)^T$.

In two-dimensional Hilbert Spaces, for single qubits, the four commonly used unitary operators are I, σ_X, σ_Z, and σ_Y.

(a) $I = |0\rangle\langle 0| + |1\rangle\langle 1|$—$I$ keeps the quantum state unchanged, and its matrix form is,

$$I = \begin{pmatrix} 1 & 0 \\ 0 & 1 \end{pmatrix} \tag{2.8}$$

(b) $\sigma_X = |0\rangle\langle 1| + |1\rangle\langle 0|$—this operator can achieve a bit flip of a quantum state, such as $\sigma_X|0\rangle = |1\rangle$ and $\sigma_X|1\rangle = |0\rangle$, and its matrix form is

$$\sigma_X = \begin{pmatrix} 0 & 1 \\ 1 & 0 \end{pmatrix} \tag{2.9}$$

(c) $\sigma_Z = |0\rangle\langle 0| - |1\rangle\langle 1|$—the operator can flip the phase of the quantum state, $\sigma_Z|0\rangle = |0\rangle$ and $\sigma_Z|1\rangle = -|1\rangle$, and its matrix form is

$$\sigma_Z = \begin{pmatrix} 1 & 0 \\ 0 & -1 \end{pmatrix} \tag{2.10}$$

(d) $\sigma_Y = |1\rangle\langle 0| - |0\rangle\langle 1|$—the operator can achieve bit flip and phase reverse at the same time, $\sigma_X|0\rangle = |1\rangle$ and $\sigma_X|1\rangle = -|0\rangle$, and its matrix form is

$$\sigma_Y = \begin{pmatrix} 0 & -1 \\ 1 & 0 \end{pmatrix} \tag{2.11}$$

In addition to the above four operators, the Hadamard operator is also one of the most commonly used single quantum gates in quantum communication, which can be expressed as

$$H = \frac{1}{\sqrt{2}} \begin{pmatrix} 1 & 1 \\ 1 & -1 \end{pmatrix} \tag{2.12}$$

The operator H can achieve the flip of the measurement basis, such as $H|0\rangle = |+\rangle$ and $H|1\rangle = -|-\rangle$.

When operating on two qubits, there is the famous CNOT gate:

$$\begin{pmatrix} 1 & 0 & 0 & 0 \\ 0 & 1 & 0 & 0 \\ 0 & 0 & 1 & 0 \\ 0 & 0 & 0 & 1 \end{pmatrix} \tag{2.13}$$

When the CNOT gate acts on two qubits, one qubit is the control bit and the other qubit is the target bit, and the target bit is controlled by the control bit, while the control bit itself remains unchanged.

2.1.4 *Measurement*

In quantum mechanics, the quantum state will collapse to an eigenstate of the measurement operator after being measured, and the measurement result is probabilistic. The two most important measurements in quantum information are Von Neumann measurement (i.e., projective measurement) and positive operator valued measurement (POVM).

Next, we define quantum measurements using the set of measurement operators $\{M_i\}$, where the subscript i represents the first kind of measurement result i and the corresponding eigenstate is $|\phi_i\rangle$. Suppose that the system is in the initial state $|\phi\rangle$, and the probability that the result is $|\phi_i\rangle$ after the action of the measurement operator M_i is

$$p(i) = \langle\phi|M_i^\dagger M_i|\phi\rangle \tag{2.14}$$

The measured state collapse is

$$|\phi'\rangle = \frac{M_i|\phi\rangle}{\sqrt{p(i)}} \tag{2.15}$$

In addition, in order to ensure the completeness of the measurement, it is required that

$$\sum_i M_i^\dagger M_i = I, \tag{2.16}$$

where I represents the unit operator. This causes (i) to satisfy the normalization condition:

$$\sum_i p(i) = \langle\phi|M_i^\dagger M_i|\phi\rangle = 1 \tag{2.17}$$

We perform the projection measurement on $|\psi\rangle = \alpha|0\rangle + \beta|1\rangle$, the measurement of operator set for $\{M_i\} = \{M_0, M_1\}$, defined as

$$M_0 = |0\rangle\langle 0| \tag{2.18}$$

$$M_1 = |1\rangle\langle 1| \tag{2.19}$$

and satisfy

$$M_0^\dagger M_0 = M_0 \tag{2.20}$$

$$M_1^\dagger M_1 = M_1 \tag{2.21}$$

The probability of $|\psi\rangle = \alpha|0\rangle + \beta|1\rangle$ collapsing to $|0\rangle$ and $|1\rangle$ is measured as

$$p(0) = \langle\psi|M_0^\dagger M_0|\psi\rangle = |\alpha|^2 \tag{2.22}$$

$$p(1) = \langle\psi|M_1^\dagger M_1|\psi\rangle = |\alpha|^2 \tag{2.23}$$

1. **Projective Measurement**
The projection measurement can be described by an observable Hermitian operator M on the state space of a quantum system, and the lapse space of $|\phi\rangle$ can be projected onto the eigenspace of P_i using the Hermitian operator P_i. The spectral decomposition form of the observed quantity M is

$$M = \sum_m mP_m \tag{2.24}$$

where P_i is a projection of the eigenspace M and has an eigenvalue m, and the measurement results correspond to the eigenvalue M of the observed quantity m. When the state of the measured quantum system is $|\phi\rangle$, the probability of obtaining m can be expressed as

$$p(m) = \langle\phi|P_m|\phi\rangle \tag{2.25}$$

According to the measurement result of m, the measurement result of the measured quantum system can be obtained as

$$\frac{P_m|\psi\rangle}{\sqrt{P(m)}} \tag{2.26}$$

Calculate the mean value of M according to the definition:

$$\begin{aligned} E(m) &= \sum_m mp(m) \\ &= \sum_m m\langle\phi|P_m|\phi\rangle \\ &= \left\langle\phi\right|\left(\sum_m mP_m\right)\left|\phi\right\rangle \\ &= \langle\phi|M|\phi\rangle \end{aligned} \tag{2.27}$$

2. **POVM Measurement**
If there is a quantum system in the state $|\phi\rangle$, the operator set $E_i = M_i M_i^*$ of the measurement operation used in it if the condition for completeness of the measurement is satisfied, where

$$\sum_i E_i = I \tag{2.28}$$

If E_i is a positive definite operator, then the measurement is said to be a POVM measurement, where the possible probability of the result m_i is

$$P_{m_i} = \langle\phi|E_i|\phi\rangle \tag{2.29}$$

2.1.5 Composite Systems

Postulate The quantum states of a composite system are tensor products of the states of each subsystem. Suppose there are n systems, the system is in state $|\psi_i\rangle$, $i \in \{1, 2, \cdots, n\}$, is the composite of the whole system state $|\psi_1\rangle \otimes |\psi_2\rangle \cdots \otimes |\psi_n\rangle$.

From the postulator, we can see that a $2n$-dimensional quantum system obtained by the composition of n qubits, whose orthonormal basis is $\{|i_1 i_2 \cdots i_n\rangle, i_1, i_2, \cdots, i_n \in \{0, 1\}\}$ and whose corresponding decimal representation is $\{|k\rangle, k = 0, 1, \cdots, 2^n - 1\}$. This state is a set of linear combinations,

$$|\psi\rangle = \sum_{k=0}^{2^n-1} a_k |k\rangle, \text{where} \sum_{k=0}^{2^n-1} |a_k|^2 = 1 \tag{2.30}$$

2.2 Density Operator

2.2.1 Density Operator

Density operator, whose matrix form is also called density matrix, and state vector are both important tools to describe quantum mechanics. It is mathematically equivalent to the state vector method. Density operators provide a convenient way to describe quantum systems whose states are not completely known. If a quantum system happens to be in one of a set of states $|\psi_i\rangle$ with probability p_i and i is an index, then $\{p_i, |\psi_i\rangle\}$ can be called an ensemble of pure states, and the density operator of the system can be defined as

$$\rho = \sum_i p_i |\psi\rangle\langle\psi| \tag{2.31}$$

Then when a quantum system has an exact known state, it is in a pure state, in which case, its density operator

$$\rho = |\psi\rangle\langle\psi| \tag{2.32}$$

Otherwise, it is in a mixed state, that is, the system consists of a set of subsystems described by the state vector $|\psi_i\rangle$, $i \in \{1, 2, \cdots, n\}$, and subsystem i will appear in the whole system with probability p_i, the corresponding density operator is $\rho = \sum_i p_i |\psi_i\rangle\langle\psi_i|$, $\sum_i p_i = 1$.

2.2.2 *Reduced Density Operator*

The reduced density operator is an important tool for analyzing subsystems. Suppose there are physical systems A and B whose states can be represented by the density operator ρ^{AB}, then the reduced density operator for system A can be defined as

$$\rho^A = tr_B\left(\rho^{AB}\right) \tag{2.33}$$

where tr_B is called the partial trace on system B and is defined as

$$tr_B(|a_1\rangle\langle a_2|\otimes|b_1\rangle\langle b_2|) = |a_1\rangle\langle a_2|tr(|b_1\rangle\langle b_2|) \tag{2.34}$$

Here $|a_1\rangle$, $|a_2\rangle$ and $|b_1\rangle$, $|b_2\rangle$ are two vectors of the state space A and B, respectively.

2.3 Quantum Entanglement

The study of quantum entanglement not only has important scientific significance for understanding the entanglement phenomenon and further studying the basic theory of quantum mechanics but also has important practical significance because of the great application value of quantum entanglement in quantum information technology.

Quantum entanglement has three interesting basic properties:

(a) Entanglement, especially multibody entanglement, has a complex mathematical structure.
(b) Easy to interact with the environment, resulting in decoherence, so very fragile.
(c) Local operation and classical communication cannot increase entanglement.

The basic problems to be solved by quantum entanglement theory include:

(a) How to characterize, control, and measure entanglement
(b) How best to detect entanglement theoretically and experimentally
(c) How to maintain or recover entanglement in unavoidable interactions with the environment

2.3.1 *Entanglement State*

The definition of entanglement is given below.

For a two-body pure state $|\Psi\rangle_{AB}$, it can always be represented as follows:

$$|\Psi\rangle_{AB} = \sum_i \lambda_i |i_A\rangle |i_B\rangle, \tag{2.35}$$

where $| i_A\rangle$ and $| i_B\rangle$ are the subsets of the system A and the system B, respectively. nonnegative real number λ_i called Schmidt coefficient and $\sum_i \lambda_i^2 = 1$. The number of nonzero Schmidt coefficients is called the Schmidt number, and Eq. 2.35 is called the Schmidt decomposition of $| \Psi\rangle_{AB}$.

For two-qubit pure states, the pure state is entangled if the Schmidt number of the system is greater than 1. That is, if a two-qubit pure state cannot be represented as a direct product of a pure state of two subsets, then the pure state is entangled.

For example, the four Bell states mentioned above are typical entangled states of a two-qubit system.

$$| \phi^{\pm}\rangle = \frac{1}{\sqrt{2}}(|00\rangle \pm |11\rangle) \tag{2.36}$$

$$| \psi^{\pm}\rangle = \frac{1}{\sqrt{2}}(|01\rangle \pm |10\rangle) \tag{2.37}$$

For a two-qubit mixed state, if the density matrix of the mixed state ρ_{AB} cannot be expressed as a separable state,

$$\rho_{AB} = \sum_i p_i \rho_A^{(i)} \otimes \rho_B^{(i)}, \sum_i p_i = 1 \tag{2.38}$$

The mixed states are entangled.

For quantum systems with three qubits and above, the typical entangled state is the GHZ state.

2.3.2 EPR Paradox and Bell Inequality

In 1935, Einstein, Podolsky, and Rosen published a joint paper questioning the nonlocal nature of quantum mechanics, later called the EPR paradox. The nonlocality shown in quantum mechanics was difficult for many physicists, such as Einstein, to accept. Einstein is best known for his statement that "God doesn't play dice."

The EPR paradox given by Einstein mainly includes local causality and physical objectivity. Local causality means that if two measurements are separated by a certain amount of time, then there is no causal relationship between the two measurements. Physical objectivity means that if the system is not disturbed, the measurement of any physical quantity should have a definite value. In 1951, Bohm presented a simplified version of the EPR paradox, considering the Bell state:

$$|\psi^{+}\rangle = \frac{1}{\sqrt{2}}(|0\rangle|1\rangle + |1\rangle|0\rangle) \tag{2.39}$$

According to local causality, when particles 1 and 2 are far enough apart that they are measured independently and the measurement times are close together, then the measurement of particle 1 will have no effect on particle 2.

However, according to quantum mechanics, the structure and collapse of entangled states are considered to be nonlocal, so it can be seen that there is a contradiction between local causality and quantum mechanics. In recent years, many experimental results have proved that local causality and physical objectivity are wrong.

In 1964, Bell further investigated the EPR paradox and proposed the Bell inequality based on the local hidden variable theory.

2.4 Properties of Quantum Information

2.4.1 Nonorthogonal Quantum States Are Indistinguishable

The indistinguishable theorem of nonorthogonal quantum states that nonorthogonal quantum states $|\psi_1\rangle$ and $|\psi_2\rangle$ cannot be precisely distinguished by measurement. Suppose the binary information "0" and "1" are encoded in the quantum states $|\psi_1\rangle$ and $|\psi_2\rangle$, respectively. After the two quantum states are transmitted to the receiver through a noiseless channel, can the receiver decode the secret message? The indistinguishable theorem states that when the two quantum states are orthogonal to each other, the receiver can successfully recover the encoded information, that is, distinguish the two states successfully. If the two quantum states are nonorthogonal, the receiver cannot recover the encoded information with 100% probability, that is, the two quantum states cannot be completely distinguished. Here, the theorem is also proved by means of proof by contradiction.

2.4.2 No-Cloning of Unknown Quantum States

In 1982, Wootters and Zurek first proposed the quantum no-cloning theorem, demonstrating that an unknown quantum state cannot be accurately copied. The quantum no-cloning theorem is an important guarantee for the security of quantum cryptography. Next, the theorem is proved mathematically by the method of proof by contradiction.

Consider a quantum system A, take any state $|\psi\rangle$ in the system. In another quantum system B, take any pure state $|S\rangle$ in the system B, if there is an unitary operation U that makes

$$U(|\psi\rangle \otimes |S\rangle) = |\psi\rangle \otimes |\psi\rangle \tag{2.40}$$

Suppose $|\phi\rangle$ is another state in the quantum system A and $|\phi\rangle \neq |\psi\rangle$, obtained by unitary operation

$$U(|\phi\rangle \otimes |S\rangle) = |\phi\rangle \otimes |\phi\rangle = |\phi, \phi\rangle \tag{2.41}$$

According to the linear transformation of the vectors, the quantum state $|\chi\rangle = \lambda|\psi\rangle + u|\phi\rangle$ is obtained in system A, where λ, u are two complex numbers. After U operation, we have

$$U(|\chi\rangle \otimes |S\rangle) = |\chi\rangle \otimes |\chi\rangle = |\chi, \chi\rangle \tag{2.42}$$

We derive the left-hand side of Eq. 2.42 and get

$$\begin{aligned} U(|\chi\rangle \otimes |S\rangle) &= U(\lambda|\psi\rangle + u|\phi\rangle) \otimes |\chi\rangle \\ &= \lambda|\psi\rangle \otimes |\psi\rangle + u|\phi\rangle \otimes |\phi\rangle \\ &= \lambda|\psi, \psi\rangle + u|\phi, \phi\rangle \end{aligned} \tag{2.43}$$

The right-hand side of the Eq. 2.42 is

$$\begin{aligned} \chi|\rangle \otimes |\chi\rangle &= (\lambda|\psi\rangle + u|\phi\rangle) \otimes (\lambda|\psi\rangle + u|\phi\rangle) \\ &= \lambda^2|\psi\rangle \otimes |\psi\rangle + \lambda u(|\psi\rangle \otimes |\phi\rangle + |\phi\rangle \otimes |\psi\rangle) + u^2|\phi\rangle \otimes |\phi\rangle \\ &= \lambda^2|\psi, \psi\rangle + \lambda u|\psi, \phi\rangle + \lambda u|\phi, \psi\rangle + u^2|\phi, \phi\rangle \end{aligned} \tag{2.44}$$

Assuming that the quantum state can be precisely cloned, the following is obtained from Eqs. 2.43 and 2.44:

$$(\lambda - 1)|\psi, \psi\rangle + \lambda u|\psi, \phi\rangle + \lambda u|\phi, \psi\rangle + u(u - 1)|\phi, \phi\rangle \equiv 0 \tag{2.45}$$

It follows from the Eq. 2.45 that it can be cloned when there is $\lambda u = 0$, that is, $|\chi\rangle = |\psi\rangle$ or $|\chi\rangle = |\phi\rangle$. For any superposition state B, it cannot be cloned precisely.

2.4.3 Heisenberg Uncertainty

The quantum uncertainty principle was proposed by the German physicist Heisenberg in 1927, so it is also known as the Heisenberg uncertainty principle. In classical physics, the coordinates and momentum of a classical particle in a given state can give precise measurements at the same time. However, in the micro domain, the coordinates and momentum of the micro particles cannot be accurately measured at the same time because of the volatility of the particles.

According to the basic properties of quantum measurement, for a quantum state $|\psi\rangle$, given two mechanical quantities $\widehat{P}$ and $\widehat{Q}$ that have a reciprocal effect on each other and a reciprocal effect on each other, the sufficient and necessary condition for accurate observation of these two mechanical quantities is that the operators of these two mechanical quantities are reciprocal, that is, $\left[\widehat{P}, \widehat{Q}\right] = \widehat{P}\widehat{Q} - \widehat{Q}\widehat{P} = 0$ is satisfied.

When two noncommutative mechanical quantities are measured for a given quantum state $|\psi\rangle$, an accurate measurement cannot be obtained at the same time.

For two mechanical quantity $\widehat{P}$ and $\widehat{Q}$, the measurement uncertainty is defined as

$$\triangle P \equiv \sqrt{\overline{\left(\widehat{P} - \overline{P}\right)^2}} \tag{2.46}$$

$$\triangle Q \equiv \sqrt{\overline{\left(\widehat{Q} - \overline{Q}\right)^2}}, \tag{2.47}$$

where $\triangle P$ represents the deviation of the mechanical quantity $\widehat{P}$ from the measurement of the quantum state and $\triangle Q$ represents the deviation of the mechanical quantity $\widehat{Q}$ from the measurement of the quantum state.

The Heisenberg uncertainty relation given as

$$\triangle P \triangle Q \geq \frac{1}{2}\left|\left\langle \psi \right| \left[\widehat{P}, \widehat{Q}\right] \left|\psi\right\rangle\right| \tag{2.48}$$

The above formula means that when two noncommutative mechanical quantities are observed in quantum states at the same time, the product of their measurement deviation is greater than a certain value. In other words, it is not possible to obtain the exact values of two mechanical quantities at the same time.

2.5 Semi-quantum Correlation Theory

2.5.1 Semi-quantum Communication Environment

"Semi-quantum" means that there is at least one communicating party with only classical capabilities. In other words, one or several parties would be limited to only a set of quantum states. Generally, all parties use quantum resources and establish a shared key. And they against adversaries who obey quantum mechanics. It is necessary for all parties to have ability of quantum? In addition, with the mature development of quantum communication networks, a semi-quantum scheme may hold the unique advantage of allowing lower-cost devices (i.e., less quantum devices) or devices that are more robust (some devices fail and can switch to a semi-quantum mode of operation). These will reduce the usage limits of the quantum communication party. The introduction of the semi-quantum concept is of great significance, which reduces the consumption of quantum resources and the difficulty of realization.

As mentioned in Sect. 1.1.2, Boyer [1] defines an environment involving two types of parties: the quantum party (Alice) obtains full quantum capabilities, and the classical party (Bob) has classical power and limited quantum capabilities. Bob can perform three operations:

(a) Measure the qubit in the classical $\{0, 1\}$ basis (i.e., Z basis)
(b) Prepare Z basis (i.e., $\{|0\rangle, |1\rangle\}$) qubits in the classical basis and send it
(c) Reorder the qubits (by using different delay lines)

The computational basis "classical" in (a) actually uses the classical notations $\{0, 1\}$ to describe the two quantum states $\{|0\rangle, |1\rangle\}$ defining this basis. If all parties were only allowed to perform three above operations, or do nothing at all. The operation of parties was similar to using qubits in the classic base and never obtained any quantum superposition of the computational basis state. The qubits can then be considered "classical bits." The Secure Quantum Communication scheme would be equivalent to the classic scheme. This is why Bob is considered a classical party.

Next, we summarize the specific semi-quantum environments as follows (see Table 2.1):

1. **Measure-resend Environment**
   In this environment, Bob is restricted to perform operations (a) and (b) or transfer the qubit back to Alice.
2. **Randomization-Based Environment**
   This environment is based on randomization. Bob is restricted to perform operations (a) and (b) or do nothing.
   After the semi-quantum environment was proposed, various kinds of semi-quantum schemes have been proposed for different security issues. Zhou will

Table 2.1 Summary of semi-quantum environment

Environment	Capabilities of classical party
Measure-resend environment	1. Generate Z basis qubits 2. Z basis measurement 3. Reflect photons without disturbance
Randomization-based environment	1. Z basis measurement 2. Reorder photons using different delay lines 3. Reflect photons without disturbance
Measurement-free environment	1. Generate Z basis qubits 2. Reorder photons using different delay lines 3. Reflect photons without disturbance
Unitary-operation-based environment	1. Z basis measurement 2. Generate Z basis qubits 3. Perform the unitary operation

further limit the capabilities of the classical party [2]. In the SQKD scheme, the classical party Bob dose not have measurement capability. This means, in SQKD, the measurement capability of the classical party is not necessary. In other words, to implement the SQKD scheme, the classical party does not need any quantum measuring device.

3. **Measurement-Free Environment**

   The classical party's measurement capability:

   (a) Generate Z basis qubits
   (b) Reorder photons using different delay lines
   (c) Reflect photons without disturbance

   It minimizes the quantum resources used by classical Bob.

   With the introduction of the mediated SQKD (MSQKD) scheme using a measurement environment, the problem is that Bell states are more difficult and expensive to generate and measure than single photons. So, the light-weight MSQKD scheme is proposed. The classical party, in semi-quantum cryptographic schemes, is based on a unitary-operation-based environment.

4. **Unitary-Operation-Based Environment**

   The classical party's measurement capability:

   (a) Z basis measurement
   (b) Generate Z basis qubits
   (c) Perform the unitary operation

2.5.2 *The Difference Between Semi-quantum and Full Quantum*

Full quantum communication often needs to make complex operation rules, higher hardware conditions, and higher technical requirements. It needs the support of a series of technologies, such as the preparation of entangled state, the storage of quantum state, anti-noise interference, and so on. Full quantum has the perfect hardware conditions and comprehensive technical support, so it can be applied to many different communication scenarios. But the use conditions are more stringent, and the required technology is more complex. Compared to all-quantum communication, semi-quantum communication usually requires fewer and simpler technologies, as long as it has the basic quantum functions of Z basis measurement, single quantum preparation, and forwarding and follows some simple rules.

Full quantum communication requires high hardware costs. Current Secure Quantum Communication schemes require users to be equipped with advanced quantum devices such as quantum generator, quantum memory, and unitary operator. However, as an ordinary user, there is no way to afford such expensive quantum devices, which limit the application of these Secure Quantum Communication schemes.

Semi-quantum communication is a new kind of quantum communication. Compared to full quantum communication, it is simple to operate and needs lower hardware condition. Unlike all-quantum communication, which requires expensive quantum equipment and complicated technical support, semi-quantum communication requires only a few simple hardware devices to communicate, thus effectively reducing the cost of quantum communication. In addition, the classical users with limited quantum capabilities will have the opportunity to participate in quantum communication, which will promote the application of the quantum key agreement scheme and the popularization of quantum communication. Therefore, although semi-quantum communication has some limitations, its advantages are obvious, with a wide range of applications.

2.6 Quantum Information Metric

2.6.1 Classical Shannon Entropy

Classical information theory uses entropy to describe the uncertainty of a random variable, a concept first proposed by Shannon, also known as Shannon entropy. According to Shannon's theory, when the information change is a continuous quantity, the symbol representing the information is a continuously distributed random variable, and its continuous entropy is defined as

$$H(X) = -\int P(x) \log P(x) dx \tag{2.49}$$

Similarly, joint entropy and conditional entropy can be defined as

$$H(X\ Y) = -\iint P(x,y) \log P(x,y) dxdy \tag{2.50}$$

$$H(X|Y) = -\iint P(x,y) \log P(x|y) dxdy \tag{2.51}$$

$$H(Y|X) = -\iint P(x,y) \log P(y|x) dxdy \tag{2.52}$$

Mutual information can be derived from the above Eqs. 2.50, 2.51, and 2.52 as follows:

$$\begin{aligned} I(X:Y) = I(Y:X) &= H(X) - H(X|Y) - H(Y|X) \\ &= H(Y) - H(Y|X) \\ &= H(X) + H(Y) - H(X,Y) \end{aligned} \tag{2.53}$$

The maximum mutual information of two communication parties is the channel capacity

$$\max I(X,Y) = \frac{1}{2}(1 + SNR) \tag{2.54}$$

2.6.2 Quantum Von Neumann Entropy

Shannon entropy, as the mathematical basis of classical information theory, can be used to describe information systems mathematically. The classical Shannon information entropy is extended to the field of quantum information, that is, quantum information entropy. Shannon entropy is based on the probability distribution of a variable and can be used to describe the uncertainty of a random variable. Quantum information entropy is based on the density matrix of quantum states and is used to measure the uncertainty of quantum states. Quantum information entropy was first proposed by von Neumann, so it is also called von Neumann entropy. Let the density matrix ρ corresponding to the quantum state, then the information carried by the quantum state can be expressed by von Neumann entropy as

$$S(\rho) = -Tr(\rho \log_2 \rho) \tag{2.55}$$

From the above Eq. 2.55, it can be seen that von Neumann entropy is represented by the quantum density matrix. Assuming that λ_i is the eigenvalue of the density matrix ρ, the von Neumann entropy can also be expressed as

$$S(\rho) = -\sum_i \lambda_i \log_2 \lambda_i \tag{2.56}$$

Von Neumann entropy has the following properties:

(a) Entropy is nonnegative. For a pure state, the entropy is 0. Assuming a pure state $|\psi\rangle = \frac{1}{\sqrt{2}}(|0\rangle + |1\rangle)$, the density matrix of the quantum state is

$$\rho = |\psi\rangle\langle\psi| = \frac{1}{2}\begin{pmatrix} 1 & 1 \\ 1 & 1 \end{pmatrix} \tag{2.57}$$

The eigenvalues of the density matrix are 0 and 1. $S(\rho) = -\ 0log_2 0 - 1log_2 1 = 0$ is calculated according to the expression of von Neumann entropy.

(b) In an n-dimensional Hilbert space, entropy is at most $log_2 n$.

(c) For a given set of positive p_i addresses, if $\sum_i p_i = 1$, $S\left(\sum_i p_i \rho_i\right) \geq \sum_i p_i S(\rho_i)$.

(d) Von Neumann entropy remains constant under unitary transformation U, $S(U\rho U^{-1}) = S(\rho)$.

Since the calculation of von Neumann entropy depends only on the eigenvalues of the density matrix, and unitary transformations do not change the eigenvalues of the matrix, von Neumann entropy remains unchanged under unitary transformations.

Similar to classical information entropy, joint entropy, relative entropy, conditional entropy, and mutual information can also be defined in quantum systems. The quantum joint entropy is defined as

$$S(\rho_{AB}) = -Tr(\rho_{AB} \log \rho_{AB}) \tag{2.58}$$

The quantum joint entropy satisfies the following properties:

(a) Von Neumann entropy is additive, and for two independent systems A and B, the density matrices are ρ_A and ρ_B:

$$S(\rho_A \otimes \rho_B) = S(\rho_A) + S(\rho_B) \tag{2.59}$$

If ρ_A and ρ_B are the reduced density matrices of ρ_{AB}, then there is

$$|S(\rho_A) - S(\rho_B)| \leq S(\rho_{AB}) \leq S(\rho_A) + S(\rho_B) \tag{2.60}$$

(b) Von Neumann entropy is strongly subadditive; given three Hilbert subspaces A, B, and C, there are

$$S(\rho_{ABC}) + S(\rho_B) \leq S(\rho_{AB}) + S(\rho_{BC}) \tag{2.61}$$

If the density matrices of two systems are ρ and σ, the relative entropy from ρ to σ is defined as

$$S(\rho \parallel \sigma) = Tr(\rho \log \rho) - Tr(\sigma \log \sigma) \tag{2.62}$$

Relative entropy has nonnegative property, when $\rho = \sigma$, $S(\rho \parallel \sigma) = 0$.

The quantum condition entropy is defined as

$$S(\rho_A | \rho_B) = S(\rho_{AB}) - S(\rho_B) \tag{2.63}$$

It is worth noting here that in classical information theory, conditional entropy cannot be negative, but in quantum information theory, conditional entropy can be negative due to the certainty of the joint system state and the complete uncertainty of the subsystem state. If quantum system A and B are completely unrelated,

$$S(\rho_{AB}) = S(\rho_A) + S(\rho_B) \tag{2.64}$$

$$S(\rho_A|\rho_B) = S(\rho_A) \tag{2.65}$$

$$S(\rho_B|\rho_A) = S(\rho_B) \tag{2.66}$$

Quantum mutual information is defined as

$$S(\rho_A : \rho_B) = S(\rho_A) + S(\rho_B) - S(\rho_{AB}) \tag{2.67}$$

When the quantum information source prepares the quantum state ρ_i with probability p_i, the density matrix of the system can be expressed as

$$\rho = \sum_i p_i \rho_i \tag{2.68}$$

2.6.3 *Quantum Fidelity*

For two quantum states, effectively measuring the distance or similarity between them is a key problem. For example, on a quantum channel with noise, how do we know how well the output of the channel can reproduce the quantum state of the original input system? Quantum fidelity is a measure used to measure the closeness of quantum states and is widely used in quantum information and quantum computing. For two quantum mixed states ρ_1 and ρ_2, the fidelity between them is defined as

$$F(\rho_1, \rho_2) = \left(tr\sqrt{\rho_1^{1/2} \rho_2 \rho_1^{1/2}} \right)^2 \tag{2.69}$$

When the quantum state is pure, it can be reduced to Hilbert-Schmidt fidelity. It can also be used to measure entanglement. Fidelity has some of the following properties:

(a) The value of the truth degree ranges from 0 to 1, that is, $0 \leq F(\rho_1, \rho_2) \leq 1$, and only if $\rho_1 = \rho_2$, $F(\rho_1, \rho_2) = 1$
(b) Under unitary transformations, fidelity remains unchanged, that is,

$$F\left(U\rho_1 U^\dagger, U\rho_2 U^\dagger\right) = F(\rho_1, \rho_2) \tag{2.70}$$

(c) Fidelity is symmetric, that is,

$$F(\rho_1, \rho_2) = F(\rho_2, \rho_1) \tag{2.71}$$

(d) Fidelity has monotonicity, and for the trace-preserving operator ε,

$$F\left(\varepsilon(\rho)_1, \varepsilon(\rho)_2\right) \geq F(\rho_1, \rho_2) \tag{2.72}$$

References

1. M. Boyer, R. Gelles, D. Kenigsberg, T. Mor, Semiquantum key distribution. Phys. Rev. A. **79**(3), 032341 (2009)
2. X. Zou, D. Qiu, S. Zhang, et al., Semiquantum key distribution without invoking the classical party's measurement capability. Quantum Inf Process **14**, 2981–2996 (2015)

Chapter 3
Quantum Key Distribution

Abstract The main purpose of quantum key distribution is to ensure that the keys generated by the two parties can be safely distributed over a quantum channel, which allows the two parties, not initially sharing secret information, to create a shared secret bit sequence. Security is based on the physical laws of quantum mechanics. As quantum key distribution technology matures, its distribution options are becoming more diverse, such as discrete variable, continuous variable, distribution phase reference, and semi-quantum.

3.1 Introduction

In recent years, with the rapid development of quantum communication, quantum-based data encryption technology has made rapid progress. Quantum key distribution (QKD) is based on the basic principle of quantum mechanics, which distributes the secret key to the two parties who need to keep the secret communication. This process guarantees that the key can be distributed between the two parties, and the unauthorized third party cannot get the contents of the key in the process of distribution so that the key can be truly done once. This ensures the absolute security of encrypted information. For the current public-key systems, such as the famous RSA scheme, its security depends on the computational complexity, that is, if the attacker does not have the computing ability of fast large-number factorization, he cannot crack the secret key in the effective time, the quantum computer can be used to decompose the large number factor quickly and effectively, thus destroying the security system of RSA. For QKD, security is not under the threat of an attacker's computing power, even if the attacker has enough and fast computing resources within the bounds of physical theory, QKD still ensures that both parties can share a set of identical and secure keys for subsequent secure communication. The key generated by QKD combined with one-time paid scheme can guarantee the secure transmission of information.

The security analysis of QKD has gone through the process from simple conceptual proof to strict mathematical proof based on physical model. However, the actual system has some defects because it cannot completely satisfy some assumptions of

R.-G. Zhou et al., *Design and Analysis of Secure Quantum Communication Schemes*,
https://doi.org/10.1007/978-3-031-78428-6_3

physical model, vulnerable to attack by eavesdroppers. After the improvement of the theoretical assumptions and the improvement of the scheme, further ensure the safety. Practicality is another important element, given its critical importance to the financial, military, aerospace and other fields and the ease with which it can be implemented in real-world systems, leading to opportunities for development and application in practice. At present, it has developed from the model of laboratory verification, short distance and low code rate to the practical system of network, long distance and high efficiency, and high code rate and high stability. At present, some companies have carried out commercial production systems and the coverage of the network is gradually expanding.

The goal of QKD technology is to establish a real-time, provable, and secure key between the two sides of communication. It overcomes the inherent security hidden trouble of classical encryption technology and can solve the problem of secret transmission of information in the application fields with high security demand, such as national defense, finance, government affairs, energy, commerce, and so on. The United States, Europe, Canada, Australia, Japan, and other middle-sized countries attach great importance to the research of the QKD technology and invest a lot of R & D efforts, which makes it a very important research hotspot of quantum information science in the world. This technology is likely to drive a major technological revolution in the information industry.

3.2 Principle and Research Progress of Quantum Key Distribution

The QKD principle is a key distribution process that allows both parties to create, transmit, and measure quantum states for information security and allows legitimate senders and receivers to share symmetric information security keys. The QKD technology makes use of the Heisenberg uncertainty principle and quantum No-cloning theorem of quantum physics, the Heisenberg uncertainty principle ensures that the eavesdropper cannot accurately measure the quantum state without knowing the sender's coding base, and the quantum noncloning theorem ensures that the eavesdropper cannot even be told that the quantum state has been measured. The same quantum state cannot be replicated, which makes it inevitable that the eavesdropper will make an obvious error in judgment, so the eavesdropper's presence can be detected by both sides of the communication.

The result of QKD is that the two parties share a set of secret keys composed of binary random sequences. To transmit secret information, the sender encrypts the secret information with the key to generate a cipher text and then sends the cipher text to the receiver. The receiver uses the key to decode the cipher text and get the secret message, which means that with the help of the quantum cryptography, the two sides have to have another classical communication to complete the secure transmission of the secret message.

QKD is the foundation of realizing a Secure Quantum Communication network. For many years, researchers have made a lot of research results on QKD technology, which can be divided into four categories: QKD based on discrete variables (DV-QKD), QKD based on continuous-variables (CV-QKD), QKD based on distribution phase reference (DPR-QKD), and semi-quantum key distribution (SQKD).

DV-QKD has achieved remarkable results. From the ground state perspective, it can be divided into two categories: One is based on single-photon state and the other is based on entanglement state. In 1984, Bennett and Brassard proposed the first QKD scheme, BB84 [1], based on four single-photon polarization states, using uncertainty principle and no-cloning theorem. The implementation is simple and has unconditional security [2]. In 1992, Bennett simplified the BB84 scheme and proposed a simpler scheme with half efficiency based on two single-photon polarization states, which is called B92 scheme [3]. Although the unconditional security of the ideal BB84 scheme has been proved, the security vulnerability caused by the nonideal physical devices in the actual QKD system will seriously threaten the security of the whole system. Most schemes require a single-photon source, but it is difficult to produce a perfect single-photon source in a practical system, so a weak coherent light source is used instead, but the problem with coherent light sources is that eavesdroppers use the multi-photon signals as a copy of the information to steal the key. For example, the ideal BB84 scheme uses a single photon as the light source, while the actual QKD system usually uses a strong attenuation laser pulse as the light source, which has a certain probability of transmitting a multi-photon signal. To address this, researchers have proposed new schemes. First, the basic vector alignment process of the BB84 scheme is modified slightly, and the SARG04 scheme is proposed [4]. Second, the decoy state is used to detect the eavesdropper‘s photon-number-splitting (PNS) attack [5]. The single-photon component can be analyzed by randomly modulating the light intensity of the light source, and the eavesdropper‘s multi-photon separation attack operation can be found. Its performance approximates single-photon source, which is widely used at present, and it is one of the most important developments of QKD technology in the last 10 years [6–10]. In 1991, Ekert proposed a QKD scheme based on Einstein-Podolsky-Rosen (EPR) entanglement, namely, the E91 scheme [11]. After the E91 scheme was proposed, Bennett, Brassard, and Mermin proposed a new BBM92 scheme based on entangled photon pairs [12] and proved that it is essentially equivalent to the BB84 scheme. Compared with the security of the light source, the detection end of the system has more and more problems. At present, many kinds of attacks against detection devices are proposed, and a Device-Independent QKD (DI-QKD) scheme is proposed [13]. DI-QKD scheme is based on the idea of the E91 scheme. The security of the DI-QKD scheme does not depend on the assumption of non-ideal devices in the real system, but only on Bell inequality test. However, this kind of scheme requires high efficiency of the system detection end, so it is not practical at present. Then in 2012, H. K. Lo first proposed the Measurement-Device-Independent QKD (MDI-QKD) scheme [14], which solves the security problem of the probe segment and has high practicability, and is the most concerned scheme in recent years. Because MDI-QKD can be immune to all kinds of detector attacks, this scheme is considered an important way to realize the security of Secure Quantum Communication in real world. On the other hand, this scheme can be combined with the decoy scheme for the source to ensure the security of the weak coherent light source. Therefore, MDI-QKD has become a research hotspot in recent years [15–24].

The principle of CV-QKD is that the key is loaded on the orthogonal components X and P of the quantum state, and the information is extracted by zero-difference detection or heterodyne detection, which has potential advantages in communication rate and system cost. tTe theoretical unconditional security typical of GG02 scheme has been strictly proved [25]. The secure bit rate of this scheme [26–32] is greatly affected by system noise and the efficiency of error correction algorithm.

The DPR-QKD has the best practicability, the simplest equipment, and good compatibility with the existing optical fiber network, and the related experimental research progresses rapidly [33], but the security of this kind of scheme has not been proven completely.

Although the QKD scheme mentioned above is widely accepted, most of them require both parties to communicate with complete quantum capabilities, and expensive quantum devices will deter many ordinary users. To maximize the advantages of quantum communication and reduce costs, the first SQKD scheme, the BKM07 scheme, was proposed by Canadian scholar Beyer and colleagues [34] in 2007. This scheme allows one party to have a limited quantum capability, called a classical party, and the other party to be an unrestricted quantum party, thus cooperating to generate a secure key string. Subsequently, Boyer et al. extended SQKD scheme in 2009 and designed a new SQKD scheme based on random sorting [35]. As the SQKD scheme can greatly reduce the equipment cost of quantum communication systems, more and more researchers are beginning to participate in the research of SQKD [36–42]. In 2017, Boyer and colleagues experimentally implemented the SQKD scheme [43].

3.3 Single-Photon Quantum Key Distribution Scheme

3.3.1 BB84 Scheme

BB84 scheme [1] uses four non-orthogonal states to represent classical bits. These four states belong to two groups of conjugate bases, each group of bases in the two states orthogonal and nonorthogonal states in different groups. Two groups of bases are mutually conjugate, which means that any base vector of one group of bases has the same projection on any base vector of the other group of bases. Taking a single photon as an example, the photon polarization forms a two-dimensional Hilbert space in which the state can be expressed by the horizontal and vertical polarization basis (denoted as the Z basis) $\{|0\rangle, |1\rangle\}$, it can also be expressed as a diagonal polarized basis (denoted as the X basis) $\{|+\rangle, |-\rangle\}$. The horizontal and vertical polarization basis and the diagonal polarization basis are mutually conjugate.

Assume Alice wants to set up a random set of keys with Bob. Alice first uses the random number dispenser to generate two random strings of binary numbers before distributing them. A set of random numbers is used to select the basis of the measurement (Z basis when the random number appears as 0, X basis when 1), and a set is used to encode the photon state. Here's how the BB84 scheme works in detail as follows:

Step 1: Alice prepares a string of photons from a string of binary numbers generated by the random number dispenser, each randomly placed in one of two sets of conjugate bases, and sends the string to Bob.

Step 2: Bob randomly selects either the Z basis or the X basis to measure the received photons. After the measurement, Bob publishes the selected measurement basis.

Step 3: Alice compares the bases it chooses to prepare photons with the bases that Bob publishes and tells Bob which bases are correct.

Step 4: Bob saves the results of the measurements with the correct measurement basis. The quantum states are encoded as binary classical bits in the following way:

1. $|0\rangle$ and $|+\rangle$ are encoded as bit 0.
2. $|1\rangle$ and $|-\rangle$ are encoded as bit 1.

Step 5: Alice randomly selects bits of information for error rate analysis. If the error rate is less than the threshold value set by both parties, the remaining bits of information can be used as the original key; otherwise, the scheme is aborted.

Step 6: Alice and Bob perform error correction and secret amplification on the original key to obtain the security key.

In the BB84 scheme, Bob has only a 50% probability of choosing the correct measurement basis (Step 3). If eavesdropper Eve randomly chooses a Z basis or X basis eavesdropping quantum channel, her eavesdropping behavior will introduce a 25% error rate in Step 5. Alice and Bob discover the eavesdropper through error rate analysis. The advantage of the BB84 scheme is that it is proven to be an unconditionally secure QKD scheme, and the preparation and measurement of quantum signals are relatively easy to implement. In order to ensure the security of the BB84 scheme in a noisy quantum channel, an ideal single-photon source is needed. If the ideal single-photon source is replaced by a weak laser pulse, there may be information leakage.

3.3.2 B92 Scheme

Usage of two different bases in the BB84 scheme is redundant. In 1992, Bennett proposed a new scheme named B92 using one non-orthogonal basis [3]. In the B92 scheme, Alice used only one non-orthogonal basis to send information. That is,

Alice represents random bits 0 and 1 by $|0\rangle$ and $|+\rangle$, respectively. The specific Steps of the B92 scheme are explained as follows:

Step 1: Alice chooses n-random bit by flipping coin. She sends $|0\rangle$ for random bit 0 and $|+\rangle$ for random bit 1.

Step 2: Bob measures received qubit in Z or X basis. Bob certain cases occur when he observes $|1\rangle$ and $|-\rangle$. Bob uncertain cases occur when he observes $|0\rangle$ and $|+\rangle$.

Step 3: Bob publicly informs Alice regarding uncertain bits and shares half of the certain bits to ensure security. The remaining half of the certain bit can be used as a secret key.

Overall, in the B92 scheme, only 25% of Bob's concepts were measured. The B92 scheme is similar to BB84 scheme in error rate analysis, error correction, and secret amplification. Its advantage is that quantum signal preparation and measurement is simpler than BB84 scheme, but the efficiency is low.

3.4 Entanglement-Photon Quantum Key Distribution Scheme

3.4.1 E91 Scheme

In the E91 scheme [11], Alice's and Bob's results are perfectly correlated or anti-correlated, which helps in identifying the eavesdropper. Entangled pairs become disentangled due to noise in the environment. Therefore, we need to compare the matching of bases as in the BB84 scheme. In the original E91 scheme [11], Ekert had considered three bases for Alice ($|0\rangle$, $|+\rangle$ and $|1\rangle$) and Bob ($|+\rangle$, $|1\rangle$ and $|-\rangle$). There is a 1/3 chance that Alice and Bob measure in compatible bases (E91 original scheme consider three bases). Alice and Bob publicly announce their bases and discard incompatible bases. In the original E91 scheme, to produce a key size of N, we need to $6\,N$ original key size as there is a 33% chance that bases are compatible and half of the key is used to check the eavesdropper. The main Steps of the E91 scheme are as follows,

Step 1: Alice and Bob receive their entangled photon from a central source using any one of four maximal entangled states. Consider source generate EPR pair $|\psi\rangle = \frac{1}{\sqrt{2}}(|10\rangle + |01\rangle)$.

Step 2: Alice chooses one of the bases from $|0\rangle$, $|+\rangle$, and $|1\rangle$ to measure her received particle from entangled pair.

Step 3: Bob chooses one of the basis from $|+\rangle$, $|1\rangle$ and $|-\rangle$ to measure his received particle from entangled pair.

Step 4: Alice and Bob publicly compare their basis using an authentic classical channel. For the same basis measurement, Alice and Bob's results are perfectly anti-correlated for $|\psi\rangle$ entangled state. Inversion or Step 1's complement of Bob's string is equal to Alice's string. Alice and Bob carry out the post-processing, which involves any error detection and correction phase.

Step 5: For different Bases, Alice and Bob announce their result publicly for checking the performance of the channel using the Bell test. This test is used to ensure whether there is a potential eavesdropper present or not.

Different from the BB84 scheme, the E91 scheme uses the distribution and measurement of entangled photon pairs to generate the key. Second, in the BB84 scheme, quantum states are measured by randomly selected bases, while the E91 scheme requires Alice and Bob to select one of three specific base vectors to measure. Finally, the security of the E91 scheme uses Bell inequality [44] or CHSH inequality [45] to detect whether there is eavesdropping, according to whether the inequality is broken, and the extent of the damage can be inferred the possibility of eavesdropping. This is also different from the security analysis of BB84 scheme.

3.4.2 BBM92 Scheme

In 1992, Bennett et al. proposed a QKD scheme without Bell inequality, which is called the BBM92 scheme [12], by using the idea of entanglement and the BB84 scheme. The BBM92 scheme requires Alice and Bob to share a pair of entangled photon states, which is similar to the E91 scheme. When Alice and Bob choose the same basis to measure the quantum state, they always get the opposite result. At this point, Bob simply flips the bit in his hand to get the same key as Alice.

In the aspect of state measurement, the BBM92 scheme uses two sets of non-orthogonal basis vectors to measure, which is different from the E91 scheme using three sets of basis vectors to measure.

In terms of security analysis, BBM92 takes a similar approach to BB84, unlike E91, which uses Bell's inequality to test for the presence of eavesdroppers, that is, to compare the bit error rate when both sides choose Z basis or X basis and then use the data post-processing to remove the error code and the possible leaked information in the original key, so as to obtain a consistent security key. Thus, the BBM92 scheme can be thought of as the entangled version of the BB84 scheme.

3.4.3 MDI-QKD Scheme

In the actual QKD system, in addition to the photon source, detectors, and other devices will also bring certain security holes. There are many kinds of attacks against these security vulnerabilities, such as time-shifting attack, pseudo-state attack, detector blinding attack, etc. Considering that most of these attacks are aimed at the detector side, in 2012, Lo et al. proposed the first MDI-QKD scheme [14]. The MDI-QKD scheme not only solves the security vulnerability of the detector but also can be implemented under the existing conditions.

The steps of the MDI-QKD scheme are as follows:

Step 1: Alice and Bob prepare phase-random weak coherent pulses (WCP) and randomly generate one of four BB84 states, namely, $|0\rangle$, $|1\rangle$, $|+\rangle$, and $|-\rangle$, by Polarization modulator (Pol-M). Where $|+\rangle=\frac{1}{\sqrt{2}}(|0\rangle+|1\rangle)|+\rangle=\frac{1}{\sqrt{2}}(|0\rangle+|1\rangle)$, $|-\rangle=\frac{1}{\sqrt{2}}(|0\rangle-|1\rangle)$, $|0\rangle$ and $|+\rangle$ represent bit 0, $|1\rangle$ and $|-\rangle$ represent bit 1, $|0\rangle$ and $|1\rangle$ represent Z bases, $|+\rangle$, and $|-\rangle$ represent X bases. The intensity modulator is then used to randomly modulate the strength of the decoy state, and the quantum state is then sent to a third party via a quantum channel for measurement.

Step 2: A third party makes a Bell state measurement of the received quantum state and announces the successful result, with Alice and Bob retaining the coded information corresponding to the successful Bell state measurement. The forms of the four polarized Bell states are as follows:

$$|\phi^{+}\rangle=\frac{1}{\sqrt{2}}(|00\rangle+|11\rangle) \tag{3.1}$$

$$|\phi^{-}\rangle=\frac{1}{\sqrt{2}}(|00\rangle-|11\rangle) \tag{3.2}$$

$$|\psi^{+}\rangle=\frac{1}{\sqrt{2}}(|10\rangle+|01\rangle) \tag{3.3}$$

$$|\psi^{-}\rangle=\frac{1}{\sqrt{2}}(|10\rangle-|01\rangle) \tag{3.4}$$

Step 3: Alice and Bob perform base operations on the open channel, keeping the prepared base to select the same data and discarding the base to select different data. In order for Alice and Bob to form the correct related keys, Bob needs to manipulate his data based on the prepared base and the result of Bell State Measurement, as shown in Table 3.1. For example, when Alice makes $|0\rangle$, Bob makes $|1\rangle$, and the third-party measurement is $|\psi^{-}\rangle$, Bob needs to flip the bit 1 in his hand and take the same bit 0 as Alice as the original key.

Step 4: Alice and Bob repeat the above steps until they have enough of the original key.

Step 5: Alice and Bob retrieve part of the original key to detect the QBER. If the QBER exceeds the bit error rate threshold, it indicates eavesdropping and the communication is abandoned; otherwise, the remaining key is retained and the next step is taken.

Step 6: Alice and Bob perform error correction and private key amplification to obtain an unconditionally secure key.

Table 3.1 Bob's postselection operation

Alice and Bob's measurement basis	$\|\psi^{-}\rangle$	$\|\psi^{+}\rangle$
Z basis	Bit flip	Bit flip
X basis	Bit flip	No change

The proposed QKD scheme greatly promotes the implementation of the QKD scheme, making it possible to guarantee the security of the scheme without a perfect single-photon source and a detector with security holes, and the secure transmission distance of the traditional QKD system is more than doubled.

3.5 Continuous-Variable Quantum Key Distribution Scheme

In CV-QKD, the carrier of information no longer exists in the form of single-photon polarization but is the continuous orthogonal component of the light field in the phase space. The CV-QKD based on the coherent state uses the off-the-shelf weak laser as the light source and loads the information into the coherent state by means of amplitude modulation and phase modulation. CV-QKD is an effective alternative to DV-QKD because of its easy realization and high detection rate.

In 2002, Grosshans and Grangier proposed the Gauss modulation coherent state quantum cryptography scheme, called GG02 scheme [25], the main flow is as follows:

Step 1: Alice randomly generates two Gaussian distribution random numbers whose mean is 0 and variance is V_A and then generates coherent states based on them. After repeating this process N times, we get N coherent states, and then Alice sends them to Bob via the quantum channel.

Step 2: After Bob receives these coherent states, he randomly selects an orthogonal component to carry out zero-difference detection, at this point he gets N measurements;

Step 3: Bob announces to Alice the basis of his randomly selected measurements. After Alice receives this set of measurement bases, the corresponding measurement results are retained, and the remaining data are discarded.

Step 4: Alice randomly selects some preserved data for parameter estimation, and calculates the noise value from the data, the noise value is used to decide whether to terminate the communication.

Step 5: Alice and Bob process the remaining data after Step 4, including data coordination and security enhancement, and finally extract the security key.

3.6 Semi-quantum Key Distribution Scheme

The BKM07 scheme is the first semi-quantum key distribution scheme proposed by Boyer [34]. This section describes them in detail.

In the BKM07 scheme, Alice is a quantum party with strong quantum operation ability, and Bob is a semi-quantum party with limited operation ability. Secure key distribution between the two can be achieved through the following scheme Steps.

Step 1: Alice randomly chooses to prepare N qubits with Z or X basis and sends them to Bob.

Step 2: After receiving the qubit, Bob randomly selects either the measurement resend operation (SIFT) or the return directly operation (CTRL). In the measurement resend operation, Bob makes a Z basis measurement of the qubit and produces a particle that matches the measurement and sends it back to Alice. In the return directly operation, Bob does nothing with the particle and returns it directly to Alice.

Step 3: After receiving the particle sent by Bob, Alice will select the correct measurement basis to make quantum measurement according to the initial preparation basis.

Step 4: Alice and Bob publish their operation information separately. Alice announces the location of the Z basis particle at the time of the initial transmission, and Bob announces the location of the particle at which the measurement is repeated.

According to the random selection of Alice's initial preparative basis and Bob's particle operation, the particle is in one of the following four states:

1. First, the preparative basis selected by Alice is Z basis, and the particle operation selected by Bob is CTRL, these particles are denoted as Z-CTRL particles.
2. Second, Alice chooses a Z basis, Bob chooses a particle operation SIFT, and denotes these particles as Z-SIFT particles.
3. Third, Alice chooses an X basis, Bob's chosen particle operation is CTRL, and these particles are remembered as X-CTRL particles.
4. Fourth, Alice's chosen preparation base is X basis, and Bob's chosen particle operation is SIFT, and these particles are remembered as X-SIFT particles.

Step 5: Alice checks the error rates of the Z-CTRL and X-CTRL particles. If the particle error rate is higher than the predefined threshold, the scheme is terminated.

Step 6: Alice randomly selects n TEST particles from the Z-SIFT particles. Alice publishes the selected particle, and Bob publishes the measurements for the TEST particle. Alice checks the error rate on the TEST qubit, and if it is above a predefined threshold, the scheme aborts. For X-SIFT particles, nothing is done.

Step 7: Alice and Bob select the remaining Z-sift particles as the sequence of INFO information. Alice and Bob extract the final key from the INFO information sequence using error-correcting encoding and privacy enhancement.

Boyer also proves the robustness of the BKM07 scheme in detail. Based on secure quantum cryptography schemes, quantum, and semi-quantum parties can successfully obtain the key, which can play an important role in the subsequent series of semi-quantum communications.

3.7 Brief Summary

This chapter provides a detailed study and classification of quantum key distribution and a brief introduction to the typical schemes in each classification. The quantum key distribution is the most fundamental and important part of the field of Secure Quantum Communications, it can provide strong theoretical support and experimental basis for other branches in the field of Secure Quantum Communication.

References

1. C.H. Bennett, G. Brassard Quantum *Cryptography: Public Key Distribution and Coin Tossing*, in Proceedings of the IEEE International Conference on Computers, Systems, and Signal Processing, Bangalore, India (IEEE, New York, December 1984), pp. 175–179
2. P.W. Shor, J. Preskill, Simple proof of security of the BB84 quantum key distribution scheme. Phys. Rev. Lett. **85**, 441–444 (2000)
3. C. Bennett, Quantum cryptography using any two non-orthogonal states. Phys. Rev. Lett. **68**, 3121–3124 (1992)
4. V. Scarani, A. Acin, G. Ribordy, N. Gisin, Quantum cryptography schemes robust against photon number splitting attacks for weak laser pulse implementations. Phys. Rev. Lett. **92**, 057901 (2004)
5. W.Y. Hwang, Quantum key distribution with high loss: Toward global secure communication. Phys. Rev. Lett. **91**, 057901 (2003)
6. X.B. Wang, Beating the photon-number-splitting attack in practical quantum cryptography. Phys. Rev. Lett. **94**, 230503 (2005)
7. X.F. Ma, B. Qi, Y. Zh, et al., Practical decoy state for quantum key distribution. Phys. Rev. A **72**, 012326 (2005)
8. X.F. Ma, C.H. Fred Fung, F. Dupuis, et al., Decoy-state quantum key distribution with two-way classical post- processing. Phys. Rev. A **74**, 032330 (2006)
9. W. Mauerer, C. Silberhorn, Quantum key distribution with passive decoy stateselection. Phys. Rev. A **75**, 050305 (2007)
10. Z.Q. Yin, Y.B. Zhao, Z.W. Zhou, et al., Decoy states for quantum key distribution based on decoherence-free subspaces. Phys. Rev. A **77**, 062326 (2008)
11. A. Ekert, Quantum cryptography based on Bell's theorem. Phys. Rev. Lett. **67**, 661–663 (1991)
12. C.H. Bennett, G. Brassard, N.D. Mermin, Quantum cryptography without Bell's theorem. Phys. Rev. Lett. **68**, 557–559 (1992)
13. A. Acin, N. Brunner, N. Gisin, S. Massar, S. Pironio, V. Scarani, Device-independent security of quantum cryptography against collective attacks. Phys. Rev. Lett. **98**, 230501-1–230501-4 (2007)
14. H.K. Lo, M. Curty, B. Qi, Measurement-device-independent quantum key distribution. Phys. Rev. Lett. **108**(13), 130503 (2012)
15. F. Xu, M. Curty, B. Qi, H.K. Lo, Measurement-device independent quantum cryptography. IEEE J. Sel. Top Quantum Electron. **21**(3), 148–158 (2015)
16. Z. Tang, K. Wei, O. Bedroya, L. Qian, H.K. Lo, Experimental measurement-device-independent quantum key distribution with imperfect sources. Phys. Rev. A **93**, 042308 (2016)
17. R. Valivarthi, P. Umesh, C. John, K.A. Owen, V.B. Verma, S.W. Nam, D. Oblak, Q. Zhou, W. Tittel, Measurement-device independent quantum key distribution coexisting with classical communication. Quantum Sci. Technol. **4**(4), 045002 (2019)
18. F. Xu, M. Curty, B. Qi, H.K. Lo, Practical aspects of measurement-device-independent quantum key distribution. New J. Phys. **15**, 113007 (2013)

19. G.L. Roberts, M. Lucamarini, Z.L. Yuan, J.F. Dynes, L.C. Comandar, A.W. Sharpe, A.J. Shields, M. Curty, I.V. Puthoor, E. Andersson, Experimental measurement-device-independent quantum digital signatures. Nat. Commun. **8**, 1098 (2017)
20. X.L. Hu, Y. Cao, Z.W. Yu, X.B. Wang, Measurement-device independent quantum key distribution over asymmetric channel and unstable channel. Sci. Rep. **8**, 17634 (2018)
21. Y. Qiao, G. Wang, Z. Li, B. Xu, H. Guo, Monitoring an untrusted light source with single-photon detectors in measurement-device-independent quantum key distribution. Phys. Rev. A **99**(5), 052302 (2019)
22. Z.X. Cui, W. Zhong, L. Zhou, Y.B. Sheng, Measurement device-independent quantum key distribution with hyperencoding. Sci. China Phys. Mech. Astron. **62**, 110311 (2019)
23. L. Dellantonio, A.S. Sorensen, D. Bacco, High-dimensional measurement-device-independent quantum key distribution on two-dimensional subspaces. Phys. Rev. A **98**, 062301 (2018)
24. M. Pawlowski, N. Brunner, Semi-device-independent security of one-way quantum key distribution. Phys. Rev. A **84**(1), 010302 (2011)
25. F. Grosshans, P. Grangier, Continuous variable quantum cryptography using coherent states. Phys. Rev. Lett. **88**, 057902 (2002)
26. F. Grosshans, G.V. Assche, J. Wenger, R. Brouri, N.J. Cerf, P. Grangier, Quantum key distribution using gaussian-modulated coherent states. Nature **421**, 238–241 (2003)
27. J. Lodewyck, T. Debuisschert, R. Tualle-Brouri, P. Grangier, Controlling excess noise in fiber optics continuous variables quantum key distribution. Phys. Rev. A **72**, 050303 (2005)
28. C Griffet, From discrete-to continuous-variable schemes for quantum key distribution, Master Thesis, Universite Libre De Bruxelles (2019)
29. T.C. Ralph, Continuous variable quantum cryptography. Phys. Rev. A **61**, 010303 (1999)
30. M.D. Reid, Quantum cryptography with a predetermined key, using continuous variable Einstein–Podolsky–Rosen correlations. Phys. Rev. A **62**(6), 062308-1–062308-6 (2000)
31. M. Hillery, Quantum cryptography with squeezed states. Phys. Rev. A **61**, 02230 (2000)
32. P. Papanastasiou, S. Pirandola, Continuous-variable quantum cryptography with discrete alphabets: composable security under collective gaussian attacks, pp. 1–6. arXiv:1912.11418 (2020)
33. Bacco D, Christensen J B, Castaneda M, Ding Y et al.Two-dimensional distributed-phase-reference scheme for quantum key distribution [J]. Sci. Rep., 2016, 6: 36756
34. M. Boyer, D. Kenigsberg, T. Mor, Quantum key distribution with classical Bob. Phys. Rev. Lett. **99**(14), 140501 (2007)
35. M. Boyer, R. Gelles, D. Kenigsberg, T. Mor, Semi quantum key distribution. Phys. Rev. A **79**, 032341 (2009)
36. W.O. Krawec, Restricted attacks on semi-quantum key distribution schemes. Quantum Inf. Process **13**, 2417–2436 (2014)
37. W.O. Krawec, Mediated semi-quantum key distribution. Phys. Rev. A **91**, 032323 (2015)
38. X. Zou, D. Qiu, L. Li, L. Wu, L. Li, Semi quantum-key distribution using less than four quantum states. Phys. Rev. A **79**, 0522312 (2009)
39. H. Lu, Q.Y. Cai, Quantum key distribution with classical Alice. Int. J. Quantum Inf. **6**(6), 1195–1202 (2008)
40. K.F. Yu, C.W. Yang, C.H. Liao, T. Hwang, Authenticated semiquantum key distribution scheme using Bell states. Quantum Inf. Process **13**, 1457–1465 (2014)
41. Y.P. Luo, T. Hwang, Authenticated semi-quantum direct communication schemes using Bell states. Quantum Inf. Process **15**, 947–958 (2015)
42. X. Zou, D. Qiu, S. Zhang, P. Mateus, Semi quantum key distribution without invoking the classical party's measurement capability. Quantum Inf. Process **14**, 2981–2996 (2015)
43. M. Boyer, M. Katz, R. Liss, T. Mor, Experimentally feasible scheme for semi quantum key distribution. Phys. Rev. A **96**(6), 062335-1–062335-6 (2017)
44. J.S. Bell, On the Einstein-Podolsky-Rosen paradox[J]. Physics **1**, 195 (1964)
45. J.F. Clauser, M.A. Horne, A. Shimony, et al., Proposed experiment to test local hidden variable theories[J]. Phys. Rev. Lett. **23**(15), 880 (1969)

Chapter 4
Quantum Secret Sharing

Abstract Quantum secret sharing is the extension and development of classical secret sharing in the area of quantum mechanics. It provides a new way and idea for the secure transmission of information. Using quantum states as information carrier, quantum secret sharing technology can realize quantum network identity and large-capacity information transmission. This technology has the advantages of high security, high reliability, saving resources, and reducing the complexity of communication. At present, quantum secret sharing technology has occupied a leading position in the quantum communication technology, which is of great significance not only in the theoretical research. Once realized, it will have an inestimable application prospect in the fields of quantum communication and quantum computing.

4.1 Introduction

Secret sharing, one of the important cryptographic primitives in the field of secure multiparty computation, was invented independently by Shamir [1] and Blakley in 1979 [2], which aims to protect highly confidential information. Without any question, it has been naturally extended into quantum versions. As an important branch of quantum cryptography, quantum secret sharing (QSS) is growing rapidly. It can manage the key effectively, and its basic principle is to divide the secret into several parts so that each participant can obtain the secret only when they work together; otherwise, no single participant could successfully deduce the secret information.

With the development of quantum key distribution (QKD), based on the idea that it can enable two remote users to achieve unconditional secure communication, the researchers extended secret sharing to quantum scenarios, known as quantum secret sharing. It divides the shared information into several parts, and no single part can effectively get the information. Only through cooperation between the relevant parts can the original information be finally obtained, and in the process, if someone eavesdrops or cheats among the participants, the behaviors will be discovered.

R.-G. Zhou et al., *Design and Analysis of Secure Quantum Communication Schemes*,
https://doi.org/10.1007/978-3-031-78428-6_4

Compared to classical secret sharing, QSS can share not only classical information but also unknown quantum information.

QSS is different from QKD. It is not the security that can't be guaranteed by classical communication, but the unknown quantum state can be shared safely by QSS. QSS will play an important role in distributed quantum computing.

4.2 Principle and Research Progress of Quantum Secret Sharing

QSS is an extension of classical cryptography by quantum mechanics. It uses the characteristics of quantum mechanics to enhance the security and efficiency of secret sharing. QSS mainly consists of secret distributor, agent, access structure, sub-secret space, secret space, secret distribution algorithm, and secret recovery algorithm.

QSS combines the principles of classical secret sharing and quantum mechanics. Its core idea is that the sender will first break the message into several pieces and send it to a number of legitimate participants to manage the secret message together. The original secret information cannot be recovered by any one legal participant alone, and the complete secret information can be recovered only when several legal participants cooperate together, so as to achieve the goal of protecting the security of secret information.

Unlike classical secret sharing, the shared information in a quantum secret sharing (QSS) scheme can be classical or quantum. Accordingly, quantum secret sharing schemes are usually divided into two different categories: (1) when the shared information is an unknown quantum state, it is called quantum state sharing (QSTS); (2) it is generally named quantum secret sharing (QSS) when it shares the classical information. Up to now, the researches on quantum state sharing and quantum secret sharing have been extensively studied theoretically and also verified via experiments.

Actually, quantum state sharing is a controlled multiparty quantum teleportation process. It plays an important role in building quantum networks and is an essential part of quantum communication. Nowadays, there exist fruitful quantum state sharing schemes [3–8], and researchers have begun to pay more attention to the efficiency of quantum state sharing schemes [9–13]. Besides, cryptanalysis is an important and interesting work in cryptography, which finds potential loopholes and tries to overcome security flaws [14, 15]

Quantum secret sharing means that a dealer distributes the classical secret among several agents based on quantum mechanics in such a way that the secret can be reconstructed only when the legitimate agents cooperate together. The pioneering work of quantum secret sharing was proposed by Hillery [16] and Cleve [17] in 1999. In the past two decades, quantum secret sharing has been extensively studied theoretically in ideal situations with noise-free quantum channel [18–33], such as multiparty to multiparty quantum secret sharing schemes [18, 19], multiparty quantum secret sharing schemes [16, 20, 21, 34], threshold quantum secret sharing schemes [22–24], access structure-based quantum secret sharing schemes [25, 26], and dynamic quantum secret sharing (DQSS) schemes [27–33, 35–37]. It is worth noting that all the abovementioned DQSS schemes [27–33, 35–37] use a great deal of detecting particles inserted into quantum channels for eavesdropping check. However, the detecting particles increase dramatically as the number of agents increases, which makes their schemes have lower qubit efficiency. Furthermore, most dynamic quantum secret sharing schemes [27–33, 37] can only share the random bits instead of predetermined bits in [36] and random dits in [35]. What's more important is that all the existing dynamic quantum secret sharing schemes lack the secret sharing rate analysis in noise environments because the transmitted particles in quantum channel will be inevitably interacted with outer environment and affected by channel noises.

4.3 Quantum State Sharing Scheme

4.3.1 MA's Scheme

Suppose Alice is a dealer who possesses an unknown two-qubit quantum state described as

$$|\Phi\rangle_{xy} = (\alpha|00\rangle + \beta|01\rangle + \gamma|10\rangle + \delta|11\rangle)_{xy}, \tag{4.1}$$

where α, β, γ, δ are four complex numbers and satisfy the normalization condition $|\alpha|^2 + |\beta|^2 + |\gamma|^2 + |\delta|^2 = 1$.

Assume Bob_1, Bob_2, …, Bob_{n-1} are $(n-1)$ agents of Alice, and she wants to distribute her unknown two-qubit state $|\Phi\rangle_{xy}$ among these agents. Without loss of generality, let's suppose that Bob_1 is the secret state receiver, and other $(n-2)$ agents Bob_2, …, Bob_{n-1} are the controllers. Then MA's scheme [38] can be explicated as follows:

Step 1: Alice prepares n particle-pairs $(1,2), (3,4), \ldots, (2n-1, 2n)$. These pairs are all in Bell state $|\Phi^+\rangle = \frac{1}{\sqrt{2}}(|00\rangle + |11\rangle)$.

Step 2: Alice only retains the first and last particles $(1, 2n)$ and then sends the remaining two particles $(2,3)$ to Bob_1, $(4,5)$ to Bob_2, …, $(2n-2, 2n-1)$ to Bob_{n-1}. At the same time, Bob_1, Bob_2, …, Bob_{n-1} publicly announce that they have received their two particles.

Step 3: Alice performs BM on particles $(x, 1)$ and $(y, 2n)$ and then sends the outcome to Bob_1 by classical channel. So far, the unknown two-qubit state $|\Phi\rangle_{xy}$ that Alice wants to share is collapsed to quantum system consisting of particles $(2, 3, 4, \ldots, 2n-1)$.

Step 4: Bob_2, …, Bob_{n-1} implement BM on their own two particles $(4,5)$, …, $(2n-2,\ 2n-1)$ separately and send their corresponding measurement results to Bob_1 by classical channel.

Step 5: On the basis of Alice's, Bob_2's, ..., Bob_{n-2}'s and Bob_{n-1}'s measurement results, Bob_1 reestablishes the unknown two-qubit state $|\Phi\rangle_{xy}$ by using appropriate local unitary transformations on his two particles $(2, 3)$.

From the review above, now let's give a detailed description of the collusion attack in this n-party QSTS scheme as follows:

Since Alice distributes each two particles to all agents simultaneously, after all agents have received their two particles, the first agent Bob_1 and the last agent Bob_{n-1} can interchange one of their particles to implement the collusion attack. More precisely, there exist two different situations described as

Situation 1: Bob_1 sends his particle 3 to Bob_{n-1}, while in the meantime, Bob_{n-1} sends his particle $2n - 1$ to Bob_1; thus it leads to Bob_1 holding two particles $(2, 2n - 1)$ and Bob_{n-1} holding two particles $(3, 2n - 2)$;

Situation 2: Bob_1 sends his particle 2 to Bob_{n-1}, while in the meantime, Bob_{n-1} sends his particle $2n - 2$ to Bob_1, which leads to Bob_1 holding two particles $(3, 2n - 2)$ and Bob_{n-1} holding two particles $(2, 2n - 1)$.

Hence, in the above two different situations, when Alice performs the BM operation on particles $(x, 1)$ and $(y, 2n)$, the unknown two-qubit state $|\Phi\rangle_{xy}$ are conveyed to two particles $(2, 2n - 1)$ directly based on entanglement swapping principle of Bell states [35], and it shows no correlation to the remaining particles $(3, 4, \ldots, 2n - 3, 2n - 2)$. Thus, based on Alice's BM results, either Bob_1 (see Situation 1) or Bob_{n-1} (see Situation 2) can recover Alice's two-qubit state $|\Phi\rangle_{xy}$ on their two particles without the help of other $(n - 3)$ agents Bob_2, ..., Bob_{n-2}.

From the above analysis, we have proved that MA's scheme has the security flaw of a collusion attack, that is, the first and last agents can conspire with each other to steal the dealer's secret state without the help of other agents.

4.3.2 Two IQSTS Schemes

In view of the specific security loopholes that existed in MA's scheme [38], this section proposes two improved multiparty quantum state sharing (IQSTS) schemes for sharing an arbitrary unknown two-qubit state [39], which are briefly described as follows:

1. **First Improved Scheme**
   Instead of distributing each two particles came from the two neighbor Bell states to all agents simultaneously, each two particles are distributed sequentially after each agent has implemented a BM operation on two received particles. Hence, based on the principle of entanglement swapping, the secret state shared by the dealer has been built the specific correlation to all agents' BM results.
2. **Second Improved Scheme**
   Instead of distributing each two particles came from the two neighbor Bell states to all agents simultaneously, first, two particles in each Bell states are divided and formed into two ordered sequences, and then all particles in the first

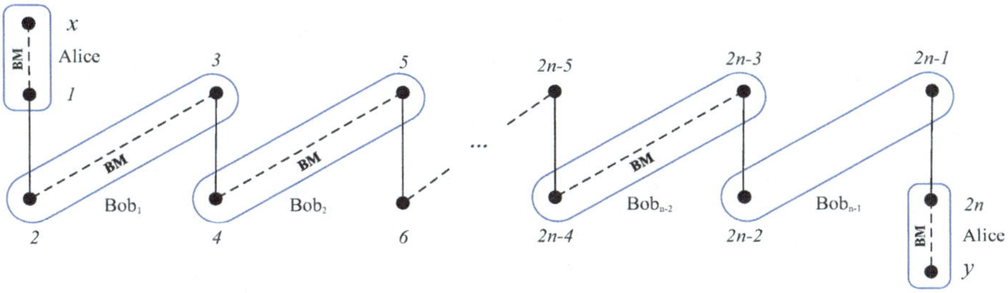

Fig. 4.1 The first IQSTS scheme among ***n***-party

sequence are disturbed by the private cyclic permutation. Following that, the dealer distributes two particles in the same position came from two ordered sequences to all agents simultaneously. Then based on the entanglement swapping of Bell states, the dealer and all agents can realize the quantum state sharing scheme among all agents safely.

On the basis of MA's scheme, two multiparty IQSTS schemes based on Bell states were proposed that can address aforementioned collusion attack.

4.3.2.1 First IQSTS Scheme

Let's also assume that Alice is a dealer who wants to share the arbitrary unknown two-qubit state $|\Phi\rangle_{xy}$ with $(n-1)$ agents $\text{Bob}_1, \text{Bob}_2, \ldots, \text{Bob}_{n-1}$, in which Bob_{n-1} is the secret receiver and other $(n-2)$ agents are the controllers. Figure 4.1 illustrates a simplified diagram of first improved scheme, and the detailed process is explained as follow:

Step 1: Alice prepares n particle-pairs $(1,2)$, $(3,4)$, ..., $(2n-1,2n)$, which are all in the Bell state $|\Phi^+\rangle = \frac{1}{\sqrt{2}}(|00\rangle + |11\rangle)$.

Step 2: Alice sends particles $(2,3)$ to Bob_1. When Bob_1 receives these two particles $(2,3)$, he performs the BM on particles $(2,3)$ and sends the measurement result to Alice via an authenticated channel. After receiving the measurement result from Bob_1, Alice performs the BM on her two particles $(1,4)$.

Step 3: Based on her measurement result, Alice can detect whether Bob makes a BM on his two particles $(2,3)$ according to the entanglement swapping principle. If Bob_1 had performed the BM, then Alice prepares a fresh two-qubit Bell state as he measured denoted as a new EPR pair $(1,4)$.

Step 4: Alice and the rest of the agents (except the receiver Bob_{n-1}) perform similar operations as Bob_1 does, that is, Alice sends two particles of $(2i, 2i+1)$ to Bob_i if and only if she has verified that Bob_{i-1} performed BM on his two particles $(2i-2, 2i-1)$.

Step 5: Alice performs BM operations on two particles $(x,1)$ and $(y,2n)$. After obtaining the successful measurement results, Alice sends two particles $(2n-2, 2n-1)$ to Bob_{n-1}.

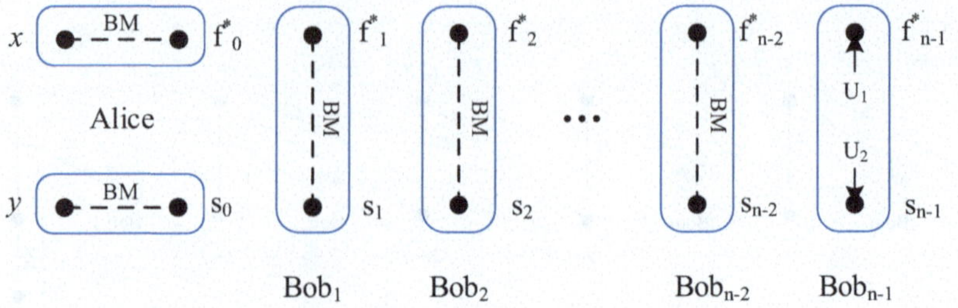

Fig. 4.2 The second IQSTS scheme among ***n***-party

Step 6: When Bob_{n-1} receives two particles $(2n-2, 2n-1)$ from Alice, he sends a confirmation to Alice. Then Alice sends the measurement results to Bob_{n-1} via an authenticated classical channel. Hence, based on Alice's and other agents Bob_1's, Bob_2's, ..., Bob_{n-2}'s measurement results, Bob_{n-1} implements the appropriate unitary transformation on his two particles $(2n-2, 2n-1)$ to recover Alice's secret state $|\Phi\rangle_{xy}$.

4.3.2.2 Second IQSTS Scheme

Instead of distributing two particles to each agent in order, this subsection proposes the second IQSTS scheme based on Bell states of sharing the arbitrary unknown two-qubit state $|\Phi\rangle_{xy}$ among n-party. Similar to the first IQSTS scheme, Fig. 4.2 illustrates a simplified diagram of second IQSTS scheme, in which Alice is the secret holder, Bob_{n-1} is the secret receiver, and other $(n-2)$ agents are the controllers. The detailed steps are described as follow;

Step 1: Alice prepares n EPR pairs $(1,2)$, $(3,4)$, ..., $(2n-1, 2n)$, which are all in the Bell state $|\Phi^+\rangle = \frac{1}{\sqrt{2}}(|00\rangle + |11\rangle)$, and then divides all these particles into two ordered sequences. That is, Alice takes the first particle of each EPR pair, particles $(1, 3, \ldots, 2n-1)$, to constitute the first sequence $F = (f_0, f_1, \cdots, f_{n-1}) = \{f_j\}_{j=0}^{n-1}$ and all the remaining particles, particles $(2, 4, \ldots, 2n)$, to constitute the second sequence $S = (s_0, s_1, \cdots, s_{n-1}) = \{s_j\}_{j=0}^{n-1}$.

Step 2: Alice performs a right cyclic permutation via different delay lines on first sequence F to form a new encrypted ordered sequence F^*, described as

$$F^* = \begin{cases} P_i F, \ i = 1, 2, \ldots, n-1, & \text{if } n \text{ is odd} \\ P_i F, \ i = 1, 2, \ldots, n-1, i \neq n/2, & \text{if } n \text{ is even} \end{cases} \tag{4.2}$$
$$P_i F = \{f_{(j+i) \bmod n}\}_{j=0}^{n-1} = \{f_j^*\}_{j=0}^{n-1}$$

This permutation rule defines a new ordered sequence. Take $i = 2$ and $n = 5$ for a simple example; the encrypted sequence F^* is

$$F^* = P_2F = \left\{f_{(j+2) \bmod n}\right\}_{j=0}^{4} = \{f_3, f_4, f_0, f_1, f_2\} = \{f_0^*, f_1^*, f_2^*, f_3^*, f_4^*\} = \left\{f_j^*\right\}_{j=0}^{4} \tag{4.3}$$

Thus, when Alice chooses one of the permutation regulations defined in Eq. 4.2 randomly to encrypte the first sequence F, she obtains a new ordered sequence: $F^* = \left\{f_j^*\right\}_{j=0}^{n-1} = \{f_0^*, f_1^*, \cdots, f_{n-1}^*\}$

Step 3: Alice sends two particles (f_k^*, s_k) to Bob_k, for $k = 1, 2, \ldots, n - 1$, and holds two particles (f_0^*, s_0) in local. When Bob_k receives two particles, he sends a confirmation to Alice via classical channel.

Step 4: Alice performs BM on particles (x, f_0^*) and (y, s_0) and Bob_k, for $k = 1, 2, \ldots, n - 2$, performs BM on their two particles (f_k^*, s_k). Then Alice sends his measurement results as well as the permutation rule to $\text{Bob}_{\text{n}-1}$ via an authenticated classical channel, and Bob_k ($k = 1, 2, \ldots, n - 2$) sends their measurement results to Bob_{n-1} via an authenticated classical channel either.

Step 5: After receivingd Alice's and Bob_k's ($k = 1, 2, \ldots, n - 2$) information, Bob_{n-1} performs the appropriate unitary transformation on his two particles (f_{n-1}^*, s_{n-1}) to recover Alice's two-qubit state $|\Phi\rangle_{xy}$.

4.3.2.3 Case Study

To illustrate the feasibility of two multiparty IQSTS schemes, this subsection takes four-party QSTS scheme of each IQSTS scheme as the simple example to explain the proposed IQSTS schemes in detail.

A. **First Case**

Suppose that Alice wants to send a two-qubit quantum state $|\Phi\rangle_{xy} = (\alpha|00\rangle + \beta|01\rangle + \gamma|10\rangle + \delta|11\rangle)_{xy}$ to Bob_3 with the help of Bob_1 and Bob_2. The simple diagram of first IQSTS scheme with four-party is given in Fig. 4.3, and the scheme can be described in detail as follows:

Step 1: Alice prepares 4 particle-pairs $(1, 2)$, $(3, 4)$, $(5, 6)$, $(7, 8)$, which are all in the Bell state $|\Phi^+\rangle = \frac{1}{\sqrt{2}}(|00\rangle + |11\rangle)$.

Step 2: Alice sends particles $(2, 3)$ to Bob_1. Then Bob_1 performs BM operation on particles $(2, 3)$ once he receives it and sends the corresponding measurement results to Alice via an authenticated channel. After receiving the measurement results from Bob_1, Alice performs the BM on her two particles $(1, 4)$.

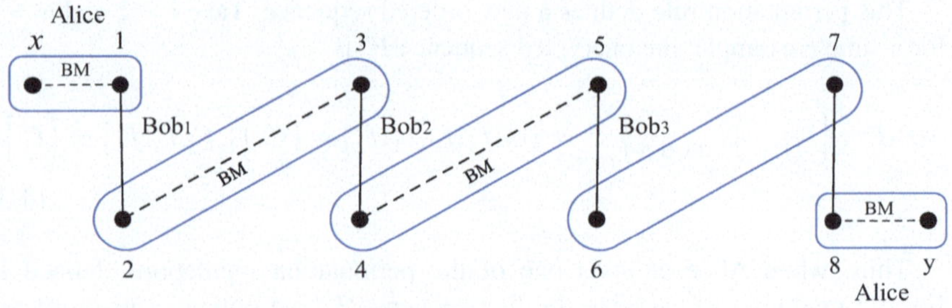

Fig. 4.3 The first IQSTS scheme of two-qubit state sharing among four-party

Step 3: Based on her own measurement result in Step 2, Alice can further detect whether Bob_1 conducts a BM on particles $(2, 3)$ according to the entanglement swapping principle. Once Alice confirms that Bob has performed BM, then she prepares a new two-qubit Bell state same as her measured denoted as the new EPR pair $(1, 4)$.

Step 4: Alice first sends two particles $(4, 5)$ to Bob_2 and then Bob_2 performs BM on two particles $(4, 5)$. After Alice has confirmed that Bob_2 truly performed the BM on his two particles $(4, 5)$, she prepares a fresh Bell state as her measured denoted as particles $(1, 6)$.

Suppose the outcomes of Bob_1 and Bob_2 are $|\Phi^-\rangle_{23} = \frac{1}{\sqrt{2}}(|00\rangle - |11\rangle)$ and $|\Phi^+\rangle_{45} = \frac{1}{\sqrt{2}}(|00\rangle + |11\rangle)$. Then, the residual system consisting of particles 1, 6, 7, 8, x, y may collapse to $|X\rangle_{1678xy}$:

$$\begin{aligned}|X\rangle_{x1y867} &= (\alpha|000000\rangle + \beta|001101\rangle - \gamma|110010\rangle - \delta|111111\rangle)_{x1y867} \\ &= \alpha|00\rangle_{x1}|00\rangle_{y8}|00\rangle_{67} + \beta|00\rangle_{x1}|11\rangle_{y8}|01\rangle_{67} - \gamma|11\rangle_{x1}|00\rangle_{y8}|10\rangle_{67} \\ &\quad - \delta|11\rangle_{x1}|11\rangle_{y8}|11\rangle_{67}\end{aligned} \tag{4.4}$$

Hence, the secret of $|\Phi\rangle_{xy} = (\alpha|00\rangle + \beta|01\rangle + \gamma|10\rangle + \delta|11\rangle)_{xy}$ is transferred to the subsystem $|X\rangle_{1678xy}$.

Step 5: Alice implements BM on particle pairs $(x, 1)$ and $(y, 8)$ separately and sends particles $(6, 7)$ to Bob_3.

Suppose the outcomes of Alice are $|\Phi^+\rangle_{x1} = \frac{1}{\sqrt{2}}(|00\rangle + |11\rangle)$ and $|\Phi^-\rangle_{y8} = \frac{1}{\sqrt{2}}(|00\rangle - |11\rangle)$. Then the residual system consisting of particles 6, 7 may collapse to $|X\rangle_{67} = \alpha|00\rangle - \beta|01\rangle - \gamma|10\rangle + \delta|11\rangle$.

Step 6: Reconstruct the initial state $|\Phi\rangle_{xy}$. Bob_3 can reconstruct Alice's secret quantum state by implementing the appropriate unitary transformation on his particles $(6, 7)$ according to the outcomes of Alice and the other two participants Bob_1 and Bob_2.

For the particle state $|X\rangle_{67}$, Bob_3 can perform local unitary transformation $\sigma_Z \otimes \sigma_Z$ to recover two-qubit state $|\Phi\rangle_{xy}$ as follows:

$$\begin{aligned}\left(\sigma_Z \otimes \sigma_Z\right) \cdot |X\rangle_{67} &= \left(\sigma_Z \otimes \sigma_Z\right) \cdot (\alpha|00\rangle - \beta|01\rangle - \gamma|10\rangle + \delta|11\rangle) \\ &= (\alpha|00\rangle + \beta|01\rangle + \gamma|10\rangle + \delta|11\rangle) = |\Phi\rangle_{xy}.\end{aligned} \tag{4.5}$$

Through the above six steps, Alice successfully shares an arbitrary unknown two-qubit state $|\Phi\rangle_{xy}$ among three agents Bob_1, Bob_2, and Bob_3. Table 4.1 (as shown in *Appendix*) illustrates the specific unitary transformation implemented by Bob_3 that corresponds to the measurement results of Alice, Bob_1, and Bob_2.

B. **Second Case**

Similar to the first IQSTS scheme, Fig. 4.4 illustrates a simplified diagram of the second IQSTS scheme, in which Alice is the secret holder, Bob_3 is the secret receiver, and the other *two* agents Bob_1 and Bob_2 are the controllers. The detailed steps are described as:

Step 1: Alice first prepares 4 EPR pairs $(1,2)$, $(3,4)$, $(5,6)$, $(7,8)$, which are all in the Bell state $|\Phi^+\rangle = \frac{1}{\sqrt{2}}(|00\rangle + |11\rangle)$. Then she divides all these particles into two ordered sequences, that is, Alice picks out the first particle in each EPR pair, the particles $(1,3,5,7)$, to form the first new particle sequence $F = (f_0, f_1, f_2, f_3) = \{f_j\}_{j=0}^{3} = \{1,3,5,7\}$. All the remaining particles form a new sequence$S = (s_0, s_1, s_2, s_3) = \{s_j\}_{j=0}^{3} = \{2,4,6,8\}$.

Step 2: Alice performs a right-shift circular permutation operation on the first particle sequence that only she knows about and forms a new encrypted sequence F^* denoted as

$$F^* = P_1 F = \left\{f_{(j+1) \bmod 4}\right\}_{j=0}^{3} = \{f_3, f_0, f_1, f_2\} = \{f_0^*, f_1^*, f_2^*, f_3^*\} = \left\{f_j^*\right\}_{j=0}^{3} \tag{4.6}$$

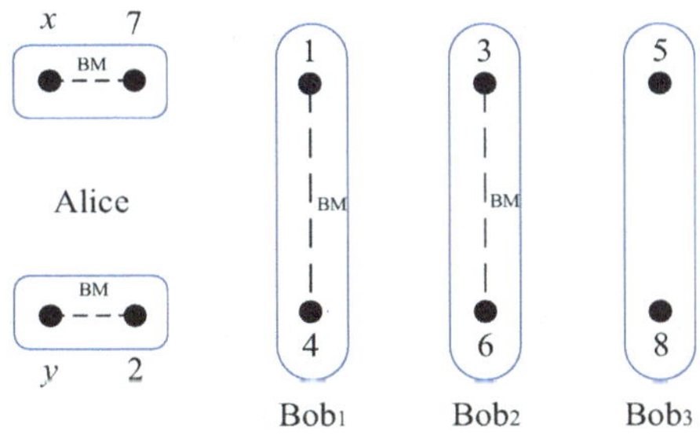

Fig. 4.4 The second IQSTS scheme of two-qubit state sharing among four-party

Go through this process, Alice now gets a fresh sequence of particles: $F^* = \{f_0^*, f_1^*, f_2^*, f_3^*\} = \{7, 1, 3, 5\}$.

Step 3: Alice sends two particles (f_k^*, s_k) to each participant (expect Alice) Bob_k simultaneously, for $k = 1, 2, 3$, and holds two particles (f_0^*, s_0) in local. Bob_k will send an acknowledgement to Alice via classical channel after he has received two particles belong to himself.

Step 4: Alice performs BM operations on her own particles $((x, f_0^*), (y, s_0))$ and Bob_k's ($k = 1, 2$) particles (f_k^*, s_k), respectively. Then Alice and Bob_k send their measurement results to Bob_3 via an authenticated classical channel. In addition, Alice will tell Bob_3 her rules of permutation together.

Suppose that Alice's BM results respectively are $|\Psi^-\rangle_{x7} = \frac{1}{\sqrt{2}}(|01\rangle - |10\rangle)$ and $|\Psi^+\rangle_{y2} = \frac{1}{\sqrt{2}}(|01\rangle + |10\rangle)$. Hence, the residual system consisting of particles 1, 3, 4, 5, 6, 8 may collapse to $|X\rangle_{134568}$.

$$
\begin{aligned}
|X\rangle_{143685} &= (\alpha|111111\rangle + \beta|000010\rangle - \gamma|111101\rangle - \delta|000000\rangle)_{143685} \\
&= \alpha|11\rangle_{14}|11\rangle_{36}|11\rangle_{85} + \beta|00\rangle_{14}|00\rangle_{36}|10\rangle_{85} - \gamma|11\rangle_{14}|11\rangle_{36}|01\rangle_{85} \\
&\quad - \delta|00\rangle_{14}|00\rangle_{36}|00\rangle_{85}
\end{aligned}
\tag{4.7}
$$

Hence, the secret of $|\Phi\rangle_{xy} = (\alpha|00\rangle + \beta|01\rangle + \gamma|10\rangle + \delta|11\rangle)_{xy}$ is transferred to the subsystem $|X\rangle_{143685}$.

And then assume the BM results of Bob_1 and Bob_2 respectively are $|\Phi^+\rangle_{14} = \frac{1}{\sqrt{2}}(|00\rangle + |11\rangle)$ and $|\Phi^-\rangle_{36} = \frac{1}{\sqrt{2}}(|00\rangle - |11\rangle)$. Hence, the remained quantum states of two qubits (8, 5) are collapsed into $|X\rangle_{85} = -\alpha|11\rangle + \beta|10\rangle + \gamma|01\rangle - \delta|00\rangle$.

Step 5: According to Alice's and Bob_k's ($k = 1, 2$) measurement results, Bob_3 performs corresponding unitary transformation on his own particles (s_3, f_3^*) to recover Alice's unknown quantum state $|\Phi\rangle_{xy}$.

Then, Bob_3 can perform local unitary transformation $i \cdot \sigma_Y \otimes i \cdot \sigma_Y$ to recover two-qubit state $|\Phi\rangle_{xy}$ as follows:

$$
\begin{aligned}
(i \cdot \sigma_Y \otimes i \cdot \sigma_Y) \cdot |X\rangle_{85} &= (i \cdot \sigma_Y \otimes i \cdot \sigma_Y) \cdot (-\alpha|11\rangle + \beta|10\rangle + \gamma|01\rangle - \delta|00\rangle) \\
&= (\alpha|00\rangle + \beta|01\rangle + \gamma|10\rangle + \delta|11\rangle) = |\Phi\rangle_{xy}
\end{aligned}
\tag{4.8}
$$

Similarly, Table 4.2 (as shown in *Appendix*) also shows that Bob_3 uses specific unitary transformations to recover Alice's master key according to the different measurement results of Alice, Bob_1 and Bob_2.

4.3.2.4 Security Analyses

Herein, let's assume that the channel is reliable and noiseless. Hence, the intercept-resend attack and measure-resend attack can be avoided in our proposed two IQSTS schemes. Furthermore, consider that an internal dishonest agent knows a part of secret share, which makes he has more substantial eavesdrop capabilities than outside eavesdroppers. Consequently, the attack behaviors of the malicious agents in two IQSTS schemes are analyzed in detail, which proves that two IQSTS schemes can resist the collusion attack and entangled probe attack.

A. **Collusion Attack**

In Sect. 4.3.1, it has proved that MA's scheme [38] exist the collusion attack, that is, the first agent Bob_1 and last agent Bob_{n-1} can cooperate with each other to steal Alice's secret quantum state without the help of other agents. In this subsection, two IQSTS schemes can avoid this attack described as follows,

1. For the first IQSTS scheme, since Alice has verified that each agent (except the secret receiver) had performed the BM on their two particles in order; thus, all agents can't implement the collusion attack by interchanging their particles to steal Alice's secret quantum state;
2. For the second IQSTS scheme, since the permutation rules make all first particles form into a new order particle sequence which only knows by Alice, thus the other agents cannot find the correct two particles that belong to one EPR pair to implement the collusion attack. Hence, the presented second IQSTS scheme also avoids this attack

B. **Entangled Probe Attack**

In addition to the abovementioned collusion attack, the dishonest agent may steal Alice's secret state via entangling the ancillary probe. Accordingly, it includes two different cases explained as follows,

1. For the first case, assume that $(n-2)$ dishonest agents $\text{Bob}_1, \ldots, \text{Bob}_{i-1}$, $\text{Bob}_{i+1}, \ldots, \text{Bob}_{n-2}, \text{Bob}_{n-1}$ want to steal Alice's secret state without the help of Bob_i via entangling the ancillary probes, where Bob_{n-1} is the secret recover and all other agents are the controllers. To fulfill this aim, $n-2$ agents $\text{Bob}_1, \ldots, \text{Bob}_{i-1}, \text{Bob}_{i+1}, \ldots, \text{Bob}_{n-2}, \text{Bob}_{n-1}$ have to know Bob_i's BM result. Suppose that one of these dishonest agents, say Bob_k, can entange two ancillary probes $\{|F\rangle, |S\rangle\}$ via using the unitary operation $\widehat{U}$ on the transmitted two particles which were sent from Alice to Bob_i.

   For single qubit $|0\rangle$ $(|1\rangle)$, the effects of unitary operation $\widehat{U}$ on two ancilla probes $\{|F\rangle, |S\rangle\}$ can be described as [36]

$$
\begin{aligned}
&\widehat{U}|0\rangle|F\rangle_E = |0\rangle|F_{00}\rangle + |1\rangle|F_{01}\rangle, \widehat{U}|1\rangle|F\rangle_E = |0\rangle|F_{10}\rangle + |1\rangle|F_{11}\rangle \\
&\widehat{U}|0\rangle|S\rangle_E = |0\rangle|S_{00}\rangle + |1\rangle|S_{01}\rangle, \widehat{U}|1\rangle|S\rangle_E = |0\rangle|S_{10}\rangle + |1\rangle|S_{11}\rangle,
\end{aligned}
\tag{4.9}
$$

where $\widehat{U} \cdot \widehat{U}^{\dagger} = \widehat{U}^{\dagger} \cdot \widehat{U} = I$, $\{|T_{00}\rangle, |T_{01}\rangle, |T_{10}\rangle, |T_{11}\rangle\}_{T \in \{F,S\}}$ are four unnormalized states that belongs to the Hilbert space of Bob_k's probe H_k, which can be distinguished by Bob_k. Obviously, to avoid introducing errors, it holds $|F_{01}\rangle = |F_{10}\rangle$, $|S_{01}\rangle = |S_{10}\rangle$. Thus, we have

$$\begin{aligned} &\widehat{U}|0\rangle|F\rangle_E = |0\rangle|F_{00}\rangle = |0\rangle|F_0\rangle, \widehat{U}|1\rangle|F\rangle_E = |1\rangle|F_{11}\rangle = |1\rangle|F_1\rangle \\ &\widehat{U}|0\rangle|S\rangle_E = |0\rangle|S_{00}\rangle = |0\rangle|S_0\rangle, \widehat{U}|1\rangle|S\rangle_E = |1\rangle|S_{11}\rangle = |1\rangle|S_1\rangle \end{aligned} \tag{4.10}$$

Now, let's suppose that Bob_i's two-qubit is in Bell state $|\Phi\rangle_{xy}$, then the entangle probe attack can be described as

$$\begin{aligned} &\widehat{U}_{1,3}\widehat{U}_{2,4}|\Phi^+\rangle_{12}|F\rangle_3|S\rangle_4 \\ &= \frac{1}{\sqrt{2}}(|00\rangle + |11\rangle)_{12}|F\rangle_3|S\rangle_4 \\ &= \frac{1}{\sqrt{2}}(|00\rangle|F_0\rangle|S_0\rangle + |11\rangle|F_1\rangle|S_1\rangle) \\ &= \frac{1}{\sqrt{2}}\left[|\Phi^+\rangle_{12}(|F_0\rangle|S_0\rangle + |F_1\rangle|S_1\rangle) + |\Phi^-\rangle_{12}(|F_0\rangle|S_0\rangle - |F_1\rangle|S_1\rangle)\right] \end{aligned} \tag{4.11}$$

Apparently, there is 1/2 probability to know the exact Bell state of Bob_i's initial Bell state and 1/2 probability to get the wrong results. Hence, the entangled probes attack of $(n - 2)$ dishonest agents $\text{Bob}_1, \ldots, \text{Bob}_{i-1}, \text{Bob}_{i+1}, \ldots, \text{Bob}_{n-2}, \text{Bob}_{n-1}$ will be failed with probability 50%.

2. For the second case, other $(n - 2)$ dishonest agents $\text{Bob}_1, \text{Bob}_2, \ldots, \text{Bob}_{n-2}$ may conspire to steal Alice's secret state without the help of secret receiver Bob_{n-1}. They can steal Alice's secret state by entangling two probes $\{|F\rangle, |S\rangle\}$ on the transmitted two particles that were sent from Alice to Bob_{n-1} via the unitary operation $\widehat{U}$ first, and then perform the appropriate local unitary transformations on two probes $\{|F\rangle, |S\rangle\}$. Without loss of generality, let's assume that state of Bob_{n-1}'s two particles is exact same as Alice's secret quantum state $|\Phi\rangle_{1,2} = (\alpha|00\rangle + \beta|01\rangle + \gamma|10\rangle + \delta|11\rangle)_{1,2}$. Then the entangled probes attack can be described as

$$\begin{aligned} \widehat{U}_{1,3}\widehat{U}_{2,4}(\alpha|00\rangle + \beta|01\rangle + \gamma|10\rangle + \delta|11\rangle)_{1,2}|F\rangle_3|S\rangle_4 = \alpha|00\rangle_{1,2}|F_0\rangle_3|S_0\rangle_4 + \beta|01\rangle_{1,2}|F_0\rangle_3|S_1\rangle_4 \\ + \gamma|10\rangle_{1,2}|F_1\rangle_3|S_0\rangle_4 + \delta|11\rangle_{1,2}|F_1\rangle_3|S_1\rangle_4 \end{aligned} \tag{4.12}$$

Obviously, the state two probes $\{|F\rangle, |S\rangle\}$ is entangled with Alice's secret state, that is, the above state in Eq. 4.12 is an unseparable state. Besides, the density matrix of two probes $\{|F\rangle, |S\rangle\}$ can be obtained via performing the trace operation on two qubits $(1, 2)$ described as

$$\begin{aligned}\rho_{3,4} &= trace_{1,2}(\alpha|00\rangle|F_0S_0\rangle + \beta|01\rangle|F_0S_1\rangle + \gamma|10\rangle|F_1S_0\rangle + \delta|11\rangle|F_1S_1\rangle)_{1,2,3,4}\\ &= \alpha a^*|F_0S_0\rangle\langle F_0S_0| + \beta\beta^*|F_0S_1\rangle\langle F_0S_1| + \gamma\gamma^*|F_1S_0\rangle\langle F_1S_0| + \delta\delta^*|F_1S_1\rangle\langle F_1S_1|\end{aligned} \tag{4.13}$$

Furthermore, the density matrix of Alice's secret quantum state $|\Phi\rangle_{1,\,2} = (\alpha|00\rangle + \beta|01\rangle + \gamma|10\rangle + \delta|11\rangle)_{1,\,2}$ can be expressed as

$$\begin{aligned}\rho_{1,2} &= |\Phi\rangle_{1,2}\langle\Phi|\\ &= (\alpha|00\rangle + \beta|01\rangle + \gamma|10\rangle + \delta|11\rangle)_{1,2}(a^*\langle 00| + \beta^*\langle 01| + \gamma^*\langle 10| + \delta^*\langle 11|)\\ &= \alpha a^*|00\rangle\langle 00| + \beta\beta^*|01\rangle\langle 01| + \gamma\gamma^*|10\rangle\langle 10| + \delta\delta^*|11\rangle\langle 11| + \alpha\beta^*|00\rangle\langle 01|\\ &\quad + \alpha\gamma^*|00\rangle\langle 10| + \alpha\delta^*|00\rangle\langle 11| + \beta a^*|01\rangle\langle 00| + \beta\gamma^*|01\rangle\langle 10| + \beta\delta^*|01\rangle\langle 11|\\ &\quad + \gamma a^*|10\rangle\langle 00| + \gamma\beta^*|10\rangle\langle 01| + \gamma\delta^*|10\rangle\langle 11| + \delta a^*|11\rangle\langle 00| + \delta\beta^*|11\rangle\langle 01|\\ &\quad + \delta\gamma^*|11\rangle\langle 10|\end{aligned} \tag{4.14}$$

By comparing the density matrices between two probes $\{|F\rangle, |S\rangle\}$ state (shown in Eq. 4.13) and Alice's secret quantum state $|\Phi\rangle_{1,\,2}$ (shown in Eq. 4.14), it is easy to find that they are quite different to each other. Hence, it can conclude that without introducing any errors to Bob_{n-1}'s two particles state, other $(n - 2)$ dishonest agents Bob_1, Bob_2, ..., Bob_{n-2} can't obtain Alice's secret quantum state. Therefore, on the basis of above analysis, our two IQSTS schemes are immune to entangled probe attack.

4.3.2.5 Efficiency Analysis

In presented two IQSTS schemes, only Bell states are used and two-qubit Bell-state measurement is needed to complete the QSTS scheme. The two photons polarization-entangled Bell states $\{|\Phi^{\pm}\rangle, |\Psi^{\pm}\rangle\}$ can be produced via using spontaneous parametric down conversion (SPDC) sources in the BBO crystal [40]. The Bell-state measurement can be realized by a resource-efficient Bell states analyzer [41], which can completely identify the four Bell states by the measurement results of the single-photon destructive and QND detectors. Thus, the successful ratio of the implementation of the proposed two IQSTS schemes based on Bell states is 100%.

In this subsection, we compare the two IQSTS schemes with the previous ones [21, 38, 42] in many aspects. Consider that a method of $\eta = \frac{q_e}{q_t}$ is used to evaluate the QSTS scheme's efficiency, where q_e stands for effective qubits have been shared and q_t denotes the total qubits used in scheme. Thus, for the first n-party IQSTS scheme, since Alice needs to prepare $2n - 2$ EPR pairs to share two-qubit unknown quantum state among $n - 1$ agents, its efficiency is $\eta_1 = \frac{2}{(2n-2)\times 2+2}$. For the second n-party IQSTS scheme, Alice prepares n EPR pairs to share two-qubit unknown quantum state among $n - 1$ agents; thus, its efficiency is $\eta_2 = \frac{2}{2n+2}$. When $n = 4$, the efficiency of first IQSTS scheme and second IQSTS scheme, respectively, are $\frac{1}{7}$ and $\frac{1}{5}$.

As shown in the figure above, we integrate efficiency and safety analysis in Table 4.3. The two IQSTS schemes we propose keep a delicate balance between efficiency and safety.

4.3.2.6 Summary

In summary, this section has proposed a special attack strategy, that is, conclusion attack, existed in MA's scheme, in which the first agent and last agent can conspire to steal the dealer's secret quantum information without the collaboration of other agents. Furthermore, based on MA's scheme, two multiparty IQSTS schemes are proposed, which makes it secure against this kind of attack. To demonstrate the feasibility of proposed two multiparty IQSTS schemes, four-party IQSTS scheme is taken as the specific example that explains proposed IQSTS scheme in detail. Finally, the performance analysis in terms of security and efficiency is demonstrated in detail, which proves that our IQSTS schemes are both safe and efficient.

4.3.2.7 Appendix

Table 4.1 Bob_3's specific unitary transformations that correspond to Bob_1's, Bob_2's and Alice's measurement results

Bob_1's measurement results of particles (2,3)	Bob_2's measurement results of particles (4,5)	Alice's measurement results		Collapsed state of particles (6,7)	Bob_3's unitary transformation
		$(x,1)$	$(y,8)$		
$\|\Phi^+\rangle$	$\|\Phi^+\rangle$	$\|\Phi^+\rangle$	$\|\Phi^+\rangle$	$\alpha\|00\rangle + \beta\|01\rangle + \gamma\|10\rangle + \delta\|11\rangle$	$I \otimes I$
			$\|\Phi^-\rangle$	$\alpha\|00\rangle - \beta\|01\rangle + \gamma\|10\rangle - \delta\|11\rangle$	$I \otimes \sigma_Z$
		$\|\Phi^-\rangle$	$\|\Phi^+\rangle$	$\alpha\|00\rangle + \beta\|01\rangle - \gamma\|10\rangle - \delta\|11\rangle$	$\sigma_Z \otimes I$
			$\|\Phi^-\rangle$	$\alpha\|00\rangle - \beta\|01\rangle - \gamma\|10\rangle + \delta\|11\rangle$	$\sigma_Z \otimes \sigma_Z$
	$\|\Phi^-\rangle$	$\|\Phi^+\rangle$	$\|\Phi^+\rangle$	$\alpha\|00\rangle + \beta\|01\rangle - \gamma\|10\rangle - \delta\|11\rangle$	$\sigma_Z \otimes I$
			$\|\Phi^-\rangle$	$\alpha\|00\rangle - \beta\|01\rangle - \gamma\|10\rangle + \delta\|11\rangle$	$\sigma_Z \otimes \sigma_Z$
		$\|\Phi^-\rangle$	$\|\Phi^+\rangle$	$\alpha\|00\rangle + \beta\|01\rangle + \gamma\|10\rangle + \delta\|11\rangle$	$I \otimes I$
			$\|\Phi^-\rangle$	$\alpha\|00\rangle - \beta\|01\rangle + \gamma\|10\rangle - \delta\|11\rangle$	$I \otimes \sigma_Z$
$\|\Phi^-\rangle$	$\|\Phi^+\rangle$	$\|\Phi^+\rangle$	$\|\Phi^+\rangle$	$\alpha\|00\rangle + \beta\|01\rangle - \gamma\|10\rangle - \delta\|11\rangle$	$\sigma_Z \otimes I$
			$\|\Phi^-\rangle$	$\alpha\|00\rangle - \beta\|01\rangle - \gamma\|10\rangle + \delta\|11\rangle$	$\sigma_Z \otimes \sigma_Z$
		$\|\Phi^-\rangle$	$\|\Phi^+\rangle$	$\alpha\|00\rangle + \beta\|01\rangle + \gamma\|10\rangle + \delta\|11\rangle$	$I \otimes I$
			$\|\Phi^-\rangle$	$\alpha\|00\rangle - \beta\|01\rangle + \gamma\|10\rangle - \delta\|11\rangle$	$I \otimes \sigma_Z$
	$\|\Phi^-\rangle$	$\|\Phi^+\rangle$	$\|\Phi^+\rangle$	$\alpha\|00\rangle + \beta\|01\rangle + \gamma\|10\rangle + \delta\|11\rangle$	$I \otimes I$
			$\|\Phi^-\rangle$	$\alpha\|00\rangle - \beta\|01\rangle + \gamma\|10\rangle - \delta\|11\rangle$	$I \otimes \sigma_Z$
		$\|\Phi^-\rangle$	$\|\Phi^+\rangle$	$\alpha\|00\rangle + \beta\|01\rangle - \gamma\|10\rangle - \delta\|11\rangle$	$\sigma_Z \otimes I$
			$\|\Phi^-\rangle$	$\alpha\|00\rangle - \beta\|01\rangle - \gamma\|10\rangle + \delta\|11\rangle$	$\sigma_Z \otimes \sigma_Z$

(continued)

Table 4.1 (continued)

Bob$_1$'s measurement results of particles (2,3)	Bob$_2$'s measurement results of particles (4,5)	Alice's measurement results		Collapsed state of particles (6,7)	Bob$_3$'s unitary transformation
		$(x,1)$	$(y,8)$		
$\|\Phi^+\rangle$	$\|\Phi^+\rangle$	$\|\Phi^+\rangle$	$\|\Psi^+\rangle$	$\alpha\|01\rangle+\beta\|00\rangle+\gamma\|11\rangle+\delta\|10\rangle$	$I\otimes\sigma_X$
			$\|\Psi^-\rangle$	$\alpha\|01\rangle-\beta\|00\rangle+\gamma\|11\rangle-\delta\|10\rangle$	$I\otimes i\cdot\sigma_Y$
		$\|\Phi^-\rangle$	$\|\Psi^+\rangle$	$\alpha\|01\rangle+\beta\|00\rangle-\gamma\|11\rangle-\delta\|10\rangle$	$\sigma_Z\otimes\sigma_X$
			$\|\Psi^-\rangle$	$\alpha\|01\rangle-\beta\|00\rangle-\gamma\|11\rangle+\delta\|10\rangle$	$\sigma_Z\otimes i\cdot\sigma_Y$
	$\|\Phi^-\rangle$	$\|\Phi^+\rangle$	$\|\Psi^+\rangle$	$\alpha\|01\rangle+\beta\|00\rangle-\gamma\|11\rangle-\delta\|10\rangle$	$\sigma_Z\otimes\sigma_X$
			$\|\Psi^-\rangle$	$\alpha\|01\rangle-\beta\|00\rangle-\gamma\|11\rangle+\delta\|10\rangle$	$\sigma_Z\otimes i\cdot\sigma_Y$
		$\|\Phi^-\rangle$	$\|\Psi^+\rangle$	$\alpha\|01\rangle+\beta\|00\rangle+\gamma\|11\rangle+\delta\|10\rangle$	$I\otimes\sigma_X$
			$\|\Psi^-\rangle$	$\alpha\|01\rangle-\beta\|00\rangle+\gamma\|11\rangle-\delta\|10\rangle$	$I\otimes i\cdot\sigma_Y$
$\|\Phi^-\rangle$	$\|\Phi^+\rangle$	$\|\Phi^+\rangle$	$\|\Psi^+\rangle$	$\alpha\|01\rangle+\beta\|00\rangle-\gamma\|11\rangle-\delta\|10\rangle$	$\sigma_Z\otimes\sigma_X$
			$\|\Psi^-\rangle$	$\alpha\|01\rangle-\beta\|00\rangle-\gamma\|11\rangle+\delta\|10\rangle$	$\sigma_Z\otimes i\cdot\sigma_Y$
		$\|\Phi^-\rangle$	$\|\Psi^+\rangle$	$\alpha\|01\rangle+\beta\|00\rangle+\gamma\|11\rangle+\delta\|10\rangle$	$I\otimes\sigma_X$
			$\|\Psi^-\rangle$	$\alpha\|01\rangle-\beta\|00\rangle+\gamma\|11\rangle-\delta\|10\rangle$	$I\otimes i\cdot\sigma_Y$
	$\|\Phi^-\rangle$	$\|\Phi^+\rangle$	$\|\Psi^+\rangle$	$\alpha\|01\rangle+\beta\|00\rangle+\gamma\|11\rangle+\delta\|10\rangle$	$I\otimes\sigma_X$
			$\|\Psi^-\rangle$	$\alpha\|01\rangle-\beta\|00\rangle+\gamma\|11\rangle-\delta\|10\rangle$	$I\otimes i\cdot\sigma_Y$
		$\|\Phi^-\rangle$	$\|\Psi^+\rangle$	$\alpha\|01\rangle+\beta\|00\rangle-\gamma\|11\rangle-\delta\|10\rangle$	$\sigma_Z\otimes\sigma_X$
			$\|\Psi^-\rangle$	$\alpha\|01\rangle-\beta\|00\rangle-\gamma\|11\rangle+\delta\|10\rangle$	$\sigma_Z\otimes i\cdot\sigma_Y$
$\|\Phi^+\rangle$	$\|\Phi^+\rangle$	$\|\Psi^+\rangle$	$\|\Phi^+\rangle$	$\alpha\|10\rangle+\beta\|11\rangle+\gamma\|00\rangle+\delta\|01\rangle$	$\sigma_X\otimes I$
			$\|\Phi^-\rangle$	$\alpha\|10\rangle-\beta\|11\rangle+\gamma\|00\rangle-\delta\|01\rangle$	$\sigma_X\otimes\sigma_Z$
		$\|\Psi^-\rangle$	$\|\Phi^+\rangle$	$\alpha\|10\rangle+\beta\|11\rangle-\gamma\|00\rangle-\delta\|01\rangle$	$i\cdot\sigma_Y\otimes I$
			$\|\Phi^-\rangle$	$\alpha\|10\rangle-\beta\|11\rangle-\gamma\|00\rangle+\delta\|01\rangle$	$i\cdot\sigma_Y\otimes\sigma_Z$
	$\|\Phi^-\rangle$	$\|\Psi^+\rangle$	$\|\Phi^+\rangle$	$-\alpha\|10\rangle-\beta\|11\rangle+\gamma\|00\rangle+\delta\|01\rangle$	$i\cdot\sigma_Y\otimes I$
			$\|\Phi^-\rangle$	$-\alpha\|10\rangle+\beta\|11\rangle+\gamma\|00\rangle-\delta\|01\rangle$	$i\cdot\sigma_Y\otimes\sigma_Z$
		$\|\Psi^-\rangle$	$\|\Phi^+\rangle$	$-\alpha\|10\rangle-\beta\|11\rangle-\gamma\|00\rangle-\delta\|01\rangle$	$\sigma_X\otimes I$
			$\|\Phi^-\rangle$	$-\alpha\|10\rangle+\beta\|11\rangle-\gamma\|00\rangle+\delta\|01\rangle$	$\sigma_X\otimes\sigma_Z$
$\|\Phi^-\rangle$	$\|\Phi^+\rangle$	$\|\Psi^+\rangle$	$\|\Phi^+\rangle$	$-\alpha\|10\rangle-\beta\|11\rangle+\gamma\|00\rangle+\delta\|01\rangle$	$i\cdot\sigma_Y\otimes I$
			$\|\Phi^-\rangle$	$-\alpha\|10\rangle+\beta\|11\rangle+\gamma\|00\rangle-\delta\|01\rangle$	$i\cdot\sigma_Y\otimes\sigma_Z$
		$\|\Psi^-\rangle$	$\|\Phi^+\rangle$	$-\alpha\|10\rangle-\beta\|11\rangle-\gamma\|00\rangle-\delta\|01\rangle$	$\sigma_X\otimes I$
			$\|\Phi^-\rangle$	$-\alpha\|10\rangle+\beta\|11\rangle-\gamma\|00\rangle+\delta\|01\rangle$	$\sigma_X\otimes\sigma_Z$
	$\|\Phi^-\rangle$	$\|\Psi^+\rangle$	$\|\Phi^+\rangle$	$\alpha\|10\rangle+\beta\|11\rangle+\gamma\|00\rangle+\delta\|01\rangle$	$\sigma_X\otimes I$
			$\|\Phi^-\rangle$	$\alpha\|10\rangle-\beta\|11\rangle+\gamma\|00\rangle-\delta\|01\rangle$	$\sigma_X\otimes\sigma_Z$
		$\|\Psi^-\rangle$	$\|\Phi^+\rangle$	$\alpha\|10\rangle+\beta\|11\rangle-\gamma\|00\rangle-\delta\|01\rangle$	$i\cdot\sigma_Y\otimes I$
			$\|\Phi^-\rangle$	$\alpha\|10\rangle-\beta\|11\rangle-\gamma\|00\rangle+\delta\|01\rangle$	$i\cdot\sigma_Y\otimes\sigma_Z$
$\|\Phi^+\rangle$	$\|\Phi^+\rangle$	$\|\Psi^+\rangle$	$\|\Psi^+\rangle$	$\alpha\|11\rangle+\beta\|10\rangle+\gamma\|01\rangle+\delta\|00\rangle$	$\sigma_X\otimes\sigma_X$
			$\|\Psi^-\rangle$	$\alpha\|11\rangle-\beta\|10\rangle+\gamma\|01\rangle-\delta\|00\rangle$	$\sigma_X\otimes i\cdot\sigma_Y$
		$\|\Psi^-\rangle$	$\|\Psi^+\rangle$	$\alpha\|11\rangle+\beta\|10\rangle-\gamma\|01\rangle-\delta\|00\rangle$	$i\cdot\sigma_Y\otimes\sigma_X$
			$\|\Psi^-\rangle$	$\alpha\|11\rangle-\beta\|10\rangle-\gamma\|01\rangle+\delta\|00\rangle$	$i\cdot\sigma_Y\otimes i\cdot\sigma_Y$
	$\|\Phi^-\rangle$	$\|\Psi^+\rangle$	$\|\Psi^+\rangle$	$-\alpha\|11\rangle-\beta\|10\rangle+\gamma\|01\rangle+\delta\|00\rangle$	$i\cdot\sigma_Y\otimes\sigma_X$
			$\|\Psi^-\rangle$	$-\alpha\|11\rangle+\beta\|10\rangle+\gamma\|01\rangle-\delta\|00\rangle$	$i\cdot\sigma_Y\otimes i\cdot\sigma_Y$
		$\|\Psi^-\rangle$	$\|\Psi^+\rangle$	$-\alpha\|11\rangle-\beta\|10\rangle-\gamma\|01\rangle-\delta\|00\rangle$	$\sigma_X\otimes\sigma_X$
			$\|\Psi^-\rangle$	$-\alpha\|11\rangle+\beta\|10\rangle-\gamma\|01\rangle+\delta\|00\rangle$	$\sigma_X\otimes i\cdot\sigma_Y$

(continued)

Table 4.1 (continued)

Bob_1's measurement results of particles (2,3)	Bob_2's measurement results of particles (4,5)	Alice's measurement results		Collapsed state of particles (6,7)	Bob_3's unitary transformation
		(*x*,1)	(*y*,8)		
$\|\Phi^-\rangle$	$\|\Phi^+\rangle$	$\|\Psi^+\rangle$	$\|\Psi^+\rangle$	$-\alpha\|11\rangle - \beta\|10\rangle + \gamma\|01\rangle + \delta\|00\rangle$	$i \cdot \sigma_Y \bigotimes \sigma_X$
			$\|\Psi^-\rangle$	$-\alpha\|11\rangle + \beta\|10\rangle + \gamma\|01\rangle - \delta\|00\rangle$	$i \cdot \sigma_Y \bigotimes i \cdot \sigma_Y$
		$\|\Psi^-\rangle$	$\|\Psi^+\rangle$	$-\alpha\|11\rangle - \beta\|10\rangle - \gamma\|01\rangle - \delta\|00\rangle$	$\sigma_X \bigotimes \sigma_X$
			$\|\Psi^-\rangle$	$-\alpha\|11\rangle + \beta\|10\rangle - \gamma\|01\rangle + \delta\|00\rangle$	$\sigma_X \bigotimes i \cdot \sigma_Y$
	$\|\Phi^-\rangle$	$\|\Psi^+\rangle$	$\|\Psi^+\rangle$	$\alpha\|11\rangle + \beta\|10\rangle + \gamma\|01\rangle + \delta\|00\rangle$	$\sigma_X \bigotimes \sigma_X$
			$\|\Psi^-\rangle$	$\alpha\|11\rangle - \beta\|10\rangle + \gamma\|01\rangle - \delta\|00\rangle$	$\sigma_X \bigotimes i \cdot \sigma_Y$
		$\|\Psi^-\rangle$	$\|\Psi^+\rangle$	$\alpha\|11\rangle + \beta\|10\rangle - \gamma\|01\rangle - \delta\|00\rangle$	$i \cdot \sigma_Y \bigotimes \sigma_X$
			$\|\Psi^-\rangle$	$\alpha\|11\rangle - \beta\|10\rangle - \gamma\|01\rangle + \delta\|00\rangle$	$i \cdot \sigma_Y \bigotimes i \cdot \sigma_Y$
$\|\Phi^+\rangle$	$\|\Psi^+\rangle$	$\|\Phi^+\rangle$	$\|\Phi^+\rangle$	$\alpha\|10\rangle + \beta\|11\rangle + \gamma\|00\rangle + \delta\|01\rangle$	$\sigma_X \bigotimes I$
			$\|\Phi^-\rangle$	$\alpha\|10\rangle - \beta\|11\rangle + \gamma\|00\rangle - \delta\|01\rangle$	$\sigma_X \bigotimes \sigma_Z$
		$\|\Phi^-\rangle$	$\|\Phi^+\rangle$	$\alpha\|10\rangle + \beta\|11\rangle - \gamma\|00\rangle - \delta\|01\rangle$	$i \cdot \sigma_Y \bigotimes I$
			$\|\Phi^-\rangle$	$\alpha\|10\rangle - \beta\|11\rangle - \gamma\|00\rangle + \delta\|01\rangle$	$i \cdot \sigma_Y \bigotimes \sigma_Z$
	$\|\Psi^-\rangle$	$\|\Phi^+\rangle$	$\|\Phi^+\rangle$	$\alpha\|10\rangle + \beta\|11\rangle - \gamma\|00\rangle - \delta\|01\rangle$	$i \cdot \sigma_Y \bigotimes I$
			$\|\Phi^-\rangle$	$\alpha\|10\rangle - \beta\|11\rangle - \gamma\|00\rangle + \delta\|01\rangle$	$i \cdot \sigma_Y \bigotimes \sigma_Z$
		$\|\Phi^-\rangle$	$\|\Phi^+\rangle$	$\alpha\|10\rangle + \beta\|11\rangle + \gamma\|00\rangle + \delta\|01\rangle$	$\sigma_X \bigotimes I$
			$\|\Phi^-\rangle$	$\alpha\|10\rangle - \beta\|11\rangle + \gamma\|00\rangle - \delta\|01\rangle$	$\sigma_X \bigotimes \sigma_Z$
$\|\Phi^-\rangle$	$\|\Psi^+\rangle$	$\|\Phi^+\rangle$	$\|\Phi^+\rangle$	$\alpha\|10\rangle + \beta\|11\rangle - \gamma\|00\rangle - \delta\|01\rangle$	$i \cdot \sigma_Y \bigotimes I$
			$\|\Phi^-\rangle$	$\alpha\|10\rangle - \beta\|11\rangle - \gamma\|00\rangle + \delta\|01\rangle$	$i \cdot \sigma_Y \bigotimes \sigma_Z$
		$\|\Phi^-\rangle$	$\|\Phi^+\rangle$	$\alpha\|10\rangle + \beta\|11\rangle + \gamma\|00\rangle + \delta\|01\rangle$	$\sigma_X \bigotimes I$
			$\|\Phi^-\rangle$	$\alpha\|10\rangle - \beta\|11\rangle + \gamma\|00\rangle - \delta\|01\rangle$	$\sigma_X \bigotimes \sigma_Z$
	$\|\Psi^-\rangle$	$\|\Phi^+\rangle$	$\|\Phi^+\rangle$	$\alpha\|10\rangle + \beta\|11\rangle + \gamma\|00\rangle + \delta\|01\rangle$	$\sigma_X \bigotimes I$
			$\|\Phi^-\rangle$	$\alpha\|10\rangle - \beta\|11\rangle + \gamma\|00\rangle - \delta\|01\rangle$	$\sigma_X \bigotimes \sigma_Z$
		$\|\Phi^-\rangle$	$\|\Phi^+\rangle$	$\alpha\|10\rangle + \beta\|11\rangle - \gamma\|00\rangle - \delta\|01\rangle$	$i \cdot \sigma_Y \bigotimes I$
			$\|\Phi^-\rangle$	$\alpha\|10\rangle - \beta\|11\rangle - \gamma\|00\rangle + \delta\|01\rangle$	$i \cdot \sigma_Y \bigotimes \sigma_Z$
$\|\Phi^+\rangle$	$\|\Psi^+\rangle$	$\|\Phi^+\rangle$	$\|\Psi^+\rangle$	$\alpha\|11\rangle + \beta\|10\rangle + \gamma\|01\rangle + \delta\|00\rangle$	$\sigma_X \bigotimes \sigma_X$
			$\|\Psi^-\rangle$	$\alpha\|11\rangle - \beta\|10\rangle + \gamma\|01\rangle - \delta\|00\rangle$	$\sigma_X \bigotimes i \cdot \sigma_Y$
		$\|\Phi^-\rangle$	$\|\Psi^+\rangle$	$\alpha\|11\rangle + \beta\|10\rangle - \gamma\|01\rangle - \delta\|00\rangle$	$i \cdot \sigma_Y \bigotimes \sigma_X$
			$\|\Psi^-\rangle$	$\alpha\|11\rangle - \beta\|10\rangle - \gamma\|01\rangle + \delta\|00\rangle$	$i \cdot \sigma_Y \bigotimes i \cdot \sigma_Y$
	$\|\Psi^-\rangle$	$\|\Phi^+\rangle$	$\|\Psi^+\rangle$	$\alpha\|11\rangle + \beta\|10\rangle - \gamma\|01\rangle - \delta\|00\rangle$	$i \cdot \sigma_Y \bigotimes \sigma_X$
			$\|\Psi^-\rangle$	$\alpha\|11\rangle - \beta\|10\rangle - \gamma\|01\rangle + \delta\|00\rangle$	$i \cdot \sigma_Y \bigotimes i \cdot \sigma_Y$
		$\|\Phi^-\rangle$	$\|\Psi^+\rangle$	$\alpha\|11\rangle + \beta\|10\rangle + \gamma\|01\rangle + \delta\|00\rangle$	$\sigma_X \bigotimes \sigma_X$
			$\|\Psi^-\rangle$	$\alpha\|11\rangle - \beta\|10\rangle + \gamma\|01\rangle - \delta\|00\rangle$	$\sigma_X \bigotimes i \cdot \sigma_Y$
$\|\Phi^-\rangle$	$\|\Psi^+\rangle$	$\|\Phi^+\rangle$	$\|\Psi^+\rangle$	$\alpha\|11\rangle + \beta\|10\rangle - \gamma\|01\rangle - \delta\|00\rangle$	$i \cdot \sigma_Y \bigotimes \sigma_X$
			$\|\Psi^-\rangle$	$\alpha\|11\rangle - \beta\|10\rangle - \gamma\|01\rangle + \delta\|00\rangle$	$i \cdot \sigma_Y \bigotimes i \cdot \sigma_Y$
		$\|\Phi^-\rangle$	$\|\Psi^+\rangle$	$\alpha\|11\rangle + \beta\|10\rangle + \gamma\|01\rangle + \delta\|00\rangle$	$\sigma_X \bigotimes \sigma_X$
			$\|\Psi^-\rangle$	$\alpha\|11\rangle - \beta\|10\rangle + \gamma\|01\rangle - \delta\|00\rangle$	$\sigma_X \bigotimes i \cdot \sigma_Y$
	$\|\Psi^-\rangle$	$\|\Phi^+\rangle$	$\|\Psi^+\rangle$	$\alpha\|11\rangle + \beta\|10\rangle + \gamma\|01\rangle + \delta\|00\rangle$	$\sigma_X \bigotimes \sigma_X$
			$\|\Psi^-\rangle$	$\alpha\|11\rangle - \beta\|10\rangle + \gamma\|01\rangle - \delta\|00\rangle$	$\sigma_X \bigotimes i \cdot \sigma_Y$
		$\|\Phi^-\rangle$	$\|\Psi^+\rangle$	$\alpha\|11\rangle + \beta\|10\rangle - \gamma\|01\rangle - \delta\|00\rangle$	$i \cdot \sigma_Y \bigotimes \sigma_X$
			$\|\Psi^-\rangle$	$\alpha\|11\rangle - \beta\|10\rangle - \gamma\|01\rangle + \delta\|00\rangle$	$i \cdot \sigma_Y \bigotimes i \cdot \sigma_Y$

(continued)

Table 4.1 (continued)

Bob$_1$'s measurement results of particles (2,3)	Bob$_2$'s measurement results of particles (4,5)	Alice's measurement results		Collapsed state of particles (6,7)	Bob$_3$'s unitary transformation
		$(x,1)$	$(y,8)$		
$\|\Phi^+\rangle$	$\|\Psi^+\rangle$	$\|\Psi^+\rangle$	$\|\Phi^+\rangle$	$\alpha\|00\rangle + \beta\|01\rangle + \gamma\|10\rangle + \delta\|11\rangle$	$I \otimes I$
			$\|\Phi^-\rangle$	$\alpha\|00\rangle - \beta\|01\rangle + \gamma\|10\rangle - \delta\|11\rangle$	$I \otimes \sigma_Z$
		$\|\Psi^-\rangle$	$\|\Phi^+\rangle$	$\alpha\|00\rangle + \beta\|01\rangle - \gamma\|10\rangle - \delta\|11\rangle$	$\sigma_Z \otimes I$
			$\|\Phi^-\rangle$	$\alpha\|00\rangle - \beta\|01\rangle - \gamma\|10\rangle + \delta\|11\rangle$	$\sigma_Z \otimes \sigma_Z$
	$\|\Psi^-\rangle$	$\|\Psi^+\rangle$	$\|\Phi^+\rangle$	$-\alpha\|00\rangle - \beta\|01\rangle + \gamma\|10\rangle + \delta\|11\rangle$	$\sigma_Z \otimes I$
			$\|\Phi^-\rangle$	$-\alpha\|00\rangle + \beta\|01\rangle + \gamma\|10\rangle - \delta\|11\rangle$	$\sigma_Z \otimes \sigma_Z$
		$\|\Psi^-\rangle$	$\|\Phi^+\rangle$	$-\alpha\|00\rangle - \beta\|01\rangle - \gamma\|10\rangle - \delta\|11\rangle$	$I \otimes I$
			$\|\Phi^-\rangle$	$-\alpha\|00\rangle + \beta\|01\rangle - \gamma\|10\rangle + \delta\|11\rangle$	$I \otimes \sigma_Z$
$\|\Phi^-\rangle$	$\|\Psi^+\rangle$	$\|\Psi^+\rangle$	$\|\Phi^+\rangle$	$-\alpha\|00\rangle - \beta\|01\rangle + \gamma\|10\rangle + \delta\|11\rangle$	$\sigma_Z \otimes I$
			$\|\Phi^-\rangle$	$-\alpha\|00\rangle + \beta\|01\rangle + \gamma\|10\rangle - \delta\|11\rangle$	$\sigma_Z \otimes \sigma_Z$
		$\|\Psi^-\rangle$	$\|\Phi^+\rangle$	$-\alpha\|00\rangle - \beta\|01\rangle - \gamma\|10\rangle - \delta\|11\rangle$	$I \otimes I$
			$\|\Phi^-\rangle$	$-\alpha\|00\rangle + \beta\|01\rangle - \gamma\|10\rangle + \delta\|11\rangle$	$I \otimes \sigma_Z$
	$\|\Psi^-\rangle$	$\|\Psi^+\rangle$	$\|\Phi^+\rangle$	$\alpha\|00\rangle + \beta\|01\rangle + \gamma\|10\rangle + \delta\|11\rangle$	$I \otimes I$
			$\|\Phi^-\rangle$	$\alpha\|00\rangle - \beta\|01\rangle + \gamma\|10\rangle - \delta\|11\rangle$	$I \otimes \sigma_Z$
		$\|\Psi^-\rangle$	$\|\Phi^+\rangle$	$\alpha\|00\rangle + \beta\|01\rangle - \gamma\|10\rangle - \delta\|11\rangle$	$\sigma_Z \otimes I$
			$\|\Phi^-\rangle$	$\alpha\|00\rangle - \beta\|01\rangle - \gamma\|10\rangle + \delta\|11\rangle$	$\sigma_Z \otimes \sigma_Z$
$\|\Phi^+\rangle$	$\|\Psi^+\rangle$	$\|\Psi^+\rangle$	$\|\Psi^+\rangle$	$\alpha\|01\rangle + \beta\|00\rangle + \gamma\|11\rangle + \delta\|10\rangle$	$I \otimes \sigma_X$
			$\|\Psi^-\rangle$	$\alpha\|01\rangle - \beta\|00\rangle + \gamma\|11\rangle - \delta\|10\rangle$	$I \otimes i \cdot \sigma_Y$
		$\|\Psi^-\rangle$	$\|\Psi^+\rangle$	$\alpha\|01\rangle + \beta\|00\rangle - \gamma\|11\rangle - \delta\|10\rangle$	$\sigma_Z \otimes \sigma_X$
			$\|\Psi^-\rangle$	$\alpha\|01\rangle - \beta\|00\rangle - \gamma\|11\rangle + \delta\|10\rangle$	$\sigma_Z \otimes i \cdot \sigma_Y$
	$\|\Psi^-\rangle$	$\|\Psi^+\rangle$	$\|\Psi^+\rangle$	$-\alpha\|01\rangle - \beta\|00\rangle + \gamma\|11\rangle + \delta\|10\rangle$	$\sigma_Z \otimes \sigma_X$
			$\|\Psi^-\rangle$	$-\alpha\|01\rangle + \beta\|00\rangle + \gamma\|11\rangle - \delta\|10\rangle$	$I \otimes i \cdot \sigma_Y$
		$\|\Psi^-\rangle$	$\|\Psi^+\rangle$	$-\alpha\|01\rangle - \beta\|00\rangle - \gamma\|11\rangle - \delta\|10\rangle$	$I \otimes \sigma_X$
			$\|\Psi^-\rangle$	$-\alpha\|01\rangle + \beta\|00\rangle - \gamma\|11\rangle + \delta\|10\rangle$	$I \otimes i \cdot \sigma_Y$
$\|\Phi^-\rangle$	$\|\Psi^+\rangle$	$\|\Psi^+\rangle$	$\|\Psi^+\rangle$	$-\alpha\|01\rangle - \beta\|00\rangle + \gamma\|11\rangle + \delta\|10\rangle$	$\sigma_Z \otimes \sigma_X$
			$\|\Psi^-\rangle$	$-\alpha\|01\rangle + \beta\|00\rangle + \gamma\|11\rangle - \delta\|10\rangle$	$I \otimes i \cdot \sigma_Y$
		$\|\Psi^-\rangle$	$\|\Psi^+\rangle$	$-\alpha\|01\rangle - \beta\|00\rangle - \gamma\|11\rangle - \delta\|10\rangle$	$I \otimes \sigma_X$
			$\|\Psi^-\rangle$	$-\alpha\|01\rangle + \beta\|00\rangle - \gamma\|11\rangle + \delta\|10\rangle$	$I \otimes i \cdot \sigma_Y$
	$\|\Psi^-\rangle$	$\|\Psi^+\rangle$	$\|\Psi^+\rangle$	$\alpha\|01\rangle + \beta\|00\rangle + \gamma\|11\rangle + \delta\|10\rangle$	$I \otimes \sigma_X$
			$\|\Psi^-\rangle$	$\alpha\|01\rangle - \beta\|00\rangle + \gamma\|11\rangle - \delta\|10\rangle$	$I \otimes i \cdot \sigma_Y$
		$\|\Psi^-\rangle$	$\|\Psi^+\rangle$	$\alpha\|01\rangle + \beta\|00\rangle - \gamma\|11\rangle - \delta\|10\rangle$	$\sigma_Z \otimes \sigma_X$
			$\|\Psi^-\rangle$	$\alpha\|01\rangle - \beta\|00\rangle - \gamma\|11\rangle + \delta\|10\rangle$	$\sigma_Z \otimes i \cdot \sigma_Y$
$\|\Psi^+\rangle$	$\|\Phi^+\rangle$	$\|\Phi^+\rangle$	$\|\Phi^+\rangle$	$\alpha\|10\rangle + \beta\|11\rangle + \gamma\|00\rangle + \delta\|01\rangle$	$\sigma_X \otimes I$
			$\|\Phi^-\rangle$	$\alpha\|10\rangle - \beta\|11\rangle + \gamma\|00\rangle - \delta\|01\rangle$	$\sigma_X \otimes \sigma_Z$
		$\|\Phi^-\rangle$	$\|\Phi^+\rangle$	$\alpha\|10\rangle + \beta\|11\rangle - \gamma\|00\rangle - \delta\|01\rangle$	$i \cdot \sigma_Y \otimes I$
			$\|\Phi^-\rangle$	$\alpha\|10\rangle - \beta\|11\rangle - \gamma\|00\rangle + \delta\|01\rangle$	$i \cdot \sigma_Y \otimes \sigma_Z$
	$\|\Phi^-\rangle$	$\|\Phi^+\rangle$	$\|\Phi^+\rangle$	$-\alpha\|10\rangle - \beta\|11\rangle + \gamma\|00\rangle + \delta\|01\rangle$	$i \cdot \sigma_Y \otimes I$
			$\|\Phi^-\rangle$	$-\alpha\|10\rangle + \beta\|11\rangle + \gamma\|00\rangle - \delta\|01\rangle$	$i \cdot \sigma_Y \otimes \sigma_Z$
		$\|\Phi^-\rangle$	$\|\Phi^+\rangle$	$-\alpha\|10\rangle - \beta\|11\rangle - \gamma\|00\rangle \quad \delta\|01\rangle$	$\sigma_X \otimes I$
			$\|\Phi^-\rangle$	$-\alpha\|10\rangle + \beta\|11\rangle - \gamma\|00\rangle + \delta\|01\rangle$	$\sigma_X \otimes \sigma_Z$

(continued)

Table 4.1 (continued)

Bob$_1$'s measurement results of particles (2,3)	Bob$_2$'s measurement results of particles (4,5)	Alice's measurement results $(x,1)$	Alice's measurement results $(y,8)$	Collapsed state of particles (6,7)	Bob$_3$'s unitary transformation
$\lvert\Psi^-\rangle$	$\lvert\Phi^+\rangle$	$\lvert\Phi^+\rangle$	$\lvert\Phi^+\rangle$	$\alpha\lvert 10\rangle+\beta\lvert 11\rangle-\gamma\lvert 00\rangle-\delta\lvert 01\rangle$	$i\cdot\sigma_Y\otimes I$
			$\lvert\Phi^-\rangle$	$\alpha\lvert 10\rangle-\beta\lvert 11\rangle-\gamma\lvert 00\rangle+\delta\lvert 01\rangle$	$i\cdot\sigma_Y\otimes\sigma_Z$
		$\lvert\Phi^-\rangle$	$\lvert\Phi^+\rangle$	$\alpha\lvert 10\rangle+\beta\lvert 11\rangle+\gamma\lvert 00\rangle+\delta\lvert 01\rangle$	$\sigma_X\otimes I$
			$\lvert\Phi^-\rangle$	$\alpha\lvert 10\rangle-\beta\lvert 11\rangle+\gamma\lvert 00\rangle-\delta\lvert 01\rangle$	$\sigma_X\otimes\sigma_Z$
	$\lvert\Phi^-\rangle$	$\lvert\Phi^+\rangle$	$\lvert\Phi^+\rangle$	$-\alpha\lvert 10\rangle-\beta\lvert 11\rangle-\gamma\lvert 00\rangle-\delta\lvert 01\rangle$	$\sigma_X\otimes I$
			$\lvert\Phi^-\rangle$	$-\alpha\lvert 10\rangle+\beta\lvert 11\rangle-\gamma\lvert 00\rangle+\delta\lvert 01\rangle$	$\sigma_X\otimes\sigma_Z$
		$\lvert\Phi^-\rangle$	$\lvert\Phi^+\rangle$	$-\alpha\lvert 10\rangle-\beta\lvert 11\rangle+\gamma\lvert 00\rangle+\delta\lvert 01\rangle$	$i\cdot\sigma_Y\otimes I$
			$\lvert\Phi^-\rangle$	$-\alpha\lvert 10\rangle+\beta\lvert 11\rangle+\gamma\lvert 00\rangle-\delta\lvert 01\rangle$	$i\cdot\sigma_Y\otimes\sigma_Z$
$\lvert\Psi^+\rangle$	$\lvert\Phi^+\rangle$	$\lvert\Phi^+\rangle$	$\lvert\Psi^+\rangle$	$\alpha\lvert 11\rangle+\beta\lvert 10\rangle+\gamma\lvert 01\rangle+\delta\lvert 00\rangle$	$\sigma_X\otimes\sigma_X$
			$\lvert\Psi^-\rangle$	$\alpha\lvert 11\rangle-\beta\lvert 10\rangle+\gamma\lvert 01\rangle-\delta\lvert 00\rangle$	$\sigma_X\otimes i\cdot\sigma_Y$
		$\lvert\Phi^-\rangle$	$\lvert\Psi^+\rangle$	$\alpha\lvert 11\rangle+\beta\lvert 10\rangle-\gamma\lvert 01\rangle-\delta\lvert 00\rangle$	$i\cdot\sigma_Y\otimes\sigma_X$
			$\lvert\Psi^-\rangle$	$\alpha\lvert 11\rangle-\beta\lvert 10\rangle-\gamma\lvert 01\rangle+\delta\lvert 00\rangle$	$i\cdot\sigma_Y\otimes i\cdot\sigma_Y$
	$\lvert\Phi^-\rangle$	$\lvert\Phi^+\rangle$	$\lvert\Psi^+\rangle$	$-\alpha\lvert 11\rangle-\beta\lvert 10\rangle+\gamma\lvert 01\rangle+\delta\lvert 00\rangle$	$i\cdot\sigma_Y\otimes\sigma_X$
			$\lvert\Psi^-\rangle$	$-\alpha\lvert 11\rangle+\beta\lvert 10\rangle+\gamma\lvert 01\rangle-\delta\lvert 00\rangle$	$i\cdot\sigma_Y\otimes i\cdot\sigma_Y$
		$\lvert\Phi^-\rangle$	$\lvert\Psi^+\rangle$	$-\alpha\lvert 11\rangle-\beta\lvert 10\rangle-\gamma\lvert 01\rangle-\delta\lvert 00\rangle$	$\sigma_X\otimes\sigma_X$
			$\lvert\Psi^-\rangle$	$-\alpha\lvert 11\rangle+\beta\lvert 10\rangle-\gamma\lvert 01\rangle+\delta\lvert 00\rangle$	$\sigma_X\otimes i\cdot\sigma_Y$
$\lvert\Psi^-\rangle$	$\lvert\Phi^+\rangle$	$\lvert\Phi^+\rangle$	$\lvert\Psi^+\rangle$	$\alpha\lvert 11\rangle+\beta\lvert 10\rangle-\gamma\lvert 01\rangle-\delta\lvert 00\rangle$	$i\cdot\sigma_Y\otimes\sigma_X$
			$\lvert\Psi^-\rangle$	$\alpha\lvert 11\rangle-\beta\lvert 10\rangle-\gamma\lvert 01\rangle+\delta\lvert 00\rangle$	$i\cdot\sigma_Y\otimes i\cdot\sigma_Y$
		$\lvert\Phi^-\rangle$	$\lvert\Psi^+\rangle$	$\alpha\lvert 11\rangle+\beta\lvert 10\rangle+\gamma\lvert 01\rangle+\delta\lvert 00\rangle$	$\sigma_X\otimes\sigma_X$
			$\lvert\Psi^-\rangle$	$\alpha\lvert 11\rangle-\beta\lvert 10\rangle+\gamma\lvert 01\rangle-\delta\lvert 00\rangle$	$\sigma_X\otimes i\cdot\sigma_Y$
	$\lvert\Phi^-\rangle$	$\lvert\Phi^+\rangle$	$\lvert\Psi^+\rangle$	$-\alpha\lvert 11\rangle-\beta\lvert 10\rangle-\gamma\lvert 01\rangle-\delta\lvert 00\rangle$	$\sigma_X\otimes\sigma_X$
			$\lvert\Psi^-\rangle$	$-\alpha\lvert 11\rangle+\beta\lvert 10\rangle-\gamma\lvert 01\rangle+\delta\lvert 00\rangle$	$\sigma_X\otimes i\cdot\sigma_Y$
		$\lvert\Phi^-\rangle$	$\lvert\Psi^+\rangle$	$-\alpha\lvert 11\rangle-\beta\lvert 10\rangle+\gamma\lvert 01\rangle+\delta\lvert 00\rangle$	$i\cdot\sigma_Y\otimes\sigma_X$
			$\lvert\Psi^-\rangle$	$-\alpha\lvert 11\rangle+\beta\lvert 10\rangle+\gamma\lvert 01\rangle-\delta\lvert 00\rangle$	$i\cdot\sigma_Y\otimes i\cdot\sigma_Y$
$\lvert\Psi^+\rangle$	$\lvert\Phi^+\rangle$	$\lvert\Psi^+\rangle$	$\lvert\Phi^+\rangle$	$\alpha\lvert 00\rangle+\beta\lvert 01\rangle+\gamma\lvert 10\rangle+\delta\lvert 11\rangle$	$I\otimes I$
			$\lvert\Phi^-\rangle$	$\alpha\lvert 00\rangle-\beta\lvert 01\rangle+\gamma\lvert 10\rangle-\delta\lvert 11\rangle$	$I\otimes\sigma_Z$
		$\lvert\Psi^-\rangle$	$\lvert\Phi^+\rangle$	$\alpha\lvert 00\rangle+\beta\lvert 01\rangle-\gamma\lvert 10\rangle-\delta\lvert 11\rangle$	$\sigma_Z\otimes I$
			$\lvert\Phi^-\rangle$	$\alpha\lvert 00\rangle-\beta\lvert 01\rangle-\gamma\lvert 10\rangle+\delta\lvert 11\rangle$	$\sigma_Z\otimes\sigma_Z$
	$\lvert\Phi^-\rangle$	$\lvert\Psi^+\rangle$	$\lvert\Phi^+\rangle$	$\alpha\lvert 00\rangle+\beta\lvert 01\rangle-\gamma\lvert 10\rangle-\delta\lvert 11\rangle$	$\sigma_Z\otimes I$
			$\lvert\Phi^-\rangle$	$\alpha\lvert 00\rangle-\beta\lvert 01\rangle-\gamma\lvert 10\rangle+\delta\lvert 11\rangle$	$\sigma_Z\otimes\sigma_Z$
		$\lvert\Psi^-\rangle$	$\lvert\Phi^+\rangle$	$\alpha\lvert 00\rangle+\beta\lvert 01\rangle+\gamma\lvert 10\rangle+\delta\lvert 11\rangle$	$I\otimes I$
			$\lvert\Phi^-\rangle$	$\alpha\lvert 00\rangle-\beta\lvert 01\rangle+\gamma\lvert 10\rangle-\delta\lvert 11\rangle$	$I\otimes\sigma_Z$
$\lvert\Psi^-\rangle$	$\lvert\Phi^+\rangle$	$\lvert\Psi^+\rangle$	$\lvert\Phi^+\rangle$	$-\alpha\lvert 00\rangle-\beta\lvert 01\rangle+\gamma\lvert 10\rangle+\delta\lvert 11\rangle$	$\sigma_Z\otimes I$
			$\lvert\Phi^-\rangle$	$-\alpha\lvert 00\rangle+\beta\lvert 01\rangle+\gamma\lvert 10\rangle-\delta\lvert 11\rangle$	$\sigma_Z\otimes\sigma_Z$
		$\lvert\Psi^-\rangle$	$\lvert\Phi^+\rangle$	$-\alpha\lvert 00\rangle-\beta\lvert 01\rangle-\gamma\lvert 10\rangle-\delta\lvert 11\rangle$	$I\otimes I$
			$\lvert\Phi^-\rangle$	$-\alpha\lvert 00\rangle+\beta\lvert 01\rangle-\gamma\lvert 10\rangle+\delta\lvert 11\rangle$	$I\otimes\sigma_Z$
	$\lvert\Phi^-\rangle$	$\lvert\Psi^+\rangle$	$\lvert\Phi^+\rangle$	$-\alpha\lvert 00\rangle-\beta\lvert 01\rangle-\gamma\lvert 10\rangle-\delta\lvert 11\rangle$	$I\otimes I$
			$\lvert\Phi^-\rangle$	$-\alpha\lvert 00\rangle+\beta\lvert 01\rangle-\gamma\lvert 10\rangle+\delta\lvert 11\rangle$	$I\otimes\sigma_Z$
		$\lvert\Psi^-\rangle$	$\lvert\Phi^+\rangle$	$-\alpha\lvert 00\rangle-\beta\lvert 01\rangle+\gamma\lvert 10\rangle+\delta\lvert 11\rangle$	$\sigma_Z\otimes I$
			$\lvert\Phi^-\rangle$	$-\alpha\lvert 00\rangle+\beta\lvert 01\rangle+\gamma\lvert 10\rangle-\delta\lvert 11\rangle$	$\sigma_Z\otimes\sigma_Z$

(continued)

Table 4.1 (continued)

Bob_1's measurement results of particles (2,3)	Bob_2's measurement results of particles (4,5)	Alice's measurement results		Collapsed state of particles (6,7)	Bob_3's unitary transformation
		$(x,1)$	$(y,8)$		
$\lvert\Psi^+\rangle$	$\lvert\Phi^+\rangle$	$\lvert\Psi^+\rangle$	$\lvert\Psi^+\rangle$	$\alpha\lvert 01\rangle + \beta\lvert 00\rangle + \gamma\lvert 11\rangle + \delta\lvert 10\rangle$	$I \otimes \sigma_X$
			$\lvert\Psi^-\rangle$	$\alpha\lvert 01\rangle - \beta\lvert 00\rangle + \gamma\lvert 11\rangle - \delta\lvert 10\rangle$	$I \otimes i \cdot \sigma_Y$
		$\lvert\Psi^-\rangle$	$\lvert\Psi^+\rangle$	$\alpha\lvert 01\rangle + \beta\lvert 00\rangle - \gamma\lvert 11\rangle - \delta\lvert 10\rangle$	$\sigma_Z \otimes \sigma_X$
			$\lvert\Psi^-\rangle$	$\alpha\lvert 01\rangle - \beta\lvert 00\rangle - \gamma\lvert 11\rangle + \delta\lvert 10\rangle$	$\sigma_Z \otimes i \cdot \sigma_Y$
	$\lvert\Phi^-\rangle$	$\lvert\Psi^+\rangle$	$\lvert\Psi^+\rangle$	$\alpha\lvert 01\rangle + \beta\lvert 00\rangle - \gamma\lvert 11\rangle - \delta\lvert 10\rangle$	$\sigma_Z \otimes \sigma_X$
			$\lvert\Psi^-\rangle$	$\alpha\lvert 01\rangle - \beta\lvert 00\rangle - \gamma\lvert 11\rangle + \delta\lvert 10\rangle$	$\sigma_Z \otimes i \cdot \sigma_Y$
		$\lvert\Psi^-\rangle$	$\lvert\Psi^+\rangle$	$\alpha\lvert 01\rangle + \beta\lvert 00\rangle + \gamma\lvert 11\rangle + \delta\lvert 10\rangle$	$I \otimes \sigma_X$
			$\lvert\Psi^-\rangle$	$\alpha\lvert 01\rangle - \beta\lvert 00\rangle + \gamma\lvert 11\rangle - \delta\lvert 10\rangle$	$I \otimes i \cdot \sigma_Y$
$\lvert\Psi^-\rangle$	$\lvert\Phi^+\rangle$	$\lvert\Psi^+\rangle$	$\lvert\Psi^+\rangle$	$-\alpha\lvert 01\rangle - \beta\lvert 00\rangle + \gamma\lvert 11\rangle + \delta\lvert 10\rangle$	$\sigma_Z \otimes \sigma_X$
			$\lvert\Psi^-\rangle$	$-\alpha\lvert 01\rangle + \beta\lvert 00\rangle + \gamma\lvert 11\rangle - \delta\lvert 10\rangle$	$\sigma_Z \otimes i \cdot \sigma_Y$
		$\lvert\Psi^-\rangle$	$\lvert\Psi^+\rangle$	$-\alpha\lvert 01\rangle - \beta\lvert 00\rangle - \gamma\lvert 11\rangle - \delta\lvert 10\rangle$	$I \otimes \sigma_X$
			$\lvert\Psi^-\rangle$	$-\alpha\lvert 01\rangle + \beta\lvert 00\rangle - \gamma\lvert 11\rangle + \delta\lvert 10\rangle$	$I \otimes i \cdot \sigma_Y$
	$\lvert\Phi^-\rangle$	$\lvert\Psi^+\rangle$	$\lvert\Psi^+\rangle$	$-\alpha\lvert 01\rangle - \beta\lvert 00\rangle - \gamma\lvert 11\rangle - \delta\lvert 10\rangle$	$I \otimes \sigma_X$
			$\lvert\Psi^-\rangle$	$-\alpha\lvert 01\rangle + \beta\lvert 00\rangle - \gamma\lvert 11\rangle + \delta\lvert 10\rangle$	$I \otimes i \cdot \sigma_Y$
		$\lvert\Psi^-\rangle$	$\lvert\Psi^+\rangle$	$-\alpha\lvert 01\rangle - \beta\lvert 00\rangle + \gamma\lvert 11\rangle + \delta\lvert 10\rangle$	$\sigma_Z \otimes \sigma_X$
			$\lvert\Psi^-\rangle$	$-\alpha\lvert 01\rangle + \beta\lvert 00\rangle + \gamma\lvert 11\rangle - \delta\lvert 10\rangle$	$\sigma_Z \otimes i \cdot \sigma_Y$
$\lvert\Psi^+\rangle$	$\lvert\Psi^+\rangle$	$\lvert\Phi^+\rangle$	$\lvert\Phi^+\rangle$	$\alpha\lvert 00\rangle + \beta\lvert 01\rangle + \gamma\lvert 10\rangle + \delta\lvert 11\rangle$	$I \otimes I$
			$\lvert\Phi^-\rangle$	$\alpha\lvert 00\rangle - \beta\lvert 01\rangle + \gamma\lvert 10\rangle - \delta\lvert 11\rangle$	$I \otimes \sigma_Z$
		$\lvert\Phi^-\rangle$	$\lvert\Phi^+\rangle$	$\alpha\lvert 00\rangle + \beta\lvert 01\rangle - \gamma\lvert 10\rangle - \delta\lvert 11\rangle$	$\sigma_Z \otimes I$
			$\lvert\Phi^-\rangle$	$\alpha\lvert 00\rangle - \beta\lvert 01\rangle - \gamma\lvert 10\rangle + \delta\lvert 11\rangle$	$\sigma_Z \otimes \sigma_Z$
	$\lvert\Psi^-\rangle$	$\lvert\Phi^+\rangle$	$\lvert\Phi^+\rangle$	$-\alpha\lvert 00\rangle - \beta\lvert 01\rangle + \gamma\lvert 10\rangle + \delta\lvert 11\rangle$	$\sigma_Z \otimes I$
			$\lvert\Phi^-\rangle$	$-\alpha\lvert 00\rangle + \beta\lvert 01\rangle + \gamma\lvert 10\rangle - \delta\lvert 11\rangle$	$\sigma_Z \otimes \sigma_Z$
		$\lvert\Phi^-\rangle$	$\lvert\Phi^+\rangle$	$-\alpha\lvert 00\rangle - \beta\lvert 01\rangle - \gamma\lvert 10\rangle - \delta\lvert 11\rangle$	$I \otimes I$
			$\lvert\Phi^-\rangle$	$-\alpha\lvert 00\rangle + \beta\lvert 01\rangle - \gamma\lvert 10\rangle + \delta\lvert 11\rangle$	$I \otimes \sigma_Z$
$\lvert\Psi^-\rangle$	$\lvert\Psi^+\rangle$	$\lvert\Phi^+\rangle$	$\lvert\Phi^+\rangle$	$\alpha\lvert 00\rangle + \beta\lvert 01\rangle - \gamma\lvert 10\rangle - \delta\lvert 11\rangle$	$\sigma_Z \otimes I$
			$\lvert\Phi^-\rangle$	$\alpha\lvert 00\rangle - \beta\lvert 01\rangle - \gamma\lvert 10\rangle + \delta\lvert 11\rangle$	$\sigma_Z \otimes \sigma_Z$
		$\lvert\Phi^-\rangle$	$\lvert\Phi^+\rangle$	$\alpha\lvert 00\rangle + \beta\lvert 01\rangle + \gamma\lvert 10\rangle + \delta\lvert 11\rangle$	$I \otimes I$
			$\lvert\Phi^-\rangle$	$\alpha\lvert 00\rangle - \beta\lvert 01\rangle + \gamma\lvert 10\rangle - \delta\lvert 11\rangle$	$I \otimes \sigma_Z$
	$\lvert\Psi^-\rangle$	$\lvert\Phi^+\rangle$	$\lvert\Phi^+\rangle$	$-\alpha\lvert 00\rangle - \beta\lvert 01\rangle - \gamma\lvert 10\rangle - \delta\lvert 11\rangle$	$I \otimes I$
			$\lvert\Phi^-\rangle$	$-\alpha\lvert 00\rangle + \beta\lvert 01\rangle - \gamma\lvert 10\rangle + \delta\lvert 11\rangle$	$I \otimes \sigma_Z$
		$\lvert\Phi^-\rangle$	$\lvert\Phi^+\rangle$	$-\alpha\lvert 00\rangle - \beta\lvert 01\rangle + \gamma\lvert 10\rangle + \delta\lvert 11\rangle$	$\sigma_Z \otimes I$
			$\lvert\Phi^-\rangle$	$-\alpha\lvert 00\rangle + \beta\lvert 01\rangle + \gamma\lvert 10\rangle - \delta\lvert 11\rangle$	$\sigma_Z \otimes \sigma_Z$
$\lvert\Psi^+\rangle$	$\lvert\Psi^+\rangle$	$\lvert\Phi^+\rangle$	$\lvert\Psi^+\rangle$	$\alpha\lvert 01\rangle + \beta\lvert 00\rangle + \gamma\lvert 11\rangle + \delta\lvert 10\rangle$	$I \otimes \sigma_X$
			$\lvert\Psi^-\rangle$	$\alpha\lvert 01\rangle - \beta\lvert 00\rangle + \gamma\lvert 11\rangle - \delta\lvert 10\rangle$	$I \otimes i \cdot \sigma_Y$
		$\lvert\Phi^-\rangle$	$\lvert\Psi^+\rangle$	$\alpha\lvert 01\rangle + \beta\lvert 00\rangle - \gamma\lvert 11\rangle - \delta\lvert 10\rangle$	$\sigma_Z \otimes \sigma_X$
			$\lvert\Psi^-\rangle$	$\alpha\lvert 01\rangle - \beta\lvert 00\rangle - \gamma\lvert 11\rangle + \delta\lvert 10\rangle$	$\sigma_Z \otimes i \cdot \sigma_Y$
	$\lvert\Psi^-\rangle$	$\lvert\Phi^+\rangle$	$\lvert\Psi^+\rangle$	$-\alpha\lvert 01\rangle - \beta\lvert 00\rangle + \gamma\lvert 11\rangle + \delta\lvert 10\rangle$	$\sigma_Z \otimes \sigma_X$
			$\lvert\Psi^-\rangle$	$-\alpha\lvert 01\rangle + \beta\lvert 00\rangle + \gamma\lvert 11\rangle - \delta\lvert 10\rangle$	$\sigma_Z \otimes i \cdot \sigma_Y$
		$\lvert\Phi^-\rangle$	$\lvert\Psi^+\rangle$	$-\alpha\lvert 01\rangle - \beta\lvert 00\rangle - \gamma\lvert 11\rangle - \delta\lvert 10\rangle$	$I \otimes \sigma_X$
			$\lvert\Psi^-\rangle$	$-\alpha\lvert 01\rangle + \beta\lvert 00\rangle - \gamma\lvert 11\rangle + \delta\lvert 10\rangle$	$I \otimes i \cdot \sigma_Y$

(continued)

Table 4.1 (continued)

Bob$_1$'s measurement results of particles (2,3)	Bob$_2$'s measurement results of particles (4,5)	Alice's measurement results		Collapsed state of particles (6,7)	Bob$_3$'s unitary transformation
		$(x,1)$	$(y,8)$		
$\|\Psi^-\rangle$	$\|\Psi^+\rangle$	$\|\Phi^+\rangle$	$\|\Psi^+\rangle$	$\alpha\|01\rangle + \beta\|00\rangle - \gamma\|11\rangle - \delta\|10\rangle$	$\sigma_Z \otimes \sigma_X$
			$\|\Psi^-\rangle$	$\alpha\|01\rangle - \beta\|00\rangle - \gamma\|11\rangle + \delta\|10\rangle$	$\sigma_Z \otimes i \cdot \sigma_Y$
		$\|\Phi^-\rangle$	$\|\Psi^+\rangle$	$\alpha\|01\rangle + \beta\|00\rangle + \gamma\|11\rangle + \delta\|10\rangle$	$I \otimes \sigma_X$
			$\|\Psi^-\rangle$	$\alpha\|01\rangle - \beta\|00\rangle + \gamma\|11\rangle - \delta\|10\rangle$	$I \otimes i \cdot \sigma_Y$
	$\|\Psi^-\rangle$	$\|\Phi^+\rangle$	$\|\Psi^+\rangle$	$-\alpha\|01\rangle - \beta\|00\rangle - \gamma\|11\rangle - \delta\|10\rangle$	$I \otimes \sigma_X$
			$\|\Psi^-\rangle$	$-\alpha\|01\rangle + \beta\|00\rangle - \gamma\|11\rangle + \delta\|10\rangle$	$I \otimes i \cdot \sigma_Y$
		$\|\Phi^-\rangle$	$\|\Psi^+\rangle$	$-\alpha\|01\rangle - \beta\|00\rangle + \gamma\|11\rangle + \delta\|10\rangle$	$\sigma_Z \otimes \sigma_X$
			$\|\Psi^-\rangle$	$-\alpha\|01\rangle + \beta\|00\rangle + \gamma\|11\rangle - \delta\|10\rangle$	$\sigma_Z \otimes i \cdot \sigma_Y$
$\|\Psi^+\rangle$	$\|\Psi^+\rangle$	$\|\Psi^+\rangle$	$\|\Phi^+\rangle$	$\alpha\|10\rangle + \beta\|11\rangle + \gamma\|00\rangle + \delta\|01\rangle$	$\sigma_X \otimes I$
			$\|\Phi^-\rangle$	$\alpha\|10\rangle - \beta\|11\rangle + \gamma\|00\rangle - \delta\|01\rangle$	$\sigma_X \otimes \sigma_Z$
		$\|\Psi^-\rangle$	$\|\Phi^+\rangle$	$\alpha\|10\rangle + \beta\|11\rangle - \gamma\|00\rangle - \delta\|01\rangle$	$i \cdot \sigma_Y \otimes I$
			$\|\Phi^-\rangle$	$\alpha\|10\rangle - \beta\|11\rangle - \gamma\|00\rangle + \delta\|01\rangle$	$i \cdot \sigma_Y \otimes \sigma_Z$
	$\|\Psi^-\rangle$	$\|\Psi^+\rangle$	$\|\Phi^+\rangle$	$\alpha\|10\rangle + \beta\|11\rangle - \gamma\|00\rangle - \delta\|01\rangle$	$i \cdot \sigma_Y \otimes I$
			$\|\Phi^-\rangle$	$\alpha\|10\rangle - \beta\|11\rangle - \gamma\|00\rangle + \delta\|01\rangle$	$i \cdot \sigma_Y \otimes \sigma_Z$
		$\|\Psi^-\rangle$	$\|\Phi^+\rangle$	$\alpha\|10\rangle + \beta\|11\rangle + \gamma\|00\rangle + \delta\|01\rangle$	$\sigma_X \otimes I$
			$\|\Phi^-\rangle$	$\alpha\|10\rangle - \beta\|11\rangle + \gamma\|00\rangle - \delta\|01\rangle$	$\sigma_X \otimes \sigma_Z$
$\|\Psi^-\rangle$	$\|\Psi^+\rangle$	$\|\Psi^+\rangle$	$\|\Phi^+\rangle$	$-\alpha\|10\rangle - \beta\|11\rangle + \gamma\|00\rangle + \delta\|01\rangle$	$i \cdot \sigma_Y \otimes I$
			$\|\Phi^-\rangle$	$-\alpha\|10\rangle + \beta\|11\rangle + \gamma\|00\rangle - \delta\|01\rangle$	$i \cdot \sigma_Y \otimes \sigma_Z$
		$\|\Psi^-\rangle$	$\|\Phi^+\rangle$	$-\alpha\|10\rangle - \beta\|11\rangle - \gamma\|00\rangle - \delta\|01\rangle$	$\sigma_X \otimes I$
			$\|\Phi^-\rangle$	$-\alpha\|10\rangle + \beta\|11\rangle - \gamma\|00\rangle + \delta\|01\rangle$	$\sigma_X \otimes \sigma_Z$
	$\|\Psi^-\rangle$	$\|\Psi^+\rangle$	$\|\Phi^+\rangle$	$-\alpha\|10\rangle - \beta\|11\rangle - \gamma\|00\rangle - \delta\|01\rangle$	$\sigma_X \otimes I$
			$\|\Phi^-\rangle$	$-\alpha\|10\rangle + \beta\|11\rangle - \gamma\|00\rangle + \delta\|01\rangle$	$\sigma_X \otimes \sigma_Z$
		$\|\Psi^-\rangle$	$\|\Phi^+\rangle$	$-\alpha\|10\rangle - \beta\|11\rangle + \gamma\|00\rangle + \delta\|01\rangle$	$i \cdot \sigma_Y \otimes I$
			$\|\Phi^-\rangle$	$-\alpha\|10\rangle + \beta\|11\rangle + \gamma\|00\rangle - \delta\|01\rangle$	$i \cdot \sigma_Y \otimes \sigma_Z$
$\|\Psi^+\rangle$	$\|\Psi^+\rangle$	$\|\Psi^+\rangle$	$\|\Psi^+\rangle$	$\alpha\|11\rangle + \beta\|10\rangle + \gamma\|01\rangle + \delta\|00\rangle$	$\sigma_X \otimes \sigma_X$
			$\|\Psi^-\rangle$	$\alpha\|11\rangle - \beta\|10\rangle + \gamma\|01\rangle - \delta\|00\rangle$	$\sigma_X \otimes i \cdot \sigma_Y$
		$\|\Psi^-\rangle$	$\|\Psi^+\rangle$	$\alpha\|11\rangle + \beta\|10\rangle - \gamma\|01\rangle - \delta\|00\rangle$	$i \cdot \sigma_Y \otimes \sigma_X$
			$\|\Psi^-\rangle$	$\alpha\|11\rangle - \beta\|10\rangle - \gamma\|01\rangle + \delta\|00\rangle$	$i \cdot \sigma_Y \otimes i \cdot \sigma_Y$
	$\|\Psi^-\rangle$	$\|\Psi^+\rangle$	$\|\Psi^+\rangle$	$\alpha\|11\rangle + \beta\|10\rangle - \gamma\|01\rangle - \delta\|00\rangle$	$i \cdot \sigma_Y \otimes \sigma_X$
			$\|\Psi^-\rangle$	$\alpha\|11\rangle - \beta\|10\rangle - \gamma\|01\rangle + \delta\|00\rangle$	$i \cdot \sigma_Y \otimes i \cdot \sigma_Y$
		$\|\Psi^-\rangle$	$\|\Psi^+\rangle$	$\alpha\|11\rangle + \beta\|10\rangle + \gamma\|01\rangle + \delta\|00\rangle$	$\sigma_X \otimes \sigma_X$
			$\|\Psi^-\rangle$	$\alpha\|11\rangle - \beta\|10\rangle + \gamma\|01\rangle - \delta\|00\rangle$	$\sigma_X \otimes i \cdot \sigma_Y$
$\|\Psi^-\rangle$	$\|\Psi^+\rangle$	$\|\Psi^+\rangle$	$\|\Psi^+\rangle$	$-\alpha\|11\rangle - \beta\|10\rangle + \gamma\|01\rangle + \delta\|00\rangle$	$i \cdot \sigma_Y \otimes \sigma_X$
			$\|\Psi^-\rangle$	$-\alpha\|11\rangle + \beta\|10\rangle + \gamma\|01\rangle - \delta\|00\rangle$	$i \cdot \sigma_Y \otimes i \cdot \sigma_Y$
		$\|\Psi^-\rangle$	$\|\Psi^+\rangle$	$-\alpha\|11\rangle - \beta\|10\rangle - \gamma\|01\rangle - \delta\|00\rangle$	$\sigma_X \otimes \sigma_X$
			$\|\Psi^-\rangle$	$-\alpha\|11\rangle + \beta\|10\rangle - \gamma\|01\rangle + \delta\|00\rangle$	$\sigma_X \otimes i \cdot \sigma_Y$
	$\|\Psi^-\rangle$	$\|\Psi^+\rangle$	$\|\Psi^+\rangle$	$-\alpha\|11\rangle - \beta\|10\rangle - \gamma\|01\rangle - \delta\|00\rangle$	$\sigma_X \otimes \sigma_X$
			$\|\Psi^-\rangle$	$-\alpha\|11\rangle + \beta\|10\rangle - \gamma\|01\rangle + \delta\|00\rangle$	$\sigma_X \otimes i \cdot \sigma_Y$
		$\|\Psi^-\rangle$	$\|\Psi^+\rangle$	$-\alpha\|11\rangle - \beta\|10\rangle + \gamma\|01\rangle + \delta\|00\rangle$	$i \cdot \sigma_Y \otimes \sigma_X$
			$\|\Psi^-\rangle$	$-\alpha\|11\rangle + \beta\|10\rangle + \gamma\|01\rangle - \delta\|00\rangle$	$i \cdot \sigma_Y \otimes i \cdot \sigma_Y$

Table 4.2 Bob_3's specific unitary transformation that correspond to Alice's, Bob_1's, and Bob_2's measurement results

Alice's measurement results		Bob_1's measurement results of particles (1,4)	Bob_2's measurement results of particles (3,6)	Collapsed state of particles (8,5)	Bob_3's unitary transformation
$(x, 7)$	$(y, 2)$				
$\|\Phi^+\rangle$	$\|\Phi^+\rangle$	$\|\Phi^+\rangle$	$\|\Phi^+\rangle$	$\alpha\|00\rangle + \beta\|01\rangle + \gamma\|10\rangle + \delta\|11\rangle$	$I \otimes I$
			$\|\Phi^-\rangle$	$\alpha\|00\rangle - \beta\|01\rangle + \gamma\|10\rangle - \delta\|11\rangle$	$I \otimes \sigma_Z$
		$\|\Phi^-\rangle$	$\|\Phi^+\rangle$	$\alpha\|00\rangle - \beta\|01\rangle + \gamma\|10\rangle - \delta\|11\rangle$	$I \otimes \sigma_Z$
			$\|\Phi^-\rangle$	$\alpha\|00\rangle + \beta\|01\rangle + \gamma\|10\rangle + \delta\|11\rangle$	$I \otimes I$
	$\|\Phi^-\rangle$	$\|\Phi^+\rangle$	$\|\Phi^+\rangle$	$\alpha\|00\rangle - \beta\|01\rangle + \gamma\|10\rangle - \delta\|11\rangle$	$I \otimes \sigma_Z$
			$\|\Phi^-\rangle$	$\alpha\|00\rangle + \beta\|01\rangle + \gamma\|10\rangle + \delta\|11\rangle$	$I \otimes I$
		$\|\Phi^-\rangle$	$\|\Phi^+\rangle$	$\alpha\|00\rangle + \beta\|01\rangle + \gamma\|10\rangle + \delta\|11\rangle$	$I \otimes I$
			$\|\Phi^-\rangle$	$\alpha\|00\rangle - \beta\|01\rangle + \gamma\|10\rangle - \delta\|11\rangle$	$I \otimes \sigma_Z$
$\|\Phi^-\rangle$	$\|\Phi^+\rangle$	$\|\Phi^+\rangle$	$\|\Phi^+\rangle$	$\alpha\|00\rangle + \beta\|01\rangle - \gamma\|10\rangle - \delta\|11\rangle$	$\sigma_Z \otimes I$
			$\|\Phi^-\rangle$	$\alpha\|00\rangle - \beta\|01\rangle - \gamma\|10\rangle + \delta\|11\rangle$	$\sigma_Z \otimes \sigma_Z$
		$\|\Phi^-\rangle$	$\|\Phi^+\rangle$	$\alpha\|00\rangle - \beta\|01\rangle - \gamma\|10\rangle + \delta\|11\rangle$	$\sigma_Z \otimes \sigma_Z$
			$\|\Phi^-\rangle$	$\alpha\|00\rangle + \beta\|01\rangle - \gamma\|10\rangle - \delta\|11\rangle$	$\sigma_Z \otimes I$
	$\|\Phi^-\rangle$	$\|\Phi^+\rangle$	$\|\Phi^+\rangle$	$\alpha\|00\rangle - \beta\|01\rangle - \gamma\|10\rangle + \delta\|11\rangle$	$\sigma_Z \otimes \sigma_Z$
			$\|\Phi^-\rangle$	$\alpha\|00\rangle + \beta\|01\rangle - \gamma\|10\rangle - \delta\|11\rangle$	$\sigma_Z \otimes I$
		$\|\Phi^-\rangle$	$\|\Phi^+\rangle$	$\alpha\|00\rangle + \beta\|01\rangle - \gamma\|10\rangle - \delta\|11\rangle$	$\sigma_Z \otimes I$
			$\|\Phi^-\rangle$	$\alpha\|00\rangle - \beta\|01\rangle - \gamma\|10\rangle + \delta\|11\rangle$	$\sigma_Z \otimes \sigma_Z$
$\|\Phi^+\rangle$	$\|\Phi^+\rangle$	$\|\Phi^+\rangle$	$\|\Psi^+\rangle$	$\alpha\|01\rangle + \beta\|00\rangle + \gamma\|11\rangle + \delta\|10\rangle$	$I \otimes \sigma_X$
			$\|\Psi^-\rangle$	$\alpha\|01\rangle - \beta\|00\rangle + \gamma\|11\rangle - \delta\|10\rangle$	$I \otimes i \cdot \sigma_Y$
		$\|\Phi^-\rangle$	$\|\Psi^+\rangle$	$\alpha\|01\rangle - \beta\|00\rangle + \gamma\|11\rangle - \delta\|10\rangle$	$I \otimes i \cdot \sigma_Y$
			$\|\Psi^-\rangle$	$\alpha\|01\rangle + \beta\|00\rangle + \gamma\|11\rangle + \delta\|10\rangle$	$I \otimes \sigma_X$
	$\|\Phi^-\rangle$	$\|\Phi^+\rangle$	$\|\Psi^+\rangle$	$\alpha\|01\rangle - \beta\|00\rangle + \gamma\|11\rangle - \delta\|10\rangle$	$I \otimes i \cdot \sigma_Y$
			$\|\Psi^-\rangle$	$\alpha\|01\rangle + \beta\|00\rangle + \gamma\|11\rangle + \delta\|10\rangle$	$I \otimes \sigma_X$
		$\|\Phi^-\rangle$	$\|\Psi^+\rangle$	$\alpha\|01\rangle + \beta\|00\rangle + \gamma\|11\rangle + \delta\|10\rangle$	$I \otimes \sigma_X$
			$\|\Psi^-\rangle$	$\alpha\|01\rangle - \beta\|00\rangle + \gamma\|11\rangle - \delta\|10\rangle$	$I \otimes i \cdot \sigma_Y$
$\|\Phi^-\rangle$	$\|\Phi^+\rangle$	$\|\Phi^+\rangle$	$\|\Psi^+\rangle$	$\alpha\|01\rangle + \beta\|00\rangle - \gamma\|11\rangle - \delta\|10\rangle$	$\sigma_Z \otimes \sigma_X$
			$\|\Psi^-\rangle$	$\alpha\|01\rangle - \beta\|00\rangle - \gamma\|11\rangle + \delta\|10\rangle$	$\sigma_Z \otimes i \cdot \sigma_Y$
		$\|\Phi^-\rangle$	$\|\Psi^+\rangle$	$\alpha\|01\rangle - \beta\|00\rangle - \gamma\|11\rangle + \delta\|10\rangle$	$\sigma_Z \otimes i \cdot \sigma_Y$
			$\|\Psi^-\rangle$	$\alpha\|01\rangle + \beta\|00\rangle - \gamma\|11\rangle - \delta\|10\rangle$	$\sigma_Z \otimes \sigma_X$
	$\|\Phi^-\rangle$	$\|\Phi^+\rangle$	$\|\Psi^+\rangle$	$\alpha\|01\rangle - \beta\|00\rangle - \gamma\|11\rangle + \delta\|10\rangle$	$\sigma_Z \otimes i \cdot \sigma_Y$
			$\|\Psi^-\rangle$	$\alpha\|01\rangle + \beta\|00\rangle - \gamma\|11\rangle - \delta\|10\rangle$	$\sigma_Z \otimes \sigma_X$
		$\|\Phi^-\rangle$	$\|\Psi^+\rangle$	$\alpha\|01\rangle + \beta\|00\rangle - \gamma\|11\rangle - \delta\|10\rangle$	$\sigma_Z \otimes \sigma_X$
			$\|\Psi^-\rangle$	$\alpha\|01\rangle - \beta\|00\rangle - \gamma\|11\rangle + \delta\|10\rangle$	$\sigma_Z \otimes i \cdot \sigma_Y$
$\|\Phi^+\rangle$	$\|\Phi^+\rangle$	$\|\Psi^+\rangle$	$\|\Phi^+\rangle$	$\alpha\|01\rangle + \beta\|00\rangle + \gamma\|11\rangle + \delta\|10\rangle$	$I \otimes \sigma_X$
			$\|\Phi^-\rangle$	$-\alpha\|01\rangle + \beta\|00\rangle - \gamma\|11\rangle + \delta\|10\rangle$	$I \otimes i \cdot \sigma_Y$
		$\|\Psi^-\rangle$	$\|\Phi^+\rangle$	$\alpha\|01\rangle - \beta\|00\rangle + \gamma\|11\rangle - \delta\|10\rangle$	$I \otimes i \cdot \sigma_Y$
			$\|\Phi^-\rangle$	$-\alpha\|01\rangle - \beta\|00\rangle - \gamma\|11\rangle - \delta\|10\rangle$	$I \otimes \sigma_X$
	$\|\Phi^-\rangle$	$\|\Psi^+\rangle$	$\|\Phi^+\rangle$	$\alpha\|01\rangle - \beta\|00\rangle + \gamma\|11\rangle - \delta\|10\rangle$	$I \otimes i \cdot \sigma_Y$
			$\|\Phi^-\rangle$	$-\alpha\|01\rangle - \beta\|00\rangle - \gamma\|11\rangle - \delta\|10\rangle$	$I \otimes \sigma_X$
		$\|\Psi^-\rangle$	$\|\Phi^+\rangle$	$\alpha\|01\rangle + \beta\|00\rangle + \gamma\|11\rangle + \delta\|10\rangle$	$I \otimes \sigma_X$
			$\|\Phi^-\rangle$	$-\alpha\|01\rangle + \beta\|00\rangle - \gamma\|11\rangle + \delta\|10\rangle$	$I \otimes i \cdot \sigma_Y$

(continued)

Table 4.2 (continued)

Alice's measurement results		Bob$_1$'s measurement results of particles (1,4)	Bob$_2$'s measurement results of particles (3,6)	Collapsed state of particles (8,5)	Bob$_3$'s unitary transformation
$(x, 7)$	$(y, 2)$				
$\|\Phi^-\rangle$	$\|\Phi^+\rangle$	$\|\Psi^+\rangle$	$\|\Phi^+\rangle$	$\alpha\|01\rangle + \beta\|00\rangle - \gamma\|11\rangle - \delta\|10\rangle$	$\sigma_Z \otimes \sigma_X$
			$\|\Phi^-\rangle$	$-\alpha\|01\rangle + \beta\|00\rangle + \gamma\|11\rangle - \delta\|10\rangle$	$\sigma_Z \otimes i \cdot \sigma_Y$
		$\|\Psi^-\rangle$	$\|\Phi^+\rangle$	$\alpha\|01\rangle - \beta\|00\rangle - \gamma\|11\rangle + \delta\|10\rangle$	$\sigma_Z \otimes i \cdot \sigma_Y$
			$\|\Phi^-\rangle$	$-\alpha\|01\rangle - \beta\|00\rangle + \gamma\|11\rangle + \delta\|10\rangle$	$\sigma_Z \otimes \sigma_X$
	$\|\Phi^-\rangle$	$\|\Psi^+\rangle$	$\|\Phi^+\rangle$	$\alpha\|01\rangle - \beta\|00\rangle - \gamma\|11\rangle + \delta\|10\rangle$	$\sigma_Z \otimes i \cdot \sigma_Y$
			$\|\Phi^-\rangle$	$-\alpha\|01\rangle - \beta\|00\rangle + \gamma\|11\rangle + \delta\|10\rangle$	$\sigma_Z \otimes \sigma_X$
		$\|\Psi^-\rangle$	$\|\Phi^+\rangle$	$\alpha\|01\rangle + \beta\|00\rangle - \gamma\|11\rangle - \delta\|10\rangle$	$\sigma_Z \otimes \sigma_X$
			$\|\Phi^-\rangle$	$-\alpha\|01\rangle + \beta\|00\rangle + \gamma\|11\rangle - \delta\|10\rangle$	$\sigma_Z \otimes i \cdot \sigma_Y$
$\|\Phi^+\rangle$	$\|\Phi^+\rangle$	$\|\Psi^+\rangle$	$\|\Psi^+\rangle$	$\alpha\|00\rangle + \beta\|01\rangle + \gamma\|10\rangle + \delta\|11\rangle$	$I \otimes I$
			$\|\Psi^-\rangle$	$-\alpha\|00\rangle + \beta\|01\rangle - \gamma\|10\rangle + \delta\|11\rangle$	$I \otimes \sigma_Z$
		$\|\Psi^-\rangle$	$\|\Psi^+\rangle$	$\alpha\|00\rangle - \beta\|01\rangle + \gamma\|10\rangle - \delta\|11\rangle$	$I \otimes \sigma_Z$
			$\|\Psi^-\rangle$	$-\alpha\|00\rangle - \beta\|01\rangle - \gamma\|10\rangle - \delta\|11\rangle$	$I \otimes I$
	$\|\Phi^-\rangle$	$\|\Psi^+\rangle$	$\|\Psi^+\rangle$	$\alpha\|00\rangle - \beta\|01\rangle + \gamma\|10\rangle - \delta\|11\rangle$	$I \otimes \sigma_Z$
			$\|\Psi^-\rangle$	$-\alpha\|00\rangle - \beta\|01\rangle - \gamma\|10\rangle - \delta\|11\rangle$	$I \otimes I$
		$\|\Psi^-\rangle$	$\|\Psi^+\rangle$	$\alpha\|00\rangle + \beta\|01\rangle + \gamma\|10\rangle + \delta\|11\rangle$	$I \otimes I$
			$\|\Psi^-\rangle$	$-\alpha\|00\rangle + \beta\|01\rangle - \gamma\|10\rangle + \delta\|11\rangle$	$I \otimes \sigma_Z$
$\|\Phi^-\rangle$	$\|\Phi^+\rangle$	$\|\Psi^+\rangle$	$\|\Psi^+\rangle$	$\alpha\|00\rangle + \beta\|01\rangle - \gamma\|10\rangle - \delta\|11\rangle$	$\sigma_Z \otimes I$
			$\|\Psi^-\rangle$	$-\alpha\|00\rangle + \beta\|01\rangle + \gamma\|10\rangle - \delta\|11\rangle$	$\sigma_Z \otimes \sigma_Z$
		$\|\Psi^-\rangle$	$\|\Psi^+\rangle$	$\alpha\|00\rangle - \beta\|01\rangle - \gamma\|10\rangle + \delta\|11\rangle$	$\sigma_Z \otimes \sigma_Z$
			$\|\Psi^-\rangle$	$-\alpha\|00\rangle - \beta\|01\rangle + \gamma\|10\rangle + \delta\|11\rangle$	$\sigma_Z \otimes I$
	$\|\Phi^-\rangle$	$\|\Psi^+\rangle$	$\|\Psi^+\rangle$	$\alpha\|00\rangle - \beta\|01\rangle - \gamma\|10\rangle + \delta\|11\rangle$	$\sigma_Z \otimes \sigma_Z$
			$\|\Psi^-\rangle$	$-\alpha\|00\rangle - \beta\|01\rangle + \gamma\|10\rangle + \delta\|11\rangle$	$\sigma_Z \otimes I$
		$\|\Psi^-\rangle$	$\|\Psi^+\rangle$	$\alpha\|00\rangle + \beta\|01\rangle - \gamma\|10\rangle - \delta\|11\rangle$	$\sigma_Z \otimes I$
			$\|\Psi^-\rangle$	$-\alpha\|00\rangle + \beta\|01\rangle + \gamma\|10\rangle - \delta\|11\rangle$	$\sigma_Z \otimes \sigma_Z$
$\|\Phi^+\rangle$	$\|\Psi^+\rangle$	$\|\Phi^+\rangle$	$\|\Phi^+\rangle$	$\alpha\|01\rangle + \beta\|00\rangle + \gamma\|11\rangle + \delta\|10\rangle$	$I \otimes \sigma_X$
			$\|\Phi^-\rangle$	$-\alpha\|01\rangle + \beta\|00\rangle - \gamma\|11\rangle + \delta\|10\rangle$	$I \otimes i \cdot \sigma_Y$
		$\|\Phi^-\rangle$	$\|\Phi^+\rangle$	$-\alpha\|01\rangle + \beta\|00\rangle - \gamma\|11\rangle + \delta\|10\rangle$	$I \otimes i \cdot \sigma_Y$
			$\|\Phi^-\rangle$	$\alpha\|01\rangle + \beta\|00\rangle + \gamma\|11\rangle + \delta\|10\rangle$	$I \otimes \sigma_X$
	$\|\Psi^-\rangle$	$\|\Phi^+\rangle$	$\|\Phi^+\rangle$	$\alpha\|01\rangle - \beta\|00\rangle + \gamma\|11\rangle - \delta\|10\rangle$	$I \otimes i \cdot \sigma_Y$
			$\|\Phi^-\rangle$	$-\alpha\|01\rangle - \beta\|00\rangle - \gamma\|11\rangle - \delta\|10\rangle$	$I \otimes \sigma_X$
		$\|\Phi^-\rangle$	$\|\Phi^+\rangle$	$-\alpha\|01\rangle - \beta\|00\rangle - \gamma\|11\rangle - \delta\|10\rangle$	$I \otimes \sigma_X$
			$\|\Phi^-\rangle$	$\alpha\|01\rangle - \beta\|00\rangle + \gamma\|11\rangle - \delta\|10\rangle$	$I \otimes i \cdot \sigma_Y$
$\|\Phi^-\rangle$	$\|\Psi^+\rangle$	$\|\Phi^+\rangle$	$\|\Phi^+\rangle$	$\alpha\|01\rangle + \beta\|00\rangle - \gamma\|11\rangle - \delta\|10\rangle$	$\sigma_Z \otimes \sigma_X$
			$\|\Phi^-\rangle$	$-\alpha\|01\rangle + \beta\|00\rangle + \gamma\|11\rangle - \delta\|10\rangle$	$\sigma_Z \otimes i \cdot \sigma_Y$
		$\|\Phi^-\rangle$	$\|\Phi^+\rangle$	$-\alpha\|01\rangle + \beta\|00\rangle + \gamma\|11\rangle - \delta\|10\rangle$	$\sigma_Z \otimes i \cdot \sigma_Y$
			$\|\Phi^-\rangle$	$\alpha\|01\rangle + \beta\|00\rangle - \gamma\|11\rangle - \delta\|10\rangle$	$\sigma_Z \otimes \sigma_X$
	$\|\Psi^-\rangle$	$\|\Phi^+\rangle$	$\|\Phi^+\rangle$	$\alpha\|01\rangle - \beta\|00\rangle - \gamma\|11\rangle + \delta\|10\rangle$	$\sigma_Z \otimes i \cdot \sigma_Y$
			$\|\Phi^-\rangle$	$-\alpha\|01\rangle - \beta\|00\rangle + \gamma\|11\rangle + \delta\|10\rangle$	$\sigma_Z \otimes \sigma_X$
		$\|\Phi^-\rangle$	$\|\Phi^+\rangle$	$-\alpha\|01\rangle - \beta\|00\rangle + \gamma\|11\rangle + \delta\|10\rangle$	$\sigma_Z \otimes \sigma_X$
			$\|\Phi^-\rangle$	$\alpha\|01\rangle - \beta\|00\rangle - \gamma\|11\rangle + \delta\|10\rangle$	$\sigma_Z \otimes i \cdot \sigma_Y$

(continued)

Table 4.2 (continued)

Alice's measurement results		Bob_1's measurement results of	Bob_2's measurement results of		Bob_3's unitary
$(x, 7)$	$(y, 2)$	particles (1,4)	particles (3,6)	Collapsed state of particles (8,5)	transformation
$\|\Phi^+\rangle$	$\|\Psi^+\rangle$	$\|\Phi^+\rangle$	$\|\Psi^+\rangle$	$\alpha\|00\rangle + \beta\|01\rangle + \gamma\|10\rangle + \delta\|11\rangle$	$I \otimes I$
			$\|\Psi^-\rangle$	$-\alpha\|00\rangle + \beta\|01\rangle - \gamma\|10\rangle + \delta\|11\rangle$	$I \otimes \sigma_Z$
		$\|\Phi^-\rangle$	$\|\Psi^+\rangle$	$-\alpha\|00\rangle + \beta\|01\rangle - \gamma\|10\rangle + \delta\|11\rangle$	$I \otimes \sigma_Z$
			$\|\Psi^-\rangle$	$\alpha\|00\rangle + \beta\|01\rangle + \gamma\|10\rangle + \delta\|11\rangle$	$I \otimes I$
	$\|\Psi^-\rangle$	$\|\Phi^+\rangle$	$\|\Psi^+\rangle$	$\alpha\|00\rangle - \beta\|01\rangle + \gamma\|10\rangle - \delta\|11\rangle$	$I \otimes \sigma_Z$
			$\|\Psi^-\rangle$	$-\alpha\|00\rangle - \beta\|01\rangle - \gamma\|10\rangle - \delta\|11\rangle$	$I \otimes I$
		$\|\Phi^-\rangle$	$\|\Psi^+\rangle$	$-\alpha\|00\rangle - \beta\|01\rangle - \gamma\|10\rangle - \delta\|11\rangle$	$I \otimes I$
			$\|\Psi^-\rangle$	$\alpha\|00\rangle - \beta\|01\rangle + \gamma\|10\rangle - \delta\|11\rangle$	$I \otimes \sigma_Z$
$\|\Phi^-\rangle$	$\|\Psi^+\rangle$	$\|\Phi^+\rangle$	$\|\Psi^+\rangle$	$\alpha\|00\rangle + \beta\|01\rangle - \gamma\|10\rangle - \delta\|11\rangle$	$\sigma_Z \otimes I$
			$\|\Psi^-\rangle$	$-\alpha\|00\rangle + \beta\|01\rangle + \gamma\|10\rangle - \delta\|11\rangle$	$\sigma_Z \otimes \sigma_Z$
		$\|\Phi^-\rangle$	$\|\Psi^+\rangle$	$-\alpha\|00\rangle + \beta\|01\rangle + \gamma\|10\rangle - \delta\|11\rangle$	$\sigma_Z \otimes \sigma_Z$
			$\|\Psi^-\rangle$	$\alpha\|00\rangle + \beta\|01\rangle - \gamma\|10\rangle - \delta\|11\rangle$	$\sigma_Z \otimes I$
	$\|\Psi^-\rangle$	$\|\Phi^+\rangle$	$\|\Psi^+\rangle$	$\alpha\|00\rangle - \beta\|01\rangle - \gamma\|10\rangle + \delta\|11\rangle$	$\sigma_Z \otimes \sigma_Z$
			$\|\Psi^-\rangle$	$-\alpha\|00\rangle - \beta\|01\rangle + \gamma\|10\rangle + \delta\|11\rangle$	$\sigma_Z \otimes I$
		$\|\Phi^-\rangle$	$\|\Psi^+\rangle$	$-\alpha\|00\rangle - \beta\|01\rangle + \gamma\|10\rangle + \delta\|11\rangle$	$\sigma_Z \otimes I$
			$\|\Psi^-\rangle$	$\alpha\|00\rangle - \beta\|01\rangle - \gamma\|10\rangle + \delta\|11\rangle$	$\sigma_Z \otimes \sigma_Z$
$\|\Phi^+\rangle$	$\|\Psi^+\rangle$	$\|\Psi^+\rangle$	$\|\Phi^+\rangle$	$\alpha\|00\rangle + \beta\|01\rangle + \gamma\|10\rangle + \delta\|11\rangle$	$I \otimes I$
			$\|\Phi^-\rangle$	$\alpha\|00\rangle - \beta\|01\rangle + \gamma\|10\rangle - \delta\|11\rangle$	$I \otimes \sigma_Z$
		$\|\Psi^-\rangle$	$\|\Phi^+\rangle$	$-\alpha\|00\rangle + \beta\|01\rangle - \gamma\|10\rangle + \delta\|11\rangle$	$I \otimes \sigma_Z$
			$\|\Phi^-\rangle$	$-\alpha\|00\rangle - \beta\|01\rangle - \gamma\|10\rangle - \delta\|11\rangle$	$I \otimes I$
	$\|\Psi^-\rangle$	$\|\Psi^+\rangle$	$\|\Phi^+\rangle$	$\alpha\|00\rangle - \beta\|01\rangle + \gamma\|10\rangle - \delta\|11\rangle$	$I \otimes \sigma_Z$
			$\|\Phi^-\rangle$	$\alpha\|00\rangle + \beta\|01\rangle + \gamma\|10\rangle + \delta\|11\rangle$	$I \otimes I$
		$\|\Psi^-\rangle$	$\|\Phi^+\rangle$	$-\alpha\|00\rangle - \beta\|01\rangle - \gamma\|10\rangle - \delta\|11\rangle$	$I \otimes I$
			$\|\Phi^-\rangle$	$-\alpha\|00\rangle + \beta\|01\rangle - \gamma\|10\rangle + \delta\|11\rangle$	$I \otimes \sigma_Z$
$\|\Phi^-\rangle$	$\|\Psi^+\rangle$	$\|\Psi^+\rangle$	$\|\Phi^+\rangle$	$\alpha\|00\rangle + \beta\|01\rangle - \gamma\|10\rangle - \delta\|11\rangle$	$\sigma_Z \otimes I$
			$\|\Phi^-\rangle$	$\alpha\|00\rangle - \beta\|01\rangle - \gamma\|10\rangle + \delta\|11\rangle$	$\sigma_Z \otimes \sigma_Z$
		$\|\Psi^-\rangle$	$\|\Phi^+\rangle$	$-\alpha\|00\rangle + \beta\|01\rangle + \gamma\|10\rangle - \delta\|11\rangle$	$\sigma_Z \otimes \sigma_Z$
			$\|\Phi^-\rangle$	$-\alpha\|00\rangle - \beta\|01\rangle + \gamma\|10\rangle + \delta\|11\rangle$	$\sigma_Z \otimes I$
	$\|\Psi^-\rangle$	$\|\Psi^+\rangle$	$\|\Phi^+\rangle$	$\alpha\|00\rangle - \beta\|01\rangle - \gamma\|10\rangle + \delta\|11\rangle$	$\sigma_Z \otimes \sigma_Z$
			$\|\Phi^-\rangle$	$\alpha\|00\rangle + \beta\|01\rangle - \gamma\|10\rangle - \delta\|11\rangle$	$\sigma_Z \otimes I$
		$\|\Psi^-\rangle$	$\|\Phi^+\rangle$	$-\alpha\|00\rangle - \beta\|01\rangle + \gamma\|10\rangle + \delta\|11\rangle$	$\sigma_Z \otimes I$
			$\|\Phi^-\rangle$	$-\alpha\|00\rangle + \beta\|01\rangle + \gamma\|10\rangle - \delta\|11\rangle$	$\sigma_Z \otimes \sigma_Z$
$\|\Phi^+\rangle$	$\|\Psi^+\rangle$	$\|\Psi^+\rangle$	$\|\Psi^+\rangle$	$\alpha\|01\rangle + \beta\|00\rangle + \gamma\|11\rangle + \delta\|10\rangle$	$I \otimes \sigma_X$
			$\|\Psi^-\rangle$	$\alpha\|01\rangle - \beta\|00\rangle + \gamma\|11\rangle - \delta\|10\rangle$	$I \otimes i \cdot \sigma_Y$
		$\|\Psi^-\rangle$	$\|\Psi^+\rangle$	$-\alpha\|01\rangle + \beta\|00\rangle - \gamma\|11\rangle + \delta\|10\rangle$	$I \otimes i \cdot \sigma_Y$
			$\|\Psi^-\rangle$	$-\alpha\|01\rangle - \beta\|00\rangle - \gamma\|11\rangle - \delta\|10\rangle$	$I \otimes \sigma_X$
	$\|\Psi^-\rangle$	$\|\Psi^+\rangle$	$\|\Psi^+\rangle$	$\alpha\|01\rangle - \beta\|00\rangle + \gamma\|11\rangle - \delta\|10\rangle$	$I \otimes i \cdot \sigma_Y$
			$\|\Psi^-\rangle$	$\alpha\|01\rangle + \beta\|00\rangle + \gamma\|11\rangle + \delta\|10\rangle$	$I \otimes \sigma_X$
		$\|\Psi^-\rangle$	$\|\Psi^+\rangle$	$\alpha\|01\rangle - \beta\|00\rangle - \gamma\|11\rangle - \delta\|10\rangle$	$I \otimes \sigma_X$
			$\|\Psi^-\rangle$	$-\alpha\|01\rangle + \beta\|00\rangle - \gamma\|11\rangle + \delta\|10\rangle$	$I \otimes i \cdot \sigma_Y$

(continued)

Table 4.2 (continued)

Alice's measurement results		Bob$_1$'s measurement results of	Bob$_2$'s measurement results of		Bob$_3$'s unitary
$(x,7)$	$(y,2)$	particles (1,4)	particles (3,6)	Collapsed state of particles (8,5)	transformation
$\|\Phi^-\rangle$	$\|\Psi^+\rangle$	$\|\Psi^+\rangle$	$\|\Psi^+\rangle$	$\alpha\|01\rangle+\beta\|00\rangle-\gamma\|11\rangle-\delta\|10\rangle$	$\sigma_Z\otimes\sigma_X$
			$\|\Psi^-\rangle$	$\alpha\|01\rangle-\beta\|00\rangle-\gamma\|11\rangle+\delta\|10\rangle$	$\sigma_Z\otimes i\cdot\sigma_Y$
		$\|\Psi^-\rangle$	$\|\Psi^+\rangle$	$-\alpha\|01\rangle+\beta\|00\rangle+\gamma\|11\rangle-\delta\|10\rangle$	$\sigma_Z\otimes i\cdot\sigma_Y$
			$\|\Psi^-\rangle$	$-\alpha\|01\rangle-\beta\|00\rangle+\gamma\|11\rangle+\delta\|10\rangle$	$\sigma_Z\otimes\sigma_X$
	$\|\Psi^-\rangle$	$\|\Psi^+\rangle$	$\|\Psi^+\rangle$	$\alpha\|01\rangle-\beta\|00\rangle-\gamma\|11\rangle+\delta\|10\rangle$	$\sigma_Z\otimes i\cdot\sigma_Y$
			$\|\Psi^-\rangle$	$\alpha\|01\rangle+\beta\|00\rangle-\gamma\|11\rangle-\delta\|10\rangle$	$\sigma_Z\otimes\sigma_X$
		$\|\Psi^-\rangle$	$\|\Psi^+\rangle$	$-\alpha\|01\rangle-\beta\|00\rangle+\gamma\|11\rangle+\delta\|10\rangle$	$\sigma_Z\otimes\sigma_X$
			$\|\Psi^-\rangle$	$-\alpha\|01\rangle+\beta\|00\rangle+\gamma\|11\rangle-\delta\|10\rangle$	$\sigma_Z\otimes i\cdot\sigma_Y$
$\|\Psi^+\rangle$	$\|\Phi^+\rangle$	$\|\Phi^+\rangle$	$\|\Phi^+\rangle$	$\alpha\|10\rangle+\beta\|11\rangle+\gamma\|00\rangle+\delta\|01\rangle$	$\sigma_X\otimes I$
			$\|\Phi^-\rangle$	$\alpha\|10\rangle-\beta\|11\rangle+\gamma\|00\rangle-\delta\|01\rangle$	$\sigma_X\otimes\sigma_Z$
		$\|\Phi^-\rangle$	$\|\Phi^+\rangle$	$\alpha\|10\rangle-\beta\|11\rangle+\gamma\|00\rangle-\delta\|01\rangle$	$\sigma_X\otimes\sigma_Z$
			$\|\Phi^-\rangle$	$\alpha\|10\rangle+\beta\|11\rangle+\gamma\|00\rangle+\delta\|01\rangle$	$\sigma_X\otimes I$
	$\|\Phi^-\rangle$	$\|\Phi^+\rangle$	$\|\Phi^+\rangle$	$\alpha\|10\rangle-\beta\|11\rangle+\gamma\|00\rangle-\delta\|01\rangle$	$\sigma_X\otimes\sigma_Z$
			$\|\Phi^-\rangle$	$\alpha\|10\rangle+\beta\|11\rangle+\gamma\|00\rangle+\delta\|01\rangle$	$\sigma_X\otimes I$
		$\|\Phi^-\rangle$	$\|\Phi^+\rangle$	$\alpha\|10\rangle+\beta\|11\rangle+\gamma\|00\rangle+\delta\|01\rangle$	$\sigma_X\otimes I$
			$\|\Phi^-\rangle$	$\alpha\|10\rangle-\beta\|11\rangle+\gamma\|00\rangle-\delta\|01\rangle$	$\sigma_X\otimes\sigma_Z$
$\|\Psi^-\rangle$	$\|\Phi^+\rangle$	$\|\Phi^+\rangle$	$\|\Phi^+\rangle$	$\alpha\|10\rangle+\beta\|11\rangle-\gamma\|00\rangle-\delta\|01\rangle$	$i\cdot\sigma_Y\otimes I$
			$\|\Phi^-\rangle$	$\alpha\|10\rangle-\beta\|11\rangle-\gamma\|00\rangle+\delta\|01\rangle$	$i\cdot\sigma_Y\otimes\sigma_Z$
		$\|\Phi^-\rangle$	$\|\Phi^+\rangle$	$\alpha\|10\rangle-\beta\|11\rangle-\gamma\|00\rangle+\delta\|01\rangle$	$i\cdot\sigma_Y\otimes\sigma_Z$
			$\|\Phi^-\rangle$	$\alpha\|10\rangle+\beta\|11\rangle-\gamma\|00\rangle-\delta\|01\rangle$	$i\cdot\sigma_Y\otimes I$
	$\|\Phi^-\rangle$	$\|\Phi^+\rangle$	$\|\Phi^+\rangle$	$\alpha\|10\rangle-\beta\|11\rangle-\gamma\|00\rangle+\delta\|01\rangle$	$i\cdot\sigma_Y\otimes\sigma_Z$
			$\|\Phi^-\rangle$	$\alpha\|10\rangle+\beta\|11\rangle-\gamma\|00\rangle-\delta\|01\rangle$	$i\cdot\sigma_Y\otimes I$
		$\|\Phi^-\rangle$	$\|\Phi^+\rangle$	$\alpha\|10\rangle+\beta\|11\rangle-\gamma\|00\rangle-\delta\|01\rangle$	$i\cdot\sigma_Y\otimes I$
			$\|\Phi^-\rangle$	$\alpha\|10\rangle-\beta\|11\rangle-\gamma\|00\rangle+\delta\|01\rangle$	$i\cdot\sigma_Y\otimes\sigma_Z$
$\|\Psi^+\rangle$	$\|\Phi^+\rangle$	$\|\Phi^+\rangle$	$\|\Psi^+\rangle$	$\alpha\|11\rangle+\beta\|10\rangle+\gamma\|01\rangle+\delta\|00\rangle$	$\sigma_X\otimes\sigma_X$
			$\|\Psi^-\rangle$	$\alpha\|11\rangle-\beta\|10\rangle+\gamma\|01\rangle-\delta\|00\rangle$	$\sigma_X\otimes i\cdot\sigma_Y$
		$\|\Phi^-\rangle$	$\|\Psi^+\rangle$	$\alpha\|11\rangle-\beta\|10\rangle+\gamma\|01\rangle-\delta\|00\rangle$	$\sigma_X\otimes i\cdot\sigma_Y$
			$\|\Psi^-\rangle$	$\alpha\|11\rangle+\beta\|10\rangle+\gamma\|01\rangle+\delta\|00\rangle$	$\sigma_X\otimes\sigma_X$
	$\|\Phi^-\rangle$	$\|\Phi^+\rangle$	$\|\Psi^+\rangle$	$\alpha\|11\rangle-\beta\|10\rangle+\gamma\|01\rangle-\delta\|00\rangle$	$\sigma_X\otimes i\cdot\sigma_Y$
			$\|\Psi^-\rangle$	$\alpha\|11\rangle+\beta\|10\rangle+\gamma\|01\rangle+\delta\|00\rangle$	$\sigma_X\otimes\sigma_X$
		$\|\Phi^-\rangle$	$\|\Psi^+\rangle$	$\alpha\|11\rangle+\beta\|10\rangle+\gamma\|01\rangle+\delta\|00\rangle$	$\sigma_X\otimes\sigma_X$
			$\|\Psi^-\rangle$	$\alpha\|11\rangle-\beta\|10\rangle+\gamma\|01\rangle-\delta\|00\rangle$	$\sigma_X\otimes i\cdot\sigma_Y$
$\|\Psi^-\rangle$	$\|\Phi^+\rangle$	$\|\Phi^+\rangle$	$\|\Psi^+\rangle$	$\alpha\|11\rangle+\beta\|10\rangle-\gamma\|01\rangle-\delta\|00\rangle$	$i\cdot\sigma_Y\otimes\sigma_X$
			$\|\Psi^-\rangle$	$\alpha\|11\rangle-\beta\|10\rangle-\gamma\|01\rangle+\delta\|00\rangle$	$i\cdot\sigma_Y\otimes i\cdot\sigma_Y$
		$\|\Phi^-\rangle$	$\|\Psi^+\rangle$	$\alpha\|11\rangle-\beta\|10\rangle-\gamma\|01\rangle+\delta\|00\rangle$	$i\cdot\sigma_Y\otimes i\cdot\sigma_Y$
			$\|\Psi^-\rangle$	$\alpha\|11\rangle+\beta\|10\rangle-\gamma\|01\rangle-\delta\|00\rangle$	$i\cdot\sigma_Y\otimes\sigma_X$
	$\|\Phi^-\rangle$	$\|\Phi^+\rangle$	$\|\Psi^+\rangle$	$\alpha\|11\rangle-\beta\|10\rangle-\gamma\|01\rangle+\delta\|00\rangle$	$i\cdot\sigma_Y\otimes i\cdot\sigma_Y$
			$\|\Psi^-\rangle$	$\alpha\|11\rangle+\beta\|10\rangle-\gamma\|01\rangle-\delta\|00\rangle$	$i\cdot\sigma_Y\otimes\sigma_X$
		$\|\Phi^-\rangle$	$\|\Psi^+\rangle$	$\alpha\|11\rangle+\beta\|10\rangle-\gamma\|01\rangle-\delta\|00\rangle$	$i\cdot\sigma_Y\otimes\sigma_X$
			$\|\Psi^-\rangle$	$\alpha\|11\rangle-\beta\|10\rangle-\gamma\|01\rangle+\delta\|00\rangle$	$i\cdot\sigma_Y\otimes i\cdot\sigma_Y$

(continued)

Table 4.2 (continued)

Alice's measurement results $(x, 7)$	Alice's measurement results $(y, 2)$	Bob_1's measurement results of particles (1,4)	Bob_2's measurement results of particles (3,6)	Collapsed state of particles (8,5)	Bob_3's unitary transformation
$\|\Psi^+\rangle$	$\|\Phi^+\rangle$	$\|\Psi^+\rangle$	$\|\Phi^+\rangle$	$\alpha\|11\rangle + \beta\|10\rangle + \gamma\|01\rangle + \delta\|00\rangle$	$\sigma_X \otimes \sigma_X$
			$\|\Phi^-\rangle$	$-\alpha\|11\rangle + \beta\|10\rangle - \gamma\|01\rangle + \delta\|00\rangle$	$\sigma_X \otimes i \cdot \sigma_Y$
		$\|\Psi^-\rangle$	$\|\Phi^+\rangle$	$\alpha\|11\rangle - \beta\|10\rangle + \gamma\|01\rangle - \delta\|00\rangle$	$\sigma_X \otimes i \cdot \sigma_Y$
			$\|\Phi^-\rangle$	$-\alpha\|11\rangle - \beta\|10\rangle - \gamma\|01\rangle - \delta\|00\rangle$	$\sigma_X \otimes \sigma_X$
	$\|\Phi^-\rangle$	$\|\Psi^+\rangle$	$\|\Phi^+\rangle$	$\alpha\|11\rangle - \beta\|10\rangle + \gamma\|01\rangle - \delta\|00\rangle$	$\sigma_X \otimes i \cdot \sigma_Y$
			$\|\Phi^-\rangle$	$-\alpha\|11\rangle - \beta\|10\rangle - \gamma\|01\rangle - \delta\|00\rangle$	$\sigma_X \otimes \sigma_X$
		$\|\Psi^-\rangle$	$\|\Phi^+\rangle$	$\alpha\|11\rangle + \beta\|10\rangle + \gamma\|01\rangle + \delta\|00\rangle$	$\sigma_X \otimes \sigma_X$
			$\|\Phi^-\rangle$	$-\alpha\|11\rangle + \beta\|10\rangle - \gamma\|01\rangle + \delta\|00\rangle$	$\sigma_X \otimes i \cdot \sigma_Y$
$\|\Psi^-\rangle$	$\|\Phi^+\rangle$	$\|\Psi^+\rangle$	$\|\Phi^+\rangle$	$\alpha\|11\rangle + \beta\|10\rangle - \gamma\|01\rangle - \delta\|00\rangle$	$i \cdot \sigma_Y \otimes \sigma_X$
			$\|\Phi^-\rangle$	$-\alpha\|11\rangle + \beta\|10\rangle + \gamma\|01\rangle - \delta\|00\rangle$	$i \cdot \sigma_Y \otimes i \cdot \sigma_Y$
		$\|\Psi^-\rangle$	$\|\Phi^+\rangle$	$\alpha\|11\rangle - \beta\|10\rangle - \gamma\|01\rangle + \delta\|00\rangle$	$i \cdot \sigma_Y \otimes i \cdot \sigma_Y$
			$\|\Phi^-\rangle$	$-\alpha\|11\rangle - \beta\|10\rangle + \gamma\|01\rangle + \delta\|00\rangle$	$i \cdot \sigma_Y \otimes \sigma_X$
	$\|\Phi^-\rangle$	$\|\Psi^+\rangle$	$\|\Phi^+\rangle$	$\alpha\|11\rangle - \beta\|10\rangle - \gamma\|01\rangle + \delta\|00\rangle$	$i \cdot \sigma_Y \otimes i \cdot \sigma_Y$
			$\|\Phi^-\rangle$	$-\alpha\|11\rangle - \beta\|10\rangle + \gamma\|01\rangle + \delta\|00\rangle$	$i \cdot \sigma_Y \otimes \sigma_X$
		$\|\Psi^-\rangle$	$\|\Phi^+\rangle$	$\alpha\|11\rangle + \beta\|10\rangle - \gamma\|01\rangle - \delta\|00\rangle$	$i \cdot \sigma_Y \otimes \sigma_X$
			$\|\Phi^-\rangle$	$-\alpha\|11\rangle + \beta\|10\rangle + \gamma\|01\rangle - \delta\|00\rangle$	$i \cdot \sigma_Y \otimes i \cdot \sigma_Y$
$\|\Psi^+\rangle$	$\|\Phi^+\rangle$	$\|\Psi^+\rangle$	$\|\Psi^+\rangle$	$\alpha\|10\rangle + \beta\|11\rangle + \gamma\|00\rangle + \delta\|01\rangle$	$\sigma_X \otimes I$
			$\|\Psi^-\rangle$	$-\alpha\|10\rangle + \beta\|11\rangle - \gamma\|00\rangle + \delta\|01\rangle$	$\sigma_X \otimes \sigma_Z$
		$\|\Psi^-\rangle$	$\|\Psi^+\rangle$	$\alpha\|10\rangle - \beta\|11\rangle + \gamma\|00\rangle - \delta\|01\rangle$	$\sigma_X \otimes \sigma_Z$
			$\|\Psi^-\rangle$	$-\alpha\|10\rangle - \beta\|11\rangle - \gamma\|00\rangle - \delta\|01\rangle$	$\sigma_X \otimes I$
	$\|\Phi^-\rangle$	$\|\Psi^+\rangle$	$\|\Psi^+\rangle$	$\alpha\|10\rangle - \beta\|11\rangle + \gamma\|00\rangle - \delta\|01\rangle$	$\sigma_X \otimes \sigma_Z$
			$\|\Psi^-\rangle$	$-\alpha\|10\rangle - \beta\|11\rangle - \gamma\|00\rangle - \delta\|01\rangle$	$\sigma_X \otimes I$
		$\|\Psi^-\rangle$	$\|\Psi^+\rangle$	$\alpha\|10\rangle + \beta\|11\rangle + \gamma\|00\rangle + \delta\|01\rangle$	$\sigma_X \otimes I$
			$\|\Psi^-\rangle$	$-\alpha\|10\rangle + \beta\|11\rangle - \gamma\|00\rangle + \delta\|01\rangle$	$\sigma_X \otimes \sigma_Z$
$\|\Psi^-\rangle$	$\|\Phi^+\rangle$	$\|\Psi^+\rangle$	$\|\Psi^+\rangle$	$\alpha\|10\rangle + \beta\|11\rangle - \gamma\|00\rangle - \delta\|01\rangle$	$i \cdot \sigma_Y \otimes I$
			$\|\Psi^-\rangle$	$-\alpha\|10\rangle + \beta\|11\rangle + \gamma\|00\rangle - \delta\|01\rangle$	$i \cdot \sigma_Y \otimes \sigma_Z$
		$\|\Psi^-\rangle$	$\|\Psi^+\rangle$	$\alpha\|10\rangle - \beta\|11\rangle - \gamma\|00\rangle + \delta\|01\rangle$	$i \cdot \sigma_Y \otimes \sigma_Z$
			$\|\Psi^-\rangle$	$-\alpha\|10\rangle - \beta\|11\rangle + \gamma\|00\rangle + \delta\|01\rangle$	$i \cdot \sigma_Y \otimes I$
	$\|\Phi^-\rangle$	$\|\Psi^+\rangle$	$\|\Psi^+\rangle$	$\alpha\|10\rangle - \beta\|11\rangle - \gamma\|00\rangle + \delta\|01\rangle$	$i \cdot \sigma_Y \otimes \sigma_Z$
			$\|\Psi^-\rangle$	$-\alpha\|10\rangle - \beta\|11\rangle + \gamma\|00\rangle + \delta\|01\rangle$	$i \cdot \sigma_Y \otimes I$
		$\|\Psi^-\rangle$	$\|\Psi^+\rangle$	$\alpha\|10\rangle + \beta\|11\rangle - \gamma\|00\rangle - \delta\|01\rangle$	$i \cdot \sigma_Y \otimes I$
			$\|\Psi^-\rangle$	$-\alpha\|10\rangle + \beta\|11\rangle + \gamma\|00\rangle - \delta\|01\rangle$	$i \cdot \sigma_Y \otimes \sigma_Z$
$\|\Psi^+\rangle$	$\|\Psi^+\rangle$	$\|\Phi^+\rangle$	$\|\Phi^+\rangle$	$\alpha\|11\rangle + \beta\|10\rangle + \gamma\|01\rangle + \delta\|00\rangle$	$\sigma_X \otimes \sigma_X$
			$\|\Phi^-\rangle$	$-\alpha\|11\rangle + \beta\|10\rangle - \gamma\|01\rangle + \delta\|00\rangle$	$\sigma_X \otimes i \cdot \sigma_Y$
		$\|\Phi^-\rangle$	$\|\Phi^+\rangle$	$-\alpha\|11\rangle + \beta\|10\rangle - \gamma\|01\rangle + \delta\|00\rangle$	$\sigma_X \otimes i \cdot \sigma_Y$
			$\|\Phi^-\rangle$	$\alpha\|11\rangle + \beta\|10\rangle + \gamma\|01\rangle + \delta\|00\rangle$	$\sigma_X \otimes \sigma_X$
	$\|\Psi^-\rangle$	$\|\Phi^+\rangle$	$\|\Phi^+\rangle$	$\alpha\|11\rangle - \beta\|10\rangle + \gamma\|01\rangle - \delta\|00\rangle$	$\sigma_X \otimes i \cdot \sigma_Y$
			$\|\Phi^-\rangle$	$-\alpha\|11\rangle - \beta\|10\rangle - \gamma\|01\rangle - \delta\|00\rangle$	$\sigma_X \otimes \sigma_X$
		$\|\Phi^-\rangle$	$\|\Phi^+\rangle$	$-\alpha\|11\rangle - \beta\|10\rangle - \gamma\|01\rangle - \delta\|00\rangle$	$\sigma_X \otimes \sigma_X$
			$\|\Phi^-\rangle$	$\alpha\|11\rangle - \beta\|10\rangle + \gamma\|01\rangle - \delta\|00\rangle$	$\sigma_X \otimes i \cdot \sigma_Y$

(continued)

Table 4.2 (continued)

Alice's measurement results (x, 7)	Alice's measurement results (y, 2)	Bob_1's measurement results of particles (1,4)	Bob_2's measurement results of particles (3,6)	Collapsed state of particles (8,5)	Bob_3's unitary transformation
$\vert\Psi^-\rangle$	$\vert\Psi^+\rangle$	$\vert\Phi^+\rangle$	$\vert\Phi^+\rangle$	$\alpha\vert 11\rangle + \beta\vert 10\rangle - \gamma\vert 01\rangle - \delta\vert 00\rangle$	$i \cdot \sigma_Y \otimes \sigma_X$
			$\vert\Phi^-\rangle$	$-\alpha\vert 11\rangle + \beta\vert 10\rangle + \gamma\vert 01\rangle - \delta\vert 00\rangle$	$i \cdot \sigma_Y \otimes i \cdot \sigma_Y$
		$\vert\Phi^-\rangle$	$\vert\Phi^+\rangle$	$-\alpha\vert 11\rangle + \beta\vert 10\rangle + \gamma\vert 01\rangle - \delta\vert 00\rangle$	$i \cdot \sigma_Y \otimes i \cdot \sigma_Y$
			$\vert\Phi^-\rangle$	$\alpha\vert 11\rangle + \beta\vert 10\rangle - \gamma\vert 01\rangle - \delta\vert 00\rangle$	$i \cdot \sigma_Y \otimes \sigma_X$
	$\vert\Psi^-\rangle$	$\vert\Phi^+\rangle$	$\vert\Phi^+\rangle$	$\alpha\vert 11\rangle - \beta\vert 10\rangle - \gamma\vert 01\rangle + \delta\vert 00\rangle$	$i \cdot \sigma_Y \otimes i \cdot \sigma_Y$
			$\vert\Phi^-\rangle$	$-\alpha\vert 11\rangle - \beta\vert 10\rangle + \gamma\vert 01\rangle + \delta\vert 00\rangle$	$i \cdot \sigma_Y \otimes \sigma_X$
		$\vert\Phi^-\rangle$	$\vert\Phi^+\rangle$	$-\alpha\vert 11\rangle - \beta\vert 10\rangle + \gamma\vert 01\rangle + \delta\vert 00\rangle$	$i \cdot \sigma_Y \otimes \sigma_X$
			$\vert\Phi^-\rangle$	$\alpha\vert 11\rangle - \beta\vert 10\rangle - \gamma\vert 01\rangle + \delta\vert 00\rangle$	$i \cdot \sigma_Y \otimes i \cdot \sigma_Y$
$\vert\Psi^+\rangle$	$\vert\Psi^+\rangle$	$\vert\Phi^+\rangle$	$\vert\Psi^+\rangle$	$\alpha\vert 10\rangle + \beta\vert 11\rangle + \gamma\vert 00\rangle + \delta\vert 01\rangle$	$\sigma_X \otimes I$
			$\vert\Psi^-\rangle$	$-\alpha\vert 10\rangle + \beta\vert 11\rangle - \gamma\vert 00\rangle + \delta\vert 01\rangle$	$\sigma_X \otimes \sigma_Z$
		$\vert\Phi^-\rangle$	$\vert\Psi^+\rangle$	$-\alpha\vert 10\rangle + \beta\vert 11\rangle - \gamma\vert 00\rangle + \delta\vert 01\rangle$	$\sigma_X \otimes \sigma_Z$
			$\vert\Psi^-\rangle$	$\alpha\vert 10\rangle + \beta\vert 11\rangle + \gamma\vert 00\rangle + \delta\vert 01\rangle$	$\sigma_X \otimes I$
	$\vert\Psi^-\rangle$	$\vert\Phi^+\rangle$	$\vert\Psi^+\rangle$	$\alpha\vert 10\rangle - \beta\vert 11\rangle + \gamma\vert 00\rangle - \delta\vert 01\rangle$	$\sigma_X \otimes \sigma_Z$
			$\vert\Psi^-\rangle$	$-\alpha\vert 10\rangle - \beta\vert 11\rangle - \gamma\vert 00\rangle - \delta\vert 01\rangle$	$\sigma_X \otimes I$
		$\vert\Phi^-\rangle$	$\vert\Psi^+\rangle$	$-\alpha\vert 10\rangle - \beta\vert 11\rangle - \gamma\vert 00\rangle - \delta\vert 01\rangle$	$\sigma_X \otimes I$
			$\vert\Psi^-\rangle$	$\alpha\vert 10\rangle - \beta\vert 11\rangle + \gamma\vert 00\rangle - \delta\vert 01\rangle$	$\sigma_X \otimes \sigma_Z$
$\vert\Psi^-\rangle$	$\vert\Psi^+\rangle$	$\vert\Phi^+\rangle$	$\vert\Psi^+\rangle$	$\alpha\vert 10\rangle + \beta\vert 11\rangle - \gamma\vert 00\rangle - \delta\vert 01\rangle$	$i \cdot \sigma_Y \otimes I$
			$\vert\Psi^-\rangle$	$-\alpha\vert 10\rangle + \beta\vert 11\rangle + \gamma\vert 00\rangle - \delta\vert 01\rangle$	$i \cdot \sigma_Y \otimes \sigma_Z$
		$\vert\Phi^-\rangle$	$\vert\Psi^+\rangle$	$-\alpha\vert 10\rangle + \beta\vert 11\rangle + \gamma\vert 00\rangle - \delta\vert 01\rangle$	$i \cdot \sigma_Y \otimes \sigma_Z$
			$\vert\Psi^-\rangle$	$\alpha\vert 10\rangle + \beta\vert 11\rangle - \gamma\vert 00\rangle - \delta\vert 01\rangle$	$i \cdot \sigma_Y \otimes I$
	$\vert\Psi^-\rangle$	$\vert\Phi^+\rangle$	$\vert\Psi^+\rangle$	$\alpha\vert 10\rangle - \beta\vert 11\rangle - \gamma\vert 00\rangle + \delta\vert 01\rangle$	$i \cdot \sigma_Y \otimes \sigma_Z$
			$\vert\Psi^-\rangle$	$-\alpha\vert 10\rangle - \beta\vert 11\rangle + \gamma\vert 00\rangle + \delta\vert 01\rangle$	$i \cdot \sigma_Y \otimes I$
		$\vert\Phi^-\rangle$	$\vert\Psi^+\rangle$	$-\alpha\vert 10\rangle - \beta\vert 11\rangle + \gamma\vert 00\rangle + \delta\vert 01\rangle$	$i \cdot \sigma_Y \otimes I$
			$\vert\Psi^-\rangle$	$\alpha\vert 10\rangle - \beta\vert 11\rangle - \gamma\vert 00\rangle + \delta\vert 01\rangle$	$i \cdot \sigma_Y \otimes \sigma_Z$
$\vert\Psi^+\rangle$	$\vert\Psi^+\rangle$	$\vert\Psi^+\rangle$	$\vert\Phi^+\rangle$	$\alpha\vert 10\rangle + \beta\vert 11\rangle + \gamma\vert 00\rangle + \delta\vert 01\rangle$	$\sigma_X \otimes I$
			$\vert\Phi^-\rangle$	$\alpha\vert 10\rangle - \beta\vert 11\rangle + \gamma\vert 00\rangle - \delta\vert 01\rangle$	$\sigma_X \otimes \sigma_Z$
		$\vert\Psi^-\rangle$	$\vert\Phi^+\rangle$	$-\alpha\vert 10\rangle + \beta\vert 11\rangle - \gamma\vert 00\rangle + \delta\vert 01\rangle$	$\sigma_X \otimes \sigma_Z$
			$\vert\Phi^-\rangle$	$-\alpha\vert 10\rangle - \beta\vert 11\rangle - \gamma\vert 00\rangle - \delta\vert 01\rangle$	$\sigma_X \otimes I$
	$\vert\Psi^-\rangle$	$\vert\Psi^+\rangle$	$\vert\Phi^+\rangle$	$\alpha\vert 10\rangle - \beta\vert 11\rangle + \gamma\vert 00\rangle - \delta\vert 01\rangle$	$\sigma_X \otimes \sigma_Z$
			$\vert\Phi^-\rangle$	$\alpha\vert 10\rangle + \beta\vert 11\rangle + \gamma\vert 00\rangle + \delta\vert 01\rangle$	$\sigma_X \otimes I$
		$\vert\Psi^-\rangle$	$\vert\Phi^+\rangle$	$-\alpha\vert 10\rangle - \beta\vert 11\rangle - \gamma\vert 00\rangle - \delta\vert 01\rangle$	$\sigma_X \otimes I$
			$\vert\Phi^-\rangle$	$-\alpha\vert 10\rangle + \beta\vert 11\rangle - \gamma\vert 00\rangle + \delta\vert 01\rangle$	$\sigma_X \otimes \sigma_Z$
$\vert\Psi^-\rangle$	$\vert\Psi^+\rangle$	$\vert\Psi^+\rangle$	$\vert\Phi^+\rangle$	$\alpha\vert 10\rangle + \beta\vert 11\rangle - \gamma\vert 00\rangle - \delta\vert 01\rangle$	$i \cdot \sigma_Y \otimes I$
			$\vert\Phi^-\rangle$	$\alpha\vert 10\rangle - \beta\vert 11\rangle - \gamma\vert 00\rangle + \delta\vert 01\rangle$	$i \cdot \sigma_Y \otimes \sigma_Z$
		$\vert\Psi^-\rangle$	$\vert\Phi^+\rangle$	$-\alpha\vert 10\rangle + \beta\vert 11\rangle + \gamma\vert 00\rangle - \delta\vert 01\rangle$	$i \cdot \sigma_Y \otimes \sigma_Z$
			$\vert\Phi^-\rangle$	$-\alpha\vert 10\rangle - \beta\vert 11\rangle + \gamma\vert 00\rangle + \delta\vert 01\rangle$	$i \cdot \sigma_Y \otimes I$
	$\vert\Psi^-\rangle$	$\vert\Psi^+\rangle$	$\vert\Phi^+\rangle$	$\alpha\vert 10\rangle - \beta\vert 11\rangle - \gamma\vert 00\rangle + \delta\vert 01\rangle$	$i \cdot \sigma_Y \otimes \sigma_Z$
			$\vert\Phi^-\rangle$	$\alpha\vert 10\rangle + \beta\vert 11\rangle - \gamma\vert 00\rangle - \delta\vert 01\rangle$	$i \cdot \sigma_Y \otimes I$
		$\vert\Psi^-\rangle$	$\vert\Phi^+\rangle$	$-\alpha\vert 10\rangle - \beta\vert 11\rangle + \gamma\vert 00\rangle + \delta\vert 01\rangle$	$i \cdot \sigma_Y \otimes I$
			$\vert\Phi^-\rangle$	$-\alpha\vert 10\rangle + \beta\vert 11\rangle + \gamma\vert 00\rangle - \delta\vert 01\rangle$	$i \cdot \sigma_Y \otimes \sigma_Z$

(continued)

Table 4.2 (continued)

Alice's measurement results $(x, 7)$	$(y, 2)$	Bob$_1$'s measurement results of particles (1,4)	Bob$_2$'s measurement results of particles (3,6)	Collapsed state of particles (8,5)	Bob$_3$'s unitary transformation
$\|\Psi^+\rangle$	$\|\Psi^+\rangle$	$\|\Psi^+\rangle$	$\|\Psi^+\rangle$	$\alpha\|11\rangle + \beta\|10\rangle + \gamma\|01\rangle + \delta\|00\rangle$	$\sigma_X \otimes \sigma_X$
			$\|\Psi^-\rangle$	$\alpha\|11\rangle - \beta\|10\rangle + \gamma\|01\rangle - \delta\|00\rangle$	$\sigma_X \otimes i \cdot \sigma_Y$
		$\|\Psi^-\rangle$	$\|\Psi^+\rangle$	$-\alpha\|11\rangle + \beta\|10\rangle - \gamma\|01\rangle + \delta\|00\rangle$	$\sigma_X \otimes i \cdot \sigma_Y$
			$\|\Psi^-\rangle$	$-\alpha\|11\rangle - \beta\|10\rangle - \gamma\|01\rangle - \delta\|00\rangle$	$\sigma_X \otimes \sigma_X$
	$\|\Psi^-\rangle$	$\|\Psi^+\rangle$	$\|\Psi^+\rangle$	$\alpha\|11\rangle - \beta\|10\rangle + \gamma\|01\rangle - \delta\|00\rangle$	$\sigma_X \otimes i \cdot \sigma_Y$
			$\|\Psi^-\rangle$	$\alpha\|11\rangle + \beta\|10\rangle + \gamma\|01\rangle + \delta\|00\rangle$	$\sigma_X \otimes \sigma_X$
		$\|\Psi^-\rangle$	$\|\Psi^+\rangle$	$-\alpha\|11\rangle - \beta\|10\rangle - \gamma\|01\rangle - \delta\|00\rangle$	$\sigma_X \otimes \sigma_X$
			$\|\Psi^-\rangle$	$-\alpha\|11\rangle + \beta\|10\rangle - \gamma\|01\rangle + \delta\|00\rangle$	$\sigma_X \otimes i \cdot \sigma_Y$
$\|\Psi^-\rangle$	$\|\Psi^+\rangle$	$\|\Psi^+\rangle$	$\|\Psi^+\rangle$	$\alpha\|11\rangle + \beta\|10\rangle - \gamma\|01\rangle - \delta\|00\rangle$	$i \cdot \sigma_Y \otimes \sigma_X$
			$\|\Psi^-\rangle$	$\alpha\|11\rangle - \beta\|10\rangle - \gamma\|01\rangle + \delta\|00\rangle$	$i \cdot \sigma_Y \otimes i \cdot \sigma_Y$
		$\|\Psi^-\rangle$	$\|\Psi^+\rangle$	$-\alpha\|11\rangle + \beta\|10\rangle + \gamma\|01\rangle - \delta\|00\rangle$	$i \cdot \sigma_Y \otimes i \cdot \sigma_Y$
			$\|\Psi^-\rangle$	$-\alpha\|11\rangle - \beta\|10\rangle + \gamma\|01\rangle + \delta\|00\rangle$	$i \cdot \sigma_Y \otimes \sigma_X$
	$\|\Psi^-\rangle$	$\|\Psi^+\rangle$	$\|\Psi^+\rangle$	$\alpha\|11\rangle - \beta\|10\rangle - \gamma\|01\rangle + \delta\|00\rangle$	$i \cdot \sigma_Y \otimes i \cdot \sigma_Y$
			$\|\Psi^-\rangle$	$\alpha\|11\rangle + \beta\|10\rangle - \gamma\|01\rangle - \delta\|00\rangle$	$i \cdot \sigma_Y \otimes \sigma_X$
		$\|\Psi^-\rangle$	$\|\Psi^+\rangle$	$-\alpha\|11\rangle - \beta\|10\rangle + \gamma\|01\rangle + \delta\|00\rangle$	$i \cdot \sigma_Y \otimes \sigma_X$
			$\|\Psi^-\rangle$	$-\alpha\|11\rangle + \beta\|10\rangle + \gamma\|01\rangle - \delta\|00\rangle$	$i \cdot \sigma_Y \otimes i \cdot \sigma_Y$

Table 4.3 The comparison between the schemes in Refs. [21, 38, 42] and ours

Scheme	Ref. [21]	Ref. [38]	Ref. [42]	Our first	Our second
The number of entangled particles	Four-particle	Eight-particle	Bell states	Eight-particle	Eight-particle
Extend to tripartite	*14.3%*	*25%*	*16.6%*	*14.3%*	*20%*
Extend to multiparty	$\frac{1}{2n-1}$	$\frac{1}{n}$	$\frac{1}{2(n-1)}$	$\frac{1}{2n-1}$	$\frac{1}{n+1}$
Resist collusive attacks	No	No	Yes	Yes	Yes
Resist entangle-measure attacks	No	No	Yes	Yes	Yes

The symbols contained in following two tables are explained as follows:

1. $|\Phi^+\rangle$, $|\Phi^-\rangle$, $|\Psi^+\rangle$, $|\Psi^-\rangle$ are four different Bell states, that is, $|\Phi^\pm\rangle = \frac{1}{\sqrt{2}}(|00\rangle \pm |11\rangle), |\Psi^\pm\rangle = \frac{1}{\sqrt{2}}(|01\rangle \pm |10\rangle)$;
2. I, σ_x, σ_y, σ_z are four different Pauli operators, and the corresponding matrices are defined as: $I = \begin{pmatrix} 1 & 0 \\ 0 & 1 \end{pmatrix}, \sigma_x = \begin{pmatrix} 0 & 1 \\ 1 & 0 \end{pmatrix}, \sigma_y = \begin{pmatrix} 0 & -i \\ i & 0 \end{pmatrix}, \sigma_z = \begin{pmatrix} 1 & 0 \\ 0 & -1 \end{pmatrix}$.

4.4 Quantum Secret Sharing Scheme

4.4.1 HBB Scheme

In 1999, the first QSS scheme was adopted by M. Hillery et al. [16]. The transport process of this scheme consists of three parties, Alice as the sender of the message and Bob and Charlie as the other two participants, which are usually called agents. The three parties share the classical secret message. Any agent cannot achieve the goal of restoring the secret message. The two parties can work together to restore the message.

Step 1: Alice randomly generates a list of three-particle entangled states (GHZ states) $|\psi_{ABC}\rangle = \frac{1}{\sqrt{2}}(|000\rangle + |111\rangle)_{ABC}$ and then distributes particle B and particle C, which are sent particle B to Bob and particle C to Charlie.

Step 2: Alice, Bob, and Charlie all measure their particles, and for the basis of their measurements, they randomly select from $X = \{|{+x}\rangle, |{-x}\rangle\}$, $Y = \{|{+y}\rangle, |{-y}\rangle\}$, where:

$$|{+x}\rangle = \frac{1}{\sqrt{2}}(|0\rangle + |1\rangle), |-x\rangle = \frac{1}{\sqrt{2}}(|0\rangle - |1\rangle) \tag{4.15}$$

$$|{+y}\rangle = \frac{1}{\sqrt{2}}(|0\rangle + i|1\rangle), |-y\rangle = \frac{1}{\sqrt{2}}(|0\rangle - i|1\rangle) \tag{4.16}$$

For the three-particle entangled state (GHZ state), expand it with the eigenstates of X and Y, that is,

$$\begin{aligned} |\psi\rangle &= \frac{1}{2}[(|{+x}\rangle_A|{+x}\rangle_B + |-x\rangle_A|-x\rangle_B)|{+x}\rangle_C \\ &\quad + (|{+x}\rangle_A|-x\rangle_B + |-x\rangle_A|{+x}\rangle_B)|-x\rangle_C] \\ &= \frac{1}{2}[(|{+y}\rangle_A|-y\rangle_B + |-y\rangle_A|{+y}\rangle_B)|{+x}\rangle_C \\ &\quad + (|{+y}\rangle_A|{+y}\rangle_B + |-y\rangle_A|-y\rangle_B)|-x\rangle_C] \end{aligned} \tag{4.17}$$

Step 3: At the request of Alice, the secret sender, Bob and Charlie need to disclose the measurement base for each particle. Bob can make a statement, or Charlie can make a statement, and then make a series of comparisons between the measurements of Alice, Bob, and Charlie, and get a concrete result. From that result, Alice can judge the validity and invalidity of the data. The other two agents are then informed of the valid data, from which Bob and Charlie discard the remaining particles.

Step 4: In Alice's eavesdropping data, which we can randomly select, in each of the eavesdropping particles, at Alice's request, Bob and Charlie need to make their measurements public, Alice takes her measurements, combines them with Bob and Charlie's statements, and computes the correlations. Alice calculates the bit error rate and compares it to the threshold set previously.

Step 5: Alice, Bob, and Charlie encode their reserved measurements and each gets a binary key string. For Bob and Charlie's measurements, $\{|+x\rangle, |+y\rangle\}$ stands for 0, $\{|-x\rangle|-y\rangle\}$ stands for 1. For Alice, $\{|+x\rangle, |+y\rangle\}$ represents 1, $\{|-x\rangle|-y\rangle\}$ represents 0 when the measurement bases selected by Alice, Bob, and Charlie are valid combinations of the other three. This allows Alice, Bob, and Charlie to get a set of keys, labeled k^rA, k^rB, k^rC, respectively. The above encoding method can ensure that the three keys satisfy the $k^rA = k^rB \oplus k^rC$. Alice, Bob, and Charlie correct and secretly enlarge the key to get the final key, which is named k_A, k_B, k_C and satisfies the relation $k_A = k_B \oplus k_C$.

4.4.2 DQSS Scheme

In view of most above QSS schemes are designed in two-dimensional quantum system (i.e., qubit) with multi-qubit maximally entangled quantum state. However, with the increasing maturity of quantum technology, it is more significant to design the QSS scheme based on high-dimensional quantum system (i.e., qudit) because the QSS scheme designed in the context of qudits would have higher resource capacity and better security than the QSS scheme designed in the context of qubits. From this, it can be concluded that the QSS schemes designed in the context of qudits can be more meaningful and practical than qubits in the future. Some existing QSS schemes designed in the context of qudits have already been reported in Refs [35, 43–48]. Herein, inspired by the previous (t, n) threshold QSS scheme [24], the author utilized the transmitted particles and local unitary operators to share the multi-particle entangled quantum state between dealer and agents. This section proposes a novel dynamic quantum secret sharing (DQSS) scheme based on GHZ state in a similar way in the context of qudits [49]. In the present DQSS scheme, each agent first prepares his own sequence based on transmitted sequence via local unitary operation and then implements local unitary operation and single-particle measurement operation on their sequence to obtain his personal shares. To enhance scheme's security, two steps eavesdropping check are employed based on the shared entangled GHZ state. In addition to above, since the transmitted particles in quantum channel will inevitably interact with the outer environment and be affected by channel noises, we demonstrate the efficiency analysis of the present DQSS scheme in four types of noise, that is, dit-flip, d-phase-flip, amplitude-damping, and depolarizing, which are usually encountered in the real world, via quantum fidelity. In summary, compared with the previous existing DQSS scheme, the proposed DQSS scheme poses some advantages described as follows:

1. The proposed DQSS scheme is designed in the context of qudits and shares the predetermined dits.
2. The transmitted particles have no information about the dealer's secret as well as no correlation to the dealer's secret.

3. The proposed scheme does not require lots of detected particles for eavesdropping check to guarantee security.
4. Both dealer and agents only need to perform the single-particle measurement operations.
5. It is more efficient when dynamically adds or revokes an agent.
6. The scheme's efficiency in noise environments is deduced based on quantum fidelity.

In this section, the proposed (n+1)-party DQSS scheme based on d-level GHZ state is illustrated in detail, which allows the dealer (Alice) to share the predetermined m-dit secrets with n agents (Bob$_1$, Bob$_2$, . . ., Bob$_n$).

4.4.2.1 Qudit and GHZ state

For a single qudit, an unknown quantum state can be written as

$$|\psi\rangle = \sum_{j=0}^{d-1} a_j|j\rangle, \tag{4.18}$$

where a_j is a complex number and satisfies the normalization condition $\sum_{j=0}^{d-1}|a_j|^2 = 1$, and $|j\rangle$ is called the computational basis state.

Similar to Bell states in the context of qubits, the generalized Bell states in the context of qudits is defined as [50]

$$|\psi(m,n)\rangle = \frac{1}{\sqrt{d}}\sum_{k=0}^{d-1} w^{mk}|k, k+n\rangle,\ \ 0\le m,n\le d-1, \tag{4.19}$$

where $w = e^{2\pi i/d}$ is a primitive d-th root of unity, and here and hereafter, all the addition '+' defines add modulo d.

Similarly, the generalized GHZ state in the context of qudits is written as [43]

$$|\psi(0,0,\cdots,0)\rangle = \frac{1}{\sqrt{d}}\sum_{k=0}^{d-1}|k,k,\cdots,k\rangle \tag{4.20}$$

Moreover, two mutually unbiased measurement bases for single qudit, thst id, Z basis and X basis, are defined as [51, 52]

$$\begin{aligned} &Z\ basis:\ |Z_j\rangle = |j\rangle,\\ &X\ basis:\ |X_j\rangle = \frac{1}{\sqrt{d}}\sum_{k=0}^{d-1} w^{jk}|k\rangle\\ &j=0,1,\cdots,d-1. \end{aligned} \tag{4.21}$$

The reason why the Z basis and X basis are recognized as mutually unbiased bases is because of $|\langle Z_j| X_k\rangle|^2 = 1/d$ for any $j,\ k \in \{0,\ 1,\ \cdots,\ d-1\}$. It means that when a single qudit stays in the eigenstate of Z basis (X basis) but is measured by X basis (Z basis), the measurement result is random and only can get the correct result with probability $1/d$.

4.4.2.2 Basic Unitary Operation

Basic unitary operations in the context of qudits are considered as the generalization of their counterparts in the context of qubits, such as Hadamard transform, Pauli matrix, and Controlled-NOT (CNOT) operations.

For single qudit, the generalized Hadamard transform [44, 50] is defined as

$$H_d = \frac{1}{\sqrt{d}} \sum_{i,j=0}^{d-1} w^{i \cdot j} |i\rangle\langle j|,\ \ [H_d]^{\dagger} = \frac{1}{\sqrt{d}} \sum_{i,j=0}^{d-1} w^{-i \cdot j} |j\rangle\langle i|, \tag{4.22}$$

where sign † denotes the Hermite conjugate.

For single qudit, the generalized Pauli operation [50] is defined as

$$P_{a,b} = \sum_{j=0}^{d-1} w^{jb} |j+a\rangle\langle j| \tag{4.23}$$

The generalized NOT operation for qudit is renamed as Right-Shift (RS) operation, and its inverse operation is renamed as Left-Shift (LS) operation, defined as [44, 50]

$$RS = \sum\nolimits_{j=0}^{d-1} |j+1\rangle\langle j|,\ \ LS = \sum\nolimits_{j=0}^{d-1} |j\rangle\langle j+1| \tag{4.24}$$

Obviously, RS and LS operations, respectively, transform any computational basis state $|j\rangle$ into $|j+1\rangle$ and $|j-1\rangle$, where the sign '−' defines subtract modulo d here and hereafter. In particular, the Controlled-RS (RS_C) operation plays the role of CNOT gate in the context of qudits written as

$$RS_C(|i\rangle, |j\rangle) = |i\rangle|i+j\rangle, \tag{4.25}$$

where $|i\rangle$ and $|j\rangle$, respectively, denote the controlled qudit and the target qudit. And the inverse operation RS_C is represented by Controlled-Left-Shift (LS_C) operation and written as

$$RS_C{}^{-1}(|i\rangle, |j\rangle) = LS_C(|i\rangle, |j\rangle) = |i\rangle|j-i\rangle \tag{4.26}$$

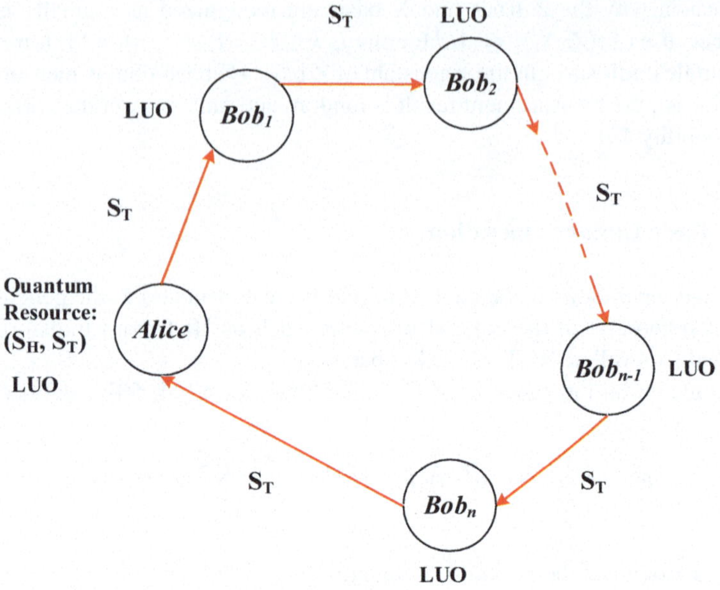

Fig. 4.5 Procedure for proposed (n+1)-party DQSS scheme. LUO means that Alice and each agent Bob$_i$ perform some local unitary operations

4.4.2.3 DQSS Scheme

Suppose that Alice wants to share the m-dit secrets $K = k_1k_2\cdots k_m$ with n agents (Bob$_1$, Bob$_2$, . . ., Bob$_n$), in such a way that the secrets can be recovered only when all agents collaborate. Figure 4.5 simply illustrates the procedure of the proposed DQSS scheme. Steps involved in the (n+1)-party DQSS scheme can be described below.

Step 1. Quantum Resources Preparation

1.1 First, Alice prepares N generalized Bell state $|\psi(0,0)\rangle$; second, she takes all the first qudit in each Bell state to form an ordered sequence $S_H=\{s_{H_i}\}, i=1,2,\cdots,N$, and the remaining qudits to form another ordered sequence $S_T=\{s_{T_i}\}$, i = 1, 2, $\cdots$, N, where the sub-subscript i of each qudit both in sequences S_H and S_T denotes that this qudit comes from the i-th Bell state; finally, she sends sequence S_T to Bob$_1$ via an authenticated quantum channel.

1.2 First, after Bob$_1$ receiving the sequence S_T, he publically announces the number of particles in sequence S_T (herein, say the number is N_1). Alice can calculate the difference $d_1 = N - N_1$. If d_1 is obviously larger than the number of loss particles due to the imperfection of quantum channel, then Alice asks Bob$_1$ to terminate the scheme and restart from the beginning. Otherwise, the scheme continues.

After that, Bob_1 publicly announces the position of those lost particles in sequence S_T first, and then Alice deletes these particles in the same position from her sequence S_H. Thus, Alice obtains the corresponding new sequence $S_H = \{s_{H_i}\}, i = 1, 2, \cdots, N_1$.

Last, Bob_1 prepares a sequence of ancillary qudits $|0\rangle^{\bigotimes N_1}$ and implements RS_C operation on each composite quantum system of two qudits described as

$$RS_C\left(s_{T_i}, |0\rangle_{1_i}\right) \to (s_{T_i}, s_{1_i}), \quad i = 1, 2, \cdots, N_1 \tag{4.27}$$

where $|0\rangle_{1_i}$ denotes i-th qudit $|0\rangle$ in ancillary qudits $|0\rangle^{\bigotimes N_1}$, and qudits s_{T_i} and $|0\rangle_{1_i}$ respectively are the control qudit and target qudit of RS_C operation. Therefore, Bob_1 can prepare his sequence $S_1 = s_{1_1} s_{1_2} \cdots s_{1_{N_1}}$ via N_1 local unitary operation RS_C. Then Bob_1 sends the sequence S_T to Bob_2 via an authenticated quantum channel.

1.3 The remaining agents (Bob_2, ..., Bob_n) perform the same operation as Bob_1 does. This process is terminated until Bob_n sends the sequence S_T back to Alice again. Herein, let us assume that the number of particles in transmitted particles sequence S_T is C when Alice finally receives it. Hence, after Alice and each agent Bob_i delete the lost particles in transmitted particles sequence S_T in the same position, each agent Bob_i holds their sequence $S_i = s_{i_1} s_{i_2} \cdots s_{i_C}$ prepared by themselves and Alice holds her two sequences $S_H = s_{H_1} s_{H_2} \cdots s_{H_C}$ and $S_T = s_{T_1} s_{T_2} \cdots s_{T_C}$.

Step 2. Eavesdropping Check

2.1 Alice performs LS_C operation on each of two qudits s_{H_j} and s_{T_j} in her sequences S_H and S_T, where each qudit s_{H_j} acts as the controlled qudit and corresponding qudit s_{T_j} acts as the target qudit. After implementing (m+h) LS_C operations, Alice measures all qudits in sequence S_T based on Z basis. Alice keeps a record of the positions of these measurement results not equal to $|0\rangle$ and calculates the corresponding rate. If the rate is higher than the threshold caused by channel noise, Alice terminates the scheme and restarts. Otherwise, Alice publicly announces the specific position of those qudit's measurement results doesn't equal to $|0\rangle$. Alice and all agents delete those qudits in their sequence in the same positions. Herein, let us assume that the number of final remained particles in Alice sequence S_H and agent Bob_i sequence S_i (for i=1 to n) both are m+h.

2.2 Alice and all agents (Bob_1, Bob_2, ..., Bob_n) perform the generalized Hadamard transform on each qudit in her/his sequences (herein, Alice's sequence is S_H and Bob_i's sequence is S_i). In order to check that all agents (Bob_1, Bob_2, ..., Bob_n) have honestly performed the scheme, Alice and all agents perform eavesdropping check as follows:

2.2.1 Alice randomly chooses h qudits (denoted as selected qudits here and thereafter) in her sequence S_H and publicly announces the specific positions of those selected qudits. Then, she asks each agent Bob_i to measure

h qudits in the same positions in their sequence S_i based on Z basis and publicly announce the corresponding measurement results in order.

2.2.2 After confirming that all agents have publicly announced their measurement results, Alice measures her h selected qudits based on Z basis and obtains the corresponding measurement results.

2.2.3 According to her personal measurement results and all agents' publicly declared measurement results, she can calculate the error rate and compare it with a pre-defined threshold decided by channel noise. If the error rate exceeds the predefined threshold, then Alice aborts the scheme; otherwise, the scheme continues.

Step 3. Secrets Encode and Recover

3.1 It is worth noting that after the above two steps, all particles in the same position in Alice's sequence S_H and agent Bob$_i$'s sequence S_i (for $i=1$ to n) form into a $(n+1)$-particle entangled state described as:

$$|\psi(0,0,0,\cdots,0)\rangle_j = \frac{1}{\sqrt{d^n}} \sum_{\substack{s_{H_j}, s_{1_j}, s_{2_j}, \cdots, s_{n_j}=0 \\ s_{H_j}+s_{1_j}+s_{2_j}+\cdots+s_{n_j}\equiv 0 (\mathrm{mod}\ d)}}^{d-1} |s_{H_j}\rangle|s_{1_j}\rangle|s_{2_j}\rangle\cdots|s_{n_j}\rangle, j=1,2,\cdots,m \tag{4.28}$$

The correctness of Eq. 4.28 is proved in *Appendix A*. To encode m-dit secrets $K = k_1k_2\cdots k_m$, Alice performs the generalized Pauli operation $P_{k_j,0}$ on the j-th qudit in sequence S_H (for $j=1$ to m), which can be described in detail:

$$\begin{aligned} &\left(P_{k_j,0} \bigotimes I_d{}^{\otimes n}\right)|\psi(0,0,0,\cdots,0)\rangle \\ &= \frac{1}{\sqrt{d^n}} \sum_{\substack{s_{H_j}, s_{1_j}, s_{2_j}, \cdots, s_{n_j}=0 \\ s_{H_j}+s_{1_j}+s_{2_j}+\cdots+s_{n_j}\equiv 0 (\mathrm{mod}\ d)}}^{d-1} \left(\sum_{s=0}^{d-1} |s+k_j\rangle\langle s|\ |s_{H_j}\rangle\right)\left(I_d|s_{1_j}\rangle\right)\left(I_d|s_{2_j}\rangle\right)\cdots\left(I_d|s_{n_j}\rangle\right) \\ &= \frac{1}{\sqrt{d^n}} \sum_{\substack{s_{H_j}, s_{1_j}, s_{2_j}, \cdots, s_{n_j}=0 \\ s_{H_j}+s_{1_j}+s_{2_j}+\cdots+s_{n_j}\equiv 0 (\mathrm{mod}\ d)}}^{d-1} |s_{H_j}+k_j\rangle|s_{1_j}\rangle|s_{2_j}\rangle\cdots|s_{n_j}\rangle \end{aligned} \tag{4.29}$$

3.2 Alice and each agent Bob$_i$ measure the remaining m qudits in their sequence based on Z basis, and Alice publicly announces the corresponding measurement results in order. Herein, let us denote that Alice's m qudits measurement results are $A = a_1a_2\cdots a_m$ and agent Bob$_i$'s m qudits measurement results are $B^i = b_1^i b_2^i \cdots b_m^i$. Then, all agents can collaboratively recover Alice's m-dit secrets $K = k_1k_2\cdots k_m$ as,

$$\begin{aligned}(A+B^1+B^2+\cdots+B^n)\bmod d&=\prod_{j=1}^{m}\left(a_j+b_j^1+b_j^2+\cdots+b_j^n\right)\bmod d\\&=\prod_{j=1}^{m}\left(s_{H_j}+k_j+s_{1_j}+s_{2_j}+\cdots+s_{n_j}\right)\bmod d=\prod_{j=1}^{m}k_j\equiv K\end{aligned}\tag{4.30}$$

4.4.2.4 Add an Agent

Suppose that a new agent Bob_{n+1} wants to join the current (n+1)-party DQSS scheme before Alice implements the Step 2 eavesdropping check in the preceding scheme. To add this new agent Bob_{n+1} into the current scheme, Alice and Bob_{n+1} can execute the following steps to form a (n+2)-party DQSS scheme:

Step A1.4 After Step 1.3 in the preceding scheme, Alice first sends her sequence S_T to Bob_{n+1} via an authenticated quantum channel first. Then. Bob_{n+1} performs the same operations as Bob_1 does, and accordingly prepares his personal sequence S_{n+1}. Finally, Bob_{n+1} sends the sequence S_T back to Alice again.

Step A2~A3 Alice and all agents (Bob_1, Bob_2, …, Bob_n, Bob_{n+1}) perform the Step 2 and Step 3 in preceding scheme in a similar way. Then, (n+1) agents (Bob_1, Bob_2, …, Bob_n, Bob_{n+1}) can collaboratively recover Alice's m-dit secrets as

$$k_j\equiv\left(a_j+b_j^1+b_j^2+\cdots+b_j^n+b_j^{n+1}\right)\bmod d,\ \ j=1,2,\cdots,m\tag{4.31}$$

4.4.2.5 Revoke an Agent

Suppose that Alice wants to revoke an agent Bob_k, $k\in\{1,\ 2,\ \ldots,\ n\}$, from the current ($n$+1)-party DQSS scheme. To achieve this aim, Alice can make minor changes to Step 2.2 and Step 3 in the preceding scheme. The details can be described as follows:

Step D2.2 After Alice and all agents (Bob_1, …, Bob_{k-1}, Bob_k, Bob_{k+1}, …, Bob_n) perform the generalized Hadamard transform on each qudit in her/his sequences. The second eavesdropping check can be described as follows:

Step D2.2.1 Alice first asks Bob_k to measure all qudits in his sequence S_k, and then Bob_k publically announces the corresponding measurement results.

Step D2.2.2 Confirming that Bob_k have publicly announced his measurement results, Alice randomly chooses h qudits (i.e., selected qudits) in his sequence S_H and publicly announces the specific positions of those selected qudits. Then, she and all the remaining agents (Bob_1, Bob_2, …, Bob_n, except Bob_k) measure those h qudits in selected positions in their sequence based on Z basis and ($n-1$) agents (Bob_1, Bob_2, …, Bob_n, except Bob_k) publicly announce the corresponding measurement results orderly.

Step D2.2.3 After that, Alice and each agent can calculate the error rate. If the error rate exceeds the predefined threshold decided by channel noise, then Alice aborts the scheme; otherwise, the scheme continues.

Step D3 It is similar to Step 3 in the preceding scheme, which can be described within the following two steps:

Step D3.1 Alice encodes her m-dit secrets $K = k_1k_2\cdots k_m$ on the remaining m qudits in sequence S_H via generalized Pauli operation $P_{k_j,0}, j = 1, 2, \cdots, m$.

Step D3.2 Alice and all the remaining agents (Bob$_1$, . . ., Bob$_{k-1}$, Bob$_{k+1}$, . . ., Bob$_n$) measure the remaining m qudits in their sequence based on Z basis and obtain the corresponding measurement results. Herein, let us denote Alice's measurement results are $A = a_1a_2\cdots a_m$, and Bob$_i$'s measurement results are $B_i = b_1^i b_2^i \cdots b_m^i$. Then according to the publicly announced measurement results by Bob$_k$, Alice updates her publicly announced measurement results A as

$$A^* = A + B_k = \prod_{j=1}^{m} \left(a_j + b_j^k\right) = \prod_{j=1}^{m} a_j^* = a_1^* a_2^* \cdots a_m^* \tag{4.32}$$

Hence, based on Alice's publicly announced measurement results, agents (Bob$_1$, . . ., Bob$_{k-1}$, Bob$_{k+1}$, . . ., Bob$_n$) can collaboratively recover Alice's secret as

$$\begin{aligned} K &\equiv \left(A^* + B^1 + \cdots + B^{k-1} + B^{k+1} + \cdots + B^n\right) \bmod d \Rightarrow \\ k_j &\equiv \left(a_j^* + b_j^1 + \cdots + b_j^{k-1} + b_j^{k+1} \cdots + b_j^n\right) \bmod d \\ &\equiv \left(a_j + b_j^k + b_j^1 + \cdots + b_j^{k-1} + b_j^{k+1} \cdots + b_j^n\right) \bmod d, \ j = 1, 2, \cdots, m \end{aligned} \tag{4.33}$$

4.4.2.6 Case Study

In this subsection, a simple example of four-party QSS scheme as well as an integrated quantum circuit illustration without consideration of dynamically adding or deleting an agent is demonstrated. Herein, it is assumed that Alice wants to share her 2-dit secrets k_1k_2 to three agents Bob$_1$, Bob$_2$, and Bob$_3$ in an ideal situation. Figure 4.6 demonstrates the simplified quantum circuit construction of four-party QSS scheme, in which three single blue lines denote three qudits in sequence S_T transmitted among Alice, Bob$_1$, Bob$_2$, and Bob$_3$. According to the illustrated quantum circuit in Fig. 4.6, we can describe the four-party QSS scheme within following three Steps in detail,

Step 1 Based on local unitary operations of H_d and RS$_C$, Alice prepares three generalized Bell states $|\psi(0,0)\rangle$. The method of Alice preparing the quantum state $|\psi_j(0,0)\rangle$, $j = 1, 2, 3$ can be described as

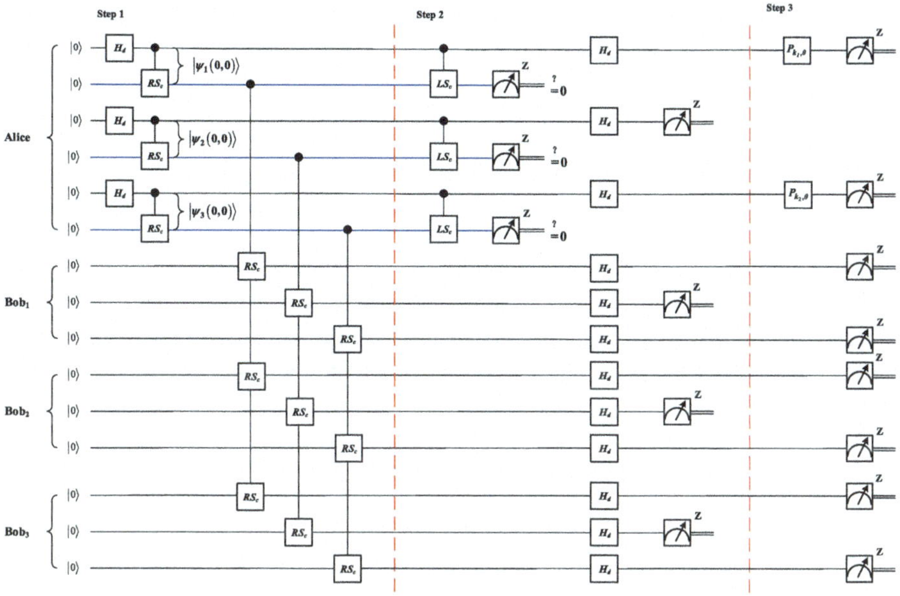

Fig. 4.6 Quantum circuit construction for our four-party QSS scheme without consideration of dynamically adding or deleting an agent. Three blue lines denote *3* qudits in sequence S_T that is transmitted among Alice, Bob$_1$, Bob$_2$, and Bob$_3$

$$|\psi_j(0,0)\rangle = RS_C(H_d|0\rangle, |0\rangle) = RS_C\left(\frac{1}{\sqrt{d}}\sum_{m=0}^{d-1}|m\rangle, |0\rangle\right) = \frac{1}{\sqrt{d}}\sum_{m=0}^{d-1}|m,m\rangle \quad (4.34)$$

Thus, Alice can divide the three pair generalized Bell states into two sequences $S_H = \{m_1^1, m_2^1, m_3^1\}$, $S_T = \{m_1^2, m_2^2, m_3^2\}$, in which the subscript j and superscript i of m_j^i respectively denote the qudit comes from the i-th qudit in j-th Bell state $|\psi_j(0,0)\rangle$.

Following that, based on the received sequence S_T, Bob$_1$, Bob$_2$, and Bob$_3$ can perform local unitary operations to prepare their personal sequence S_1, S_2, and S_3. Finally, for each generalized Bell state $|\psi_j(0,0)\rangle$, it has been transformed into *5*-qudit GHZ state written as

$$|\psi_j(0,0,0,0,0)\rangle = \frac{1}{\sqrt{d}}\sum_{m=0}^{d-1}|m,m,m,m,m\rangle, \quad j=1,2,3 \quad (4.35)$$

Accordingly, the Bob$_1$'s, Bob$_2$'s, and Bob$_3$'s sequences can be, respectively, denoted as $S_1 = \{m_1^3, m_2^3, m_3^3\}$, $S_2 = \{m_1^4, m_2^4, m_3^4\}$, and $S_3 = \{m_1^5, m_2^5, m_3^5\}$.

Step 2 First, Alice performs local unitary operation $\mathrm{LS_C}$ on each pair qudits m_j^1 and m_j^2, for *j=1* to *3*, which would transform each *5*-qudit GHZ state into the following form:

$$LS_C \bigotimes I_d^{\bigotimes 3} |\psi_j(0,0,0,0,0)\rangle = \frac{1}{\sqrt{d}} \sum_{m=0}^{d-1} |m,0,m,m,m\rangle \quad (4.36)$$

Now, it is easy to find that each qudit m_j^2 in sequence S_T being quantum state $|0\rangle$ has disentangled with former *5*-qudit GHZ state $|\psi_j(0,0,0,0,0)\rangle$. Thus, when the measurement result of each qudit in sequence S_T is not equal to *0*, Alice can detect the presence of eavesdropper and accordingly aborts the scheme; or else, the scheme continues.

Since each qudit in sequence S_T has disentangled and measured, Alice now shares three *4*-qudit GHZ states among three agents $\mathrm{Bob_1}$, $\mathrm{Bob_2}$, and $\mathrm{Bob_3}$ written as

$$|\psi_j(0,0,0,0)\rangle = \frac{1}{\sqrt{d}} \sum_{m=0}^{d-1} |m,m,m,m\rangle, \ j=1,2,3 \quad (4.37)$$

Hence, when Alice and agents ($\mathrm{Bob_1}$, $\mathrm{Bob_2}$, and $\mathrm{Bob_3}$) perform the H_d operation on each qudit in their sequence, state $|\psi_j(0,0,0,0,0)\rangle$ would be transformed into

$$\begin{aligned}
|\psi^*{}_j(0,0,0,0)\rangle &= H_d^{\bigotimes 4}(|\psi_j(0,0,0,0)\rangle) \\
&= \frac{1}{\sqrt{d}} \sum_{m=0}^{d-1} \left(\frac{1}{\sqrt{d}} \sum_{s_H=0}^{d-1} w^{ms_H} |s_H\rangle \times \frac{1}{\sqrt{d}} \sum_{s_1=0}^{d-1} w^{ms_1} |s_1\rangle \right. \\
&\quad \left. \times \frac{1}{\sqrt{d}} \sum_{s_2=0}^{d-1} w^{ms_2} |s_2\rangle \times \frac{1}{\sqrt{d}} \sum_{s_3=0}^{d-1} w^{ms_3} |s_3\rangle \right) \\
&= \frac{1}{\sqrt{d^5}} \sum_{m=0}^{d-1} \left(\sum_{s_H, s_1, s_2, s_3=0}^{d-1} w^{m(s_H+s_1+s_2+s_3)} |s_H\rangle |s_1\rangle |s_2\rangle |s_3\rangle \right) \\
&= \frac{1}{\sqrt{d^3}} \sum_{\substack{s_H, s_1, s_2, s_3=0, \\ s_H+s_1+s_2+s_3 \equiv 0 (\mathrm{mod}\ d)}}^{d-1} |s_H\rangle |s_1\rangle |s_2\rangle |s_3\rangle
\end{aligned} \quad (4.38)$$

The correctness of Eq. 4.38 can be easily proved by *Appendix A*. Herein, let's assume that the second *4*-qudit entangled quantum state $|\psi^*{}_2(0,0,0,0)\rangle$ is employed by Alice for eavesdropping check. That is, Alice measures *2*-th qudit in her sequence S_H based on Z basis and obtains the measurement result s_{H_2}, and $\mathrm{Bob_i}$ measures *2*-th qudit in his sequence S_i based on Z basis and obtains the measurement result s_{i_2}. If their measurement results do not satisfy that $s_{H_2} + s_{1_2} + s_{2_2} + s_{3_2} = 0(\mathrm{mod}\ d)$, Alice aborts the scheme; or else, the scheme continues.

Step 3 Alice encodes 2-dit secrets k_1k_2 via generalized Pauli operation $P_{k_1,0}$ and $P_{k_2,0}$ on the remaining entangled states $|\psi^*{}_1(0,0,0,0)\rangle$ and $|\psi^*{}_3(0,0,0,0)\rangle$ described as,

$$P_{k_1,0}\bigotimes I_d^{\bigotimes 3}(|\psi^*{}_1(0,0,0,0)\rangle)=\frac{1}{\sqrt{d^3}}\sum_{\substack{s_H,s_1,s_2,s_3=0,\\ s_H+s_1+s_2+s_3\equiv 0 \pmod d}}^{d-1}|s_H+k_1\rangle|s_1\rangle|s_2\rangle|s_3\rangle$$
$$P_{k_2,0}\bigotimes I_d^{\bigotimes 3}(|\psi^*{}_3(0,0,0,0)\rangle)=\frac{1}{\sqrt{d^3}}\sum_{\substack{s_H,s_1,s_2,s_3=0,\\ s_H+s_1+s_2+s_3\equiv 0 \pmod d}}^{d-1}|s_H+k_2\rangle|s_1\rangle|s_2\rangle|s_3\rangle \tag{4.39}$$

Following that, Alice and each agent measure the remaining two qudits in their sequence based on Z basis. Assume that Alice's measurement results are $A=a_1a_2$ and each agent $\mathrm{Bob_i}$'s measurement results are $B^i=b_1^ib_2^i$. To share 2-dit secrets $K=k_1k_2$, Alice publically announces the measurement results $A=a_1a_2$ first, and then three agents Bob_1, Bob_2, and Bob_3 can collaboratively recover Alice's secret as

$$\begin{aligned}K &\equiv \left(A+B^1+B^2+B^3\right)\bmod d\\ &\Rightarrow \left(a_j+b_j^1+b_j^2+b_j^3\right)\equiv k_j+s_{H_j}+s_{1_j}+s_{2_j}+s_{3_j}(\bmod\ d)\equiv k_j, j=1,2\end{aligned} \tag{4.40}$$

4.4.2.7 General Attack Analysis

This subsection proves that the proposed DQSS scheme can resist general attacks, such as intercept-resend, entangle-measure, collusion, and dishonest revoked agent attacks.

A. **Intercept-Resend Attack**

In the proposed DQSS schemes, an eavesdropper (says Eve) may try to intercept and measure all qudits in sequence S_T and then resend the qudits sequence prepared by him in advance, hoping to pass eavesdropping check. Herein, let's assume that Eve intercepts the S_T sequence transmitted from Bob_k to Bob_{k+1}. It is worth noting that after $\mathrm{Bob_k}$ has prepared his sequence S_k, qudits in the same position as Alice's sequence S_H, Bob_1's, Bob_2's, …, $\mathrm{Bob_k}$'s sequences (S_1, S_2, …, S_k) and the transmitted sequence S_T are formed into a (k+2)-qudit GHZ state written as

$$\left|\Psi\left(\underbrace{0,0,\cdots,0,0}_{k+2}\right)\right\rangle = \frac{1}{\sqrt{d}}\sum_{j=0}^{d-1}|j,j,\cdots,j,j\rangle \tag{4.41}$$

where each qudit in sequence S_T is the second qudit of each GHZ state.

However, in the proposed DQSS scheme, each qudit in transmitted sequence S_T is disentangled from the GHZ state and further measured for eavesdropping check by Alice in later, which makes each qudit in transmitted sequence S_T independent of final GHZ state (shared between Alice and n agents) and Alice's secrets (where Alice encodes her secrets on each qudit in her sequence S_H via local generalized Pauli operation after each qudit in transmitted sequence S_T has been disentangled and measured). When Eve implements the intercept-resend attack, he can't steal any information about the secrets and his existence can also be detected with the probability *100%* in Step 2.1 in ideal situation. Thus, the proposed DQSS scheme is secure under the intercept-resend attack.

B. **Entangle-Measure Attack**

The eavesdropper Eve may use a unitary operation U_e to entangle an ancillary particle $|e\rangle$ on each transmitted qudit in sequence S_T and then measure the ancillary particles to steal secret information. According to the above analysis, when Eve entangles his ancillary particle $|e\rangle$ to each qudit in transmitted sequence S_T in the process of $\text{Bob}_{\text{k-1}}$ sending the sequence S_T to Bob_{k} ($2 \le k \le n$), the density matrix of each qudit in sequence S_T can be deduced as

$$\begin{aligned}
\rho_2 &= tr_{1,\cdots,k+1,k+2}(|\psi(0,\cdots,0,0)\rangle\langle\psi(0,\cdots,0,0)|) \\
&= tr_{1,\cdots,k+1,k+2}\left(\frac{1}{\sqrt{d}}\sum_{j=0}^{d-1}|j,\cdots,j,j\rangle \times \frac{1}{\sqrt{d}}\sum_{j=0}^{d-1}\langle j,\cdots,j,j|\right) \\
&= \frac{1}{d}tr_{1,\cdots,k+1,k+2}\left(\sum_{j_1,j_2=0}^{d-1}|j_1,\cdots,j_1,j_1\rangle\langle j_2,\cdots,j_2,j_2|\right) = \frac{1}{d}\sum_{j=0}^{d-1}|j\rangle\langle j| = \frac{I_d}{d}
\end{aligned} \tag{4.42}$$

where I_d is $d \times d$-dimensional identity matrix.

Obviously, each qudit in sequence S_T is the maximally mixed state with the density matrix $\rho = I_d/d$. Furthermore, for each qudit $|i\rangle$, $i = 0, 1, \cdots, d-1$, we can deduce that Eve's unitary operation U_e [53] is expressed as

$$U_e|i\rangle|e\rangle = a_{i,0}|0\rangle|e_{i,0}\rangle + a_{i,1}|1\rangle|e_{i,1}\rangle + \cdots + a_{i,(d-1)}|d-1\rangle|e_{i,(d-1)}\rangle \tag{4.43}$$

where $|e_{i,j}\rangle$ $(i, j \in \{0, 1, \cdots, d-1\})$ are the states determined by the unitary operation U_e, and $a_{i,j}$ is the complex number that satisfies

$$|a_{i,0}|^2 + |a_{i,1}|^2 + \cdots + \left|a_{i,(d-1)}\right|^2 = 1 \tag{4.44}$$

In order to avoid introducing the disturbance to quantum system $|i\rangle$, Eve has to set

$$|a_{i,0}|^2 + \cdots + |a_{i,i-1}|^2 + |a_{i,i+1}|^2 + \cdots + \left|a_{i,(d-1)}\right|^2 = 0 \tag{4.45}$$

Accordingly, the unitary operation U_e can be simplified as $U_e|i\rangle|e\rangle = |i\rangle|e_{i,\ i}\rangle$. Thus, when $\mathrm{Bob_i}$ uses a unitary operation U_E to entangle each ancillary particle $|e\rangle$ on the transmitted qudit, the density matrix of the ancillary particle $|e\rangle$ can be described as

$$\rho_e = tr_1\left[\frac{1}{d}\sum_{j=0}^{d-1}(U_e|j\rangle|e\rangle)(\langle j|\langle e|U_e{}^{\dagger})\right] = tr_1\left(\frac{1}{d}\sum_{j=0}^{d-1}|j\rangle|e_j\rangle \times \langle j|\langle e_j|\right) = \frac{1}{d}\sum_{j=0}^{d-1}|e_j\rangle\langle e_j| \tag{4.46}$$

Obviously, it can clearly see that the ancillary particle $|e\rangle$ is also a completely mixed state. When Eve measures the ancillary particle $|e\rangle$ based on Z basis, he also gets one of measurement results $\{0, 1, \ldots, d - 1\}$ with an equal probability $1/d$. Therefore, when Eve implements the entangle-and-measure, her existence can be undetectable in Step 2.1, in which Alice obtains the corresponding measurement result $|0\rangle$. However, Eve's attack can be detected with probability $d_e = 1 - 1/d^h$ in Step 2.2, where h is the number of GHZ states selected by Alice for eavesdropping check. The reason behind this is because Eve's ancillary qubit has entangled with the shared GHZ state between Alice and n agents, and only when the measurement result of the ancillary particle being state $|0\rangle$ can he hides his existence. Thus, the probability d_e can approach to *1* arbitrarily by picking appropriate large enough h, which proves that proposed DQSS scheme is immune to this attack.

C. **Collusion Attack**

Multiple agents may also try to recover Alice's secret message without the help of some other agents. Suppose that $(n-1)$ agents (Bob_1, Bob_2, ..., Bob_{k-1}, Bob_{k+1}, ..., Bob_n) work together to recover Alice's secrets without the cooperation of $\mathrm{Bob_k}$. Since $(n-1)$ agents (Bob_1, Bob_2, ..., Bob_{k-1}, Bob_{k+1}, ..., Bob_n) can't possess $\mathrm{Bob_k}$'s sequence S_k, they have to announce the fake measurement results instead of the correct measurement results publicly announced by Bob_k. It is easy to deduce that each qudit in $\mathrm{Bob_k}$'s sequence S_k is a maximally mixed state according to former Eq. 4.53, and then the measurement result of each qudit in sequence S_k can be one of the results $\{0, 1, \ldots, d - 1\}$ with probability $1/d$. When Alice implements eavesdropping check in Step 2.2, this attack can be detected with probability $d_e = 1 - 1/d^h$. More generally speaking, if k agents $(2 \le k \le n - 1)$ conspire to implement a collusion attack, this attack can be detected with probability $d_e = 1 - (1/d^h)^{n-k}$.

D. **Dishonest Revoked Agent Attack**

In the proposed DQSS scheme, Alice may want to dynamically revoke one of agents from scheme. According to Step D2.2 in Sect. 4.4.2.5, Alice first asks the revoked agent (say $\mathrm{Bob_k}$) to publicly announce all the measurement results of each qudit in his sequence S_k. Without the prior knowledge of the specific positions of h qudits randomly selected by Alice for eavesdropping check-in later, if $\mathrm{Bob_k}$ publicly announces false measurement results, his attack can be easily detected with probability $1 - 1/d^h$ in Step D2.2. Thus, the present DQSS scheme is secure against dishonest revoked agent attack.

4.4.2.8 Efficiency Analysis in Noise Environments

Under the conditions of current mature quantum technologies, it assumes that Alice and agents can prepare the quantum system as required and implement the perfect local unitary operations. However, in the implementation of the QSS scheme, when particles are transmitted among dealer and agents via the quantum channel, the quantum system will inevitably interact with the outer environment and be affected by channel noise. Hence, this section is devoted to giving the efficiency analysis of the proposed DQSS scheme in different noise models, that is, dit-flip, d-phase-flip, amplitude-damping, and depolarizing that are usually encountered in real world.

A. **Noise Model for High-Dimensional Quantum System**

Operator sum representation is an effective approach that represents the interaction between closed quantum system and external environment [53]. Similar to quantum noise channel analysis in two-dimensional quantum system, the quantum noise channel model in high-dimensional quantum system can be defined as Kraus form [54],

$$\rho' = \varepsilon(\rho) = \sum_{m,n} E_{m,n} \rho E_{m,n}{}^{\dagger}, \tag{4.47}$$

where ε denotes the completely positive trace-preserving map, ρ and ρ' respectively represent the density matrix of input state and corresponding output state. Kraus operators $E_{m,\,n}$ are corresponding to Weyl operators $U_{m,\,n}$ [55] defined as

$$U_{m,n} = \sum_{j=0}^{d-1} w^{jm} |j\rangle \langle j+n| \tag{4.48}$$

According to Ref. [54], four types of noise—dit-flip, d-phase-flip, amplitude damping, and depolarizing channels—can be described as follows:

1. **Dit-Flip Noise**
   This kind of noise refers to turning any computational basis state $|j\rangle$ either to any other states of $|j+1\rangle$, $|j+2\rangle$, $\cdots$, or $|j+d-1\rangle$ with probability p, while keeping it unchanged with probability $1-p$. The corresponding Kraus operators are described as

$$E_{0,0}=\sqrt{1-p}\ U_{0,0},\ \ E_{0,1}=\sqrt{\frac{p}{d-1}}\ U_{0,1},\ \ \cdots,E_{0,d-1}=\sqrt{\frac{p}{d-1}}\ U_{0,d-1} \tag{4.49}$$

2. **D-Phase-Flip Noise**
   D-phase-flip noise describes the loss of quantum information without energy dissipation. This kind of noise refers to $|j\rangle$ may be with probability p suffer one out of $d-1$ phase of the form $w|j\rangle$, $w^2|j\rangle$, $\cdots$, or $w^{d-1}|j\rangle$. The Kraus operators are written as

$$E_{0,0}=\sqrt{1-p}\ U_{0,0},\ \ E_{1,0}=\sqrt{\frac{p}{d-1}}\ U_{1,0},\ \ \cdots,\ \ E_{d-1,0}=\sqrt{\frac{p}{d-1}}\ U_{d-1,0} \tag{4.50}$$

3. **Amplitude-Damping Noise**
   Amplitude damping noise describes the energy dissipation effects due to energy loss in a quantum system. The corresponding Kraus operators can be expressed as

$$E_0=|0\rangle\langle 0|+\sqrt{1-p}\ \sum_{j=1}^{d-1}|j\rangle\langle j|,\ \ E_1=\sqrt{p}\ \sum_{j=1}^{d-1}|0\rangle\langle j| \tag{4.51}$$

   It means that the effect of amplitude-damping noise will flip any basis state $|k\rangle$ (except the basis state $|0\rangle$ remains unchanged) into ground state $|0\rangle$ with probability p and with probability $1-p$ leave it unchanged.

4. **Depolarizing Noise**
   Depolarizing noise means that a quantum system initially prepared in an arbitrary state evolves to a maximally mixed state I_d/d with the probability of p. The Kraus operators for this noise are given as

$$E_{0,0}=\sqrt{1-\frac{d^2-1}{d^2}p}\ U_{0,0}\ ,\ \ E_{m,n}=\sqrt{\frac{p}{d}}\ U_{m,n},\ \ 0\leq m,n\leq d-1,(m,n)\neq(0,0) \tag{4.52}$$

Similar to the single qudit noise model in former Eq. 4.47, the noise model for n-qudit entangled quantum state via Kraus operators is defined as

$$\rho' = \varepsilon(\rho) = \sum_{k_1, k_2, \cdots, k_n} (E_{k_1} \otimes E_{k_2} \otimes \cdots \otimes E_{k_n}) * \rho^* (E_{k_1} \otimes E_{k_2} \otimes \cdots \otimes E_{k_n})^\dagger, \tag{4.53}$$

where ρ is the density matrix of input quantum state with n-qudit, E_{k_j} is the Kraus operator that represents the j-th qudit affected by channel noise.

B. Noise Models for DQSS Scheme

To simplify the analysis, the local particles held by dealer and agents are supposed to not be affected by channel noise, and in the meantime, only the same type of noise as well as identical noise parameter acts on the particles when it is transmitted via the quantum channel. In the proposed DQSS scheme, since only the qudits in sequence S_T are transmitted in the quantum channel among dealer and agents, and then the GHZ state is prepared and shared among dealers and all agents via their local unitary operations. Hence, the efficiency of the present DQSS scheme depends on how close the final shared entangled state (between Alice and all agents) to the maximally entangled GHZ state.

At the beginning of the proposed DQSS scheme, Alice first prepares the Bell state $|\psi(0,0)\rangle = 1/\sqrt{d}\sum_{k=0}^{d-1}|k,k\rangle$. Thus, when she sends the second qudit in $|\psi(0,0)\rangle$ (i.e., the transmitted particle) to agent Bob$_1$ via the quantum channel, the noise model for the whole quantum system can be described as

$$\rho_1 = \varepsilon(\rho) = \sum_{k_2} \left(I_d \otimes E_{k_2}^2\right) * \rho * \left(I_d \otimes E_{k_2}^2\right)^\dagger, \tag{4.54}$$

where ρ is the density matrix of generalized Bell state $|\psi(0,0)\rangle$ described as $\rho = 1/d\sum_{k_1,k_2=0}^{d-1}|k_1,k_1\rangle\langle k_2,k_2|$, and ρ_1 is the density matrix of the whole quantum system when Bob$_1$ receives the transmitted particle sent by Alice.

Following that, Bob$_1$ implements the local unitary operation RS$_C$ on the combined quantum system of transmitted particle and his ancillary particle $|0\rangle$. Herein, let's denote the density matrix of the whole quantum system after Bob$_1$ implementing his local unitary operation is $\rho_1{}'$. Then, it can be described as

$$\rho_1{}' = U_3\ (\rho_1 \otimes |0\rangle\langle 0|)\ U_3{}^\dagger, \tag{4.55}$$

where U_3 describes Bob$_1$'s local unitary operation RS$_C$ on the composite quantum system of pair qudits between the transmitted particle and his ancillary particle $|0\rangle$.

In a similar way, Bob$_1$ sends the transmitted particle to Bob$_2$ via quantum noise channel either. Thanks to the remaining agents (Bob$_2$, Bob$_3$, …, Bob$_n$) implement the same local unitary operation RS$_C$ as Bob$_1$ does and sends the transmitted particle to next agent via the quantum channel. Finally, Bob_{n} sends the transmitted particle back to Alice. Hence, we can easily to deduce that the general expression for the density matrices of the whole quantum system when the agent Bob_{n} receives the transmitted particle, implements his local unitary operation RS_{C}, and finally sends the transmitted particle back to Alice described as

$$\begin{aligned}
\rho_n &= \varepsilon(\rho_{n-1}{}') \\
&= \sum_{k_2} \Bigg(I_d \otimes E_{k_2}^2 \otimes \underbrace{I_d \otimes \cdots \otimes I_d}_{n-1} \Bigg) * \rho_{n-1}{}' * \Bigg(I_d \otimes E_{k_2}^2 \otimes \underbrace{I_d \otimes \cdots \otimes I_d}_{n-1} \Bigg)^{\dagger} \\
\rho_n{}' &= U_{n+2}(\rho_n \otimes |0\rangle\langle 0|)U_{n+2}{}^{\dagger} \\
\rho_A &= \varepsilon(\rho_n{}') \\
&= \sum_{k_2} \Bigg(I_d \otimes E_{k_2}^2 \otimes \underbrace{I_d \otimes \cdots \otimes I_d}_{n} \Bigg) * \rho_n{}' * \Bigg(I_d \otimes E_{k_2}^2 \otimes \underbrace{I_d \otimes \cdots \otimes I_d}_{n} \Bigg)^{\dagger},
\end{aligned} \tag{4.56}$$

where $\rho_{n-1}{}'$ denotes the instantly density matrix of the whole quantum system before $\text{Bob}_{\text{n}-1}$ sends the transmitted particles to Bob_n, ρ_n is the density matrix of whole quantum system when Bob_{n} receives the transmitted particle sent by Bob_{n-1}, U_{n+2} describes Bob_n's local unitary operation RS_{C} on the composite quantum system of each pair qudits, that is, the transmitted particle and his ancillary particle $|0\rangle$, $\rho_n{}'$ is the density matrix after Bob_{n} implementing his local unitary operation RS_{C} as Bob_1 do, and ρ_A denotes the final density matrix of whole quantum system when Alice receives the transmitted particles.

Therefore, when considers the specific noise model existed in quantum channel, the final output density matrix after Alice receiving the transmitted particles can be deduced as

$$\begin{aligned}
\rho_A &= \varepsilon^{df}(\rho_n{}') \\
&= (I_d^1 \otimes E_{0,0}^2 \otimes I_d^3 \otimes \cdots \otimes I_d^{n+2}) * \rho_n{}'^{*} (I_d^1 \otimes E_{0,0}^2 \otimes I_d^3 \otimes \cdots \otimes I_d^{n+2})^{\dagger} \\
&\quad + \sum_{j=1}^{d-1} \left(I_d^1 \otimes E_{0,j}^2 \otimes I_d^3 \otimes \cdots \otimes I_d^{n+2} \right) * \rho_n{}' * \left(I_d^1 \otimes E_{0,j}^2 \otimes I_d^3 \otimes \cdots \otimes I_d^{n+2} \right)^{\dagger} \\
&= (1-p) * (I_d^1 \otimes U_{0,0}^1 \otimes I_d^3 \otimes \cdots \otimes I_d^{n+2}) * \rho_n{}' * (I_d^1 \otimes U_{0,0}^1 \otimes I_d^3 \otimes \cdots \otimes I_d^{n+2})^{\dagger} \\
&\quad + \frac{p}{d-1} \sum_{j=1}^{d-1} \left(I_d^1 \otimes U_{0,j}^1 \otimes I_d^3 \otimes \cdots \otimes I_d^{n+2} \right) * \rho_n{}' * \left(I_d^1 \otimes U_{0,j}^1 \otimes I_d^3 \otimes \cdots \otimes I_d^{n+2} \right)^{\dagger}
\end{aligned} \tag{4.57}$$

$$
\begin{aligned}
\rho_A &= \varepsilon^{dpf}(\rho_n{}') \\
&= (I_d^1 \otimes E_{0,0}^2 \otimes I_d^3 \otimes \cdots \otimes I_d^{n+2}) * \rho_n{}'^{*} (I_d^1 \otimes E_{0,0}^2 \otimes I_d^3 \otimes \cdots \otimes I_d^{n+2})^\dagger \\
&\quad + \sum_{j=1}^{d-1} \left(I_d^1 \otimes E_{j,0}^2 \otimes I_d^3 \otimes \cdots \otimes I_d^{n+2}\right) * \rho_n{}' * \left(I_d^1 \otimes E_{j,0}^2 \otimes I_d^3 \otimes \cdots \otimes I_d^{n+2}\right)^\dagger \\
&= (1-p) * (I_d^1 \otimes U_{0,0}^2 \otimes I_d^3 \otimes \cdots \otimes I_d^{n+2}) * \rho_n{}'^{*} (I_d^1 \otimes U_{0,0}^2 \otimes I_d^3 \otimes \cdots \otimes I_d^{n+2})^\dagger \\
&\quad + \frac{p}{d-1} \sum_{j=1}^{d-1} \left(I_d^1 \otimes U_{j,0}^2 \otimes I_d^3 \otimes \cdots \otimes I_d^{n+2}\right) * \rho_{n-1}{}' * \left(I_d^1 \otimes U_{j,0}^2 \otimes I_d^3 \otimes \cdots \otimes I_d^{n+2}\right)^\dagger
\end{aligned}
\tag{4.58}
$$

$$
\begin{aligned}
\rho_A &= \varepsilon^{ad}(\rho_n{}') \\
&= (I_d^1 \otimes E_0^2 \otimes I_d^3 \otimes \cdots \otimes I_d^{n+2}) * \rho_n{}'^{*} (I_d^1 \otimes E_0^2 \otimes I_d^3 \otimes \cdots \otimes I_d^{n+2})^\dagger + \\
&\quad (I_d^1 \otimes E_1^2 \otimes I_d^3 \otimes \cdots \otimes I_d^{n+2}) * \rho_n{}' * (I_d^1 \otimes E_1^2 \otimes I_d^3 \otimes \cdots \otimes I_d^{n+2})^\dagger \\
&= (I_d^1 \otimes |0\rangle\langle 0| \otimes I_d^3 \otimes \cdots \otimes I_d^{n+2}) * \rho_n{}'^{*} (I_d^1 \otimes |0\rangle\langle 0| \otimes I_d^3 \otimes \cdots \otimes I_d^{n+2})^\dagger \\
&\quad + \sqrt{1-p}(I_d^1 \otimes |0\rangle\langle 0| \otimes I_d^3 \otimes \cdots \otimes I_d^{n+2}) * \rho_n{}'^{*} \left(I_d^1 \otimes \sum_{j=1}^{d-1} |j\rangle\langle j| \otimes I_d^3 \otimes \cdots \otimes I_d^{n+2} \right)^\dagger \\
&\quad + \sqrt{1-p}\left(I_d^1 \otimes \sum_{j=1}^{d-1} |j\rangle\langle j| \otimes I_d^3 \otimes \cdots \otimes I_d^{n+2} \right) * \rho_n{}'^{*} (I_d^1 \otimes |0\rangle\langle 0| \otimes I_d^3 \otimes \cdots \otimes I_d^{n+2})^\dagger \\
&\quad + (1-p)\left(I_d^1 \otimes \sum_{j=1}^{d-1} |j\rangle\langle j| \otimes I_d^3 \otimes \cdots \otimes I_d^{n+2} \right) * \rho_n{}'^{*} \left(I_d^1 \otimes \sum_{j=1}^{d-1} |j\rangle\langle j| \otimes I_d^3 \otimes \cdots \otimes I_d^{n+2} \right)^\dagger \\
&\quad + p\left(I_d^1 \otimes \sum_{j=1}^{d-1} |0\rangle\langle j| \otimes I_d^3 \otimes \cdots \otimes I_d^{n+2} \right) * \rho_n{}'^{*} \left(I_d^1 \otimes \sum_{j=1}^{d-1} |j\rangle\langle 0| \otimes I_d^3 \otimes \cdots \otimes I_d^{n+2} \right)^\dagger
\end{aligned}
\tag{4.59}
$$

$$
\begin{aligned}
\rho_A &= \varepsilon^{d}(\rho_n{}') \\
&= (I_d^1 \otimes E_{0,0}^2 \otimes I_d^3 \otimes \cdots \otimes I_d^{n+2}) * \rho_n{}'^{*} (I_d^1 \otimes E_{0,0}^2 \otimes I_d^3 \otimes \cdots \otimes I_d^{n+2})^\dagger + \\
&\quad \sum_{\substack{m,n=0,\\ (m,n)\neq(0,0)}}^{d-1} (I_d^1 \otimes E_{m,n}^2 \otimes I_d^3 \otimes \cdots \otimes I_d^{n+2}) * \rho_n{}' * (I_d^1 \otimes E_{m,n}^2 \otimes I_d^3 \otimes \cdots \otimes I_d^{n+2})^\dagger \\
&= \left(1 - \frac{d^2-1}{d^2}p\right) (I_d^1 \otimes U_{0,0}^2 \otimes I_d^3 \otimes \cdots \otimes I_d^{n+2}) * \rho_n{}'^{*} (I_d^1 \otimes U_{0,0}^2 \otimes I_d^3 \otimes \cdots \otimes I_d^{n+2})^\dagger + \\
&\quad \frac{p}{d^2} \sum_{\substack{m,n=0,\\ (m,n)\neq(0,0)}}^{d-1} (I_d^1 \otimes U_{m,n}^2 \otimes I_d^3 \otimes \cdots \otimes I_d^{n+2}) * \rho_n{}' * (I_d^1 \otimes U_{m,n}^2 \otimes I_d^3 \otimes \cdots \otimes I_d^{n+2})^\dagger
\end{aligned}
\tag{4.60}
$$

4.4.2.9 Efficiency Analysis

Fidelity is used to measure how close of two quantum states is [53], which can be employed here to evaluate the efficiency of the proposed DQSS scheme in noise environment. Fidelity is defined as

$$F = tr(\rho_{in}\rho_{out}), \tag{4.61}$$

where *tr*() denotes the trace of matrix, ρ_{in} and ρ_{out}, respectively, denote the density matrices of initial input quantum state and output quantum state. Moreover, when the input quantum state is a pure quantum state such as $\rho_{in} = |\varphi\rangle\langle\varphi|$, fidelity can be further defined as

$$F = \langle\varphi|\rho_{out}|\varphi\rangle \tag{4.62}$$

In ideal situation, that is, no noises exist in quantum channel, Alice and all agents (Bob_1, Bob_2, ..., Bob_n) are assumed to share the maximally entangled GHZ state written as in former Eq. 4.20. Accordingly, the fidelity between the final density matrix ρ_A (shown in above Eqs. 4.56, 4.57, 4.58, and 4.59) and maximally entangled GHZ state can be used to quantitatively evaluate the efficiency of proposed $(n+1)$-party DQSS scheme in different noises environment. However, given that the noise analysis in high-dimensional quantum systems will become extremely complicated as the number of agents and particles increase, here we only give the performance analysis in noise environment via a simple case with three-party scheme and single qudit quantum system with dimension $d=3$. According to the general expression used to calculate the density matrix shown in Eqs. 4.55, 4.56, 4.57, 4.58, and 4.59, the concrete fidelity in different noises environment is calculated by simulating the quantum system with help of MATLAB 2014a in classical computer. Table 4.4 gives the specific fidelity calculated by Alice in four different noises environment.

In order to more intuitively illustrate the fidelity in different noise models, Fig. 4.7 illustrates the relationship between fidelity and different noise parameters for above present simple case. Obviously, the efficiency of the proposed DQSS scheme is decreased with the increase of noise parameter p in four different noise models. But when noise parameter p is no greater than 0.4, the proposed DQSS scheme has better efficiency in amplitude-damping noise channel than other three noise channels of dit-flip, d-phase-flip, and depolarizing. Under the same noise parameter p, dit-flip noise channel always has the worst performance. When the noise parameter p is no less than 0.5, d-phase-flip noise channel always has the best performance than other noise channels. Similarly, when the proposed DQSS scheme is designed in a higher dimension quantum system ($d > 3$) and larger number of agents ($n > 3$), similarly conclusions can be drawn and the specific fidelity in different noise channels can be obtained. Furthermore, based on the calculated fidelity in the specific noise channel, the corresponding error rate decided by channel noise in the present DQSS scheme can be obtained, and accordingly, the minimum number of shared GHZ states

Table 4.4 Fidelity in different quantum noise channel with three-party DQSS scheme and single qudit quantum system with dimension d=3

	Noise type			
Fidelity	Dit-flip	D-phase-flip	Amplitude-damping	Depolarizing
Alice	$(1-p)^3$	$1-3p+\frac{9}{2}p^2-\frac{9}{4}p^3$	$\frac{4(1-p)^3}{9}+\frac{4\sqrt{(1-p)^3}}{9}+\frac{1}{9}$	$1-\frac{8}{3}p+\frac{82}{27}p^2-\frac{890}{729}p^3$

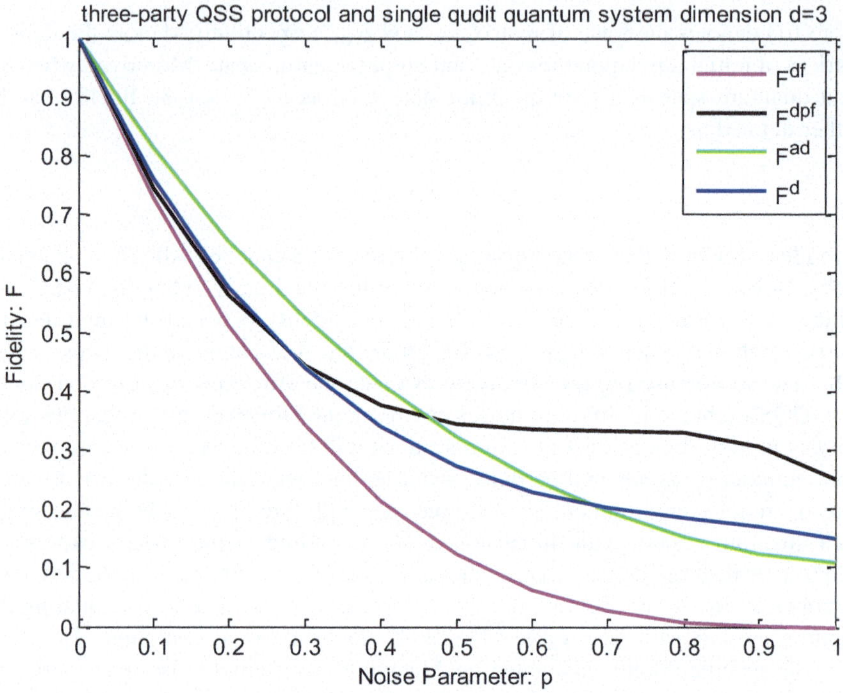

Fig. 4.7 The efficiencies of the proposed DQSS scheme with three-party in four types of noise under different noise parameters p, where F^{df} represents the mean fidelity in the dif-flip noise, F^{dpf} in the d-phase-flip noise, F^{ad} in the amplitude-damping, F^{d} in the depolarizing noise

needed for implementing eavesdropping can be determined based on this to guarantee the security of proposed DQSS scheme in noise environments.

Actually, in the proposed (n+1)-party DQSS scheme, the success of sharing the secret dit depends on the generation of (n+2)-qudit maximally entangled GHZ state (written as former Eq. 4.20) shared between Alice and n agents (Bob_1, Bob_2, ..., Bob_n). Hence, the fidelity between the (n+2)-qudit maximally entangled quantum GHZ state and (n+2)-qudit mixed entangled state generated due to the effect of noise channel can be utilized to represent the secure information sharing rate of the present DQSS scheme (i.e., efficiency, except for a little fraction of GHZ states used for eavesdropping check). Let us consider a simple DQSS scheme in which the single quantum system is three-dimensional and contains three parties, the fidelity in four types of noise environment is illustrated in Table 4.4. Thus, in four types of quantum noise environments: dit-flip, d-phase-flip, amplitude damping, and depolarizing channel, the secure information sharing rate of the present QSS scheme, respectively, are $(1 - p)^3$, $1-3p+\frac{9}{2}p^2-\frac{9}{4}p^3$, $\frac{4(1-p)^3}{9}+\frac{4\sqrt{(1-p)^3}}{9}+\frac{1}{9}$, and $1-\frac{8}{3}p+\frac{82}{27}p^2-\frac{890}{729}p^3$, where p denotes the corresponding noise parameter.

4.4.2.10 Comparisons

Given that a method of $\eta = c/q$ has been used in DQSS schemes [28, 36, 37] to evaluate the qubit efficiency, where c is the number of shared classical secret bits, and q is the number of qubits generated by the scheme. In Hsu's n-party scheme [28], the dealer must generate n Bell states (i.e., $2n$ qubits) to share a 2-bit secret. In Qin's n-party scheme [35], each (n+1)-qudit GHZ state (i.e., n+1 qudits) could be used to encode a *1*-dit secret. In Song's n-party scheme [36], the dealer must generate n-qubit GHZ state (i.e., n qubits) to share a *1*-bit secret. In Yang's n-party scheme [37], the dealer must generate $n-1$ Bell states (i.e., $2(n-1)$ qubits) to share a *1*-bit secret. In the proposed n-party scheme, GHZ state with n+1 qudits is used to share *1*-dit secret. However, it is worth noting that DQSS schemes in [28, 35–37] were inserted lots of detecting particles for eavesdropping check in quantum channels between dealers and each agent. As the number of agents increase rapidly, the qubit (qudit) efficiency of their schemes can be decreased dramatically. Compared to their schemes, the difference in the proposed DQSS scheme is that each qudit was disentangled with GHZ state and a small amount of GHZ states are used for eavesdropping check. Therefore, our scheme has high qudit efficiency. When ignores the detecting particles in the abovementioned schemes, the qubit or qudit efficiency and some important properties of DQSS schemes in [28, 35–37] and scheme as well as some important properties are illustrated in Table 4.5.

In addition to the above, the proposed DQSS scheme is constructed in the context of qudits instead of qubits in [28, 36, 37], where each shared secret is dit instead of bit that makes our scheme has higher resource capacity. Although the DQSS scheme in [35] is also constructed based on the GHZ state in the context of qudits, the proposed DQSS scheme has obvious differences and advantages, which are described as follows:

1. Dealer has to prepare the entire GHZ state and distribute qudits in GHZ state to each agent via quantum channel in [35], while in the proposed DQSS scheme, the GHZ state are prepared by the local unitary operations of dealer and agents
2. When dynamically adds or revokes agents, the proposed DQSS scheme don't needs all agents online and can share the predetermined dits rather than random dits.
3. The efficiency of proposed DQSS scheme in noises environment has been demonstrated.

Table 4.5 Comparisons between some existing schemes and the proposed DQSS scheme

	Schemes				
Items	Scheme in [28]	Scheme in [35]	Scheme in [36]	Scheme in [37]	Proposed Scheme
Dimension	Two-dimension	High-dimensional	Two-dimension	Two-dimension	High-dimensional
Quantum resources	Bell states	GHZ states	GHZ states	Bell states	GHZ state
Qubit efficiency (n-party)	$\frac{1}{n}$	$\frac{1}{n+1}$	$\frac{1}{n}$	$\frac{1}{2n-2}$	$\frac{1}{n+1}$
Resources for eavesdropping check	Decoy particles	Detecting particles	Decoy particles	Decoy particles	GHZ states
Dealer's measurement	Bell state measurement	Single-particle measurement	Single-particle measurement	GHZ state measurement	Single-particle measurement
Agent's measurement	Bell measurement	Single-particle measurement	Single-particle measurement	Single-particle measurement	Single-particle measurement
Shared secrets	Random bits	Random dits	Predetermined bits	Random bits	Predetermined dits
Dealer and all agents must be online for adding or revoking agents	Yes	Yes	No	Yes	No
Vulnerability to collusion attack	Yes	No	No	No	No
Vulnerability to dishonest revoked agent attack	Yes	Yes	No	No	No
Efficiency analysis under different quantum noise channel	No	No	No	No	Yes

4.4.2.11 Summary

A novel dynamic quantum secret sharing scheme in a high-dimensional quantum system is proposed based on transmitted particles and local unitary operations. In the scheme, dealer and each agent can perform the local unitary operations to share the multi-qudit entangled GHZ state. For secret recover, dealer and each agent only need to perform single-particle measurement operations rather than multi-particle measurement operations such as Bell states and GHZ states in previous schemes. With help of dealer, an agent can be easily added or deleted from the DQSS scheme without other agents help. To improve the qudit efficiency, detecting particles inserted into each quantum channel for eavesdropping check are omitted in proposed DQSS scheme. Furthermore, the proposed DQSS scheme can share the predetermined dits rather the random bits or dits in former researches.

Security analysis indicates that the proposed DQSS scheme is secure against intercept-and-resend, entangle-and-measure, collusion and dishonest revoked agent attacks. Furthermore, the efficiency analysis of the present DQSS scheme in different quantum noises model is deduced through quantum fidelity. The obtained results show that the present DQSS scheme has quite different efficiency in noise models of dit-flip, d-phase-flip, amplitude-damping, and depolarizing and its efficiency decreases with the increase of noise parameter in all noise models and number of agents. It is worth noting that based on the quantum fidelity calculated in specific noise channel, the threshold error rate decided by channel noise in the present DQSS scheme can be obtained, and the minimum number of shared GHZ states needed for implementing eavesdropping can be determined based on this to guarantee security of the proposed DQSS scheme in noise environment. In summary, compared with the previous existing DQSS schemes, the proposed DQSS scheme is more efficient and practical when considering implementing the DQSS scheme under actual conditions. In the future, the study of the simultaneous existence of different kinds of quantum noise in quantum channels will arouse people's interest.

4.4.2.12 Appendix

Let us assume that Alice first prepares the generalized Bell state $|\psi(0,0)\rangle = \frac{1}{\sqrt{d}} \times \sum_{m=0}^{d-1} |m,m\rangle_{12}$ and transmits the second particle among all agents (Bob_1, Bob_2, ..., Bob_n) and herself. After each agent's local unitary operation RS_c, it would form a (n +2)-particle generalized GHZ state $|\psi(0,0,0,\cdots,0)\rangle_{1,\ 2,\ 3,\ \cdots,\ n\ +\ 2}$, and the whole operation can be expressed as

$$|\psi(0,0,0,\cdots,0)\rangle_{1,2,3,\cdots,n+2} = \left[\prod_{j=3}^{n} RS_C(2,j)\right]|\psi(0,0)\rangle_{12}|0\rangle^{\otimes n} = \frac{1}{\sqrt{d}}\sum_{m=0}^{d-1}|m,m,m,\cdots,m\rangle_{1,2,3,\cdots,n+2}, \tag{A4.1}$$

where $RS_C(2,j)$ represents the implementation of local unitary operation RS_C on two qudits, and the second qudit and j-th qudit, respectively, act as the control qudit and target qudit.

When Alice performs local unitary operation LS_C on two qudits (i.e., the first qudit and second qudit), where the first qudit acts control qudit and second qudit acts the target qudit, the above (n+2)-qudit GHZ state would transform into the following form:

$$|\psi^*\rangle = \frac{1}{\sqrt{d}}\sum_{m=0}^{d-1}|m,0,m,\cdots,m\rangle_{1,2,3,\cdots,n+2} \tag{A4.2}$$

Now, it is easy to find that second qudit in the above quantum state $|\psi^*\rangle$ (i.e., particles in sequence S_T) being quantum state $|0\rangle$ has disentangled with other (n+1) qudits. Thus, when Alice measures the second qudit in Z basis, the remaining (n+1) qudits transform into the following GHZ state:

$$|\psi(0,0,\cdots,0)\rangle = \frac{1}{\sqrt{d}}\sum_{m=0}^{d-1}|m,m,\cdots,m\rangle_{1,2,3,\cdots,n+1} \tag{A4.3}$$

Hence, when the generalized Hadamard transform H_d is performed on each qudit in state $|\psi(0,0,\cdots,0)\rangle_{1,2,3,\cdots,n+1}$, Eq. A4.3 would transform into following form,

$$\begin{aligned}
|\Psi^*(0,0,\cdots,0)\rangle &= H_d^{\otimes(n+1)}|\psi(0,0,\cdots,0)\rangle \\
&= \frac{1}{\sqrt{d}}\sum_{m=0}^{d-1}\left(\frac{1}{\sqrt{d}}\sum_{s_H=0}^{d-1} w^{ms_H}|s_H\rangle \times \frac{1}{\sqrt{d}}\sum_{s_1=0}^{d-1} w^{ms_1}|s_1\rangle \times \cdots \times \frac{1}{\sqrt{d}}\sum_{s_n=0}^{d-1} w^{ms_n}|s_n\rangle\right) \\
&= \frac{1}{\sqrt{d^{n+2}}}\sum_{s_H,s_1,\cdots,s_n=0}^{d-1}\ \sum_{m=0}^{d-1} w^{m(s_H+s_1+\cdots+s_n)}|s_H\rangle|s_1\rangle\cdots|s_n\rangle \\
&= \frac{1}{\sqrt{d^n}}\sum_{\substack{s_H,s_1,\cdots,s_n=0 \\ s_H+s_1+\cdots+s_n\equiv 0(\text{mod } d)}}^{d-1}|s_H\rangle|s_1\rangle\cdots|s_n\rangle
\end{aligned} \tag{A4.4}$$

The last equation in above Eq. A4.4 can be further proved as follows:

Let us denote that $S = \sum_{m=0}^{d-1} w^{km}$ and accordingly denote $k = s_H + s_1 + \cdots + s_n$, where $w = e^{2\pi i/d}$, then we can calculate the sum S as

If $k \neq 0$ (mod d), then $w = e^{2\pi ik/d} \neq 1$. Hence, the sum S is calculated as,

$$S = \sum_{m=0}^{d-1} w^{mk} = \sum_{m=0}^{d-1} e^{\frac{2\pi ik}{d}m} = \frac{1 * \left(1 - e^{\frac{2\pi ik}{d}d}\right)}{1 - e^{\frac{2\pi ik}{d}}} = \frac{1 * \left(1 - e^{2\pi ik}\right)}{1 - e^{\frac{2\pi ik}{d}}} = 0 \tag{A4.5}$$

In other case, if $k \equiv 0 \pmod{d}$, then we have $e^{2\pi ikm/d} = e^{2\pi im} = 1$ for any $m \in \{0, d-1\}$. Thus, the sum S is calculated as

$$S = \sum_{m=0}^{d-1} w^{mk} = \sum_{m=0}^{d-1} e^{\frac{2\pi ik}{d}m} = \sum_{m=0}^{d-1} e^{2\pi im} = d \tag{A4.6}$$

In summary, we have

$$S = \sum_{m=0}^{d-1} w^{m(s_H + s_1 + \cdots + s_n)} = \begin{cases} 0, & s_H + s_1 + \cdots + s_n \neq 0 (\operatorname{mod} d) \\ d, & s_H + s_1 + \cdots + s_n \equiv 0 (\operatorname{mod} d) \end{cases} \tag{A4.7}$$

Hence, based on the above Eq. A4.7, we can verify the last equation of the abovementioned Eq. A4.4. And accordingly, the correctness of the previously mentioned Eq. 4.39 is proved.

4.4.3 MDQSS Scheme

Based on the advantages of the existing works and considering the applicability of the scheme, this section proposes a practical and effective multiparty to multiparty dynamic quantum secret sharing (MDQSS) scheme [56]. Based on the generalized GHZ state, we use the property of unbiased basis to realize secret sharing. A trusted party is needed, whose role is to prepare all quantum resources and act as an intermediary when adding and removing agents, and making the secret sharing more secure and efficient. Particularly, compared with the existing QSS schemes, the proposed scheme has the following advantages:

1. All quantum resources are prepared by the trusted party, and other participants do not need to prepare resources, which greatly reduces the cost of quantum generator.
2. The dealer sends their all particles to each agent and does not need to keep any particles. Thus, the amount of quantum memory can be saved.
3. After the quantum channel is established, the agents only need to perform single particle measurement without entanglement measurement to secret realize sharing, which further reduces the experimental difficulty and application cost.
4. The trusted party receives the particle sequence when adding or deleting agents, rather than randomly selecting agents to carry out relevant operations, which ensures the reliability of the scheme and avoids the attack of dishonest dealers.
5. When adding and deleting agents dynamically, their information is verified by the trusted party, which improves the security and avoids collusion attack.

This section will introduce a new multiparty to multiparty dynamic quantum secret sharing (MDQSS) scheme, with an application scenario of n-party to n-party. Then we will explain how to add a new agent or delete an existing agent. Here, the quantum channel is assumed to be ideal, with no channel noise and no particle loss.

4.4.3.1 Unbiased Measurement Base and Generalized GHZ State

In a quantum system, a quantum state is usually represented in a superposition state of a set of orthonormal basis state. In a two-dimensional quantum system, a common set of orthonormal basis state is $\{|Z_0\rangle, |Z_1\rangle\}$, where $|Z_0\rangle = |0\rangle$, $|Z_1\rangle = |1\rangle$, is called Z basis. According to the Fourier transform [57, 58], another orthonormal basis $\{|X_0\rangle, |X_1\rangle\}$, called the X basis, as shown in Eq. 4.63,

$$|X_j\rangle = F|Z_j\rangle = \frac{1}{\sqrt{2}}\sum_{k=0}^{1} e^{\frac{2\pi ijk}{2}}|Z_k\rangle = \frac{1}{\sqrt{2}}\sum_{k=0}^{1} e^{\pi ijk}|Z_k\rangle, \tag{4.63}$$

where $j \in \{0, 1\}$ and F is the quantum Fourier transform. Obviously, the basis vector of the X basis is the Fourier transform of the corresponding basis vector in the Z basis. Thus, it is easy to obtain that

$$|X_0\rangle = |+\rangle = \frac{1}{\sqrt{2}}(|0\rangle + |1\rangle), |X_1\rangle = |-\rangle = \frac{1}{\sqrt{2}}(|0\rangle - |1\rangle) \tag{4.64}$$

Similarly, by performing the inverse quantum Fourier transform, any basis vector $|Z_j\rangle$ $(j=0,1)$ in the Z basis can be expressed in terms of the X basis as follows:

$$|Z_j\rangle = F^{-1}|X_j\rangle = \frac{1}{\sqrt{2}}\sum_{k=0}^{1} e^{\frac{-2\pi ijk}{2}}|X_k\rangle = \frac{1}{\sqrt{2}}\sum_{k=0}^{1} e^{-\pi ijk}|X_k\rangle \tag{4.65}$$

According to the concept of complementary basis, for any j, $k \in 0, 1$, there is $|\langle Z_j|X_k\rangle| = \frac{1}{\sqrt{2}}$. Therefore, X basis and Z basis are complementary basis, which indicates that when a quantum state is in the eigenstate of Z basis (X basis) and measured by X basis (Z basis), the results obtained are completely random. Only by selecting the same measurement basis can the quantum state be recognized and the correct result can be obtained.

We introduce the generalized GHZ states of n particle entanglement, denoted as $|\Phi_0\rangle$ and $|\Phi_1\rangle$, as shown in the following form:

$$|\Phi_0\rangle = \frac{1}{\sqrt{2}}\left(|0\rangle^{\otimes n} + |1\rangle^{\otimes n}\right) \tag{4.66}$$

and

$$|\Phi_1\rangle = \frac{1}{\sqrt{2}}\left(|0\rangle^{\otimes n} - |1\rangle^{\otimes n}\right) \tag{4.67}$$

where $n \geq 2$. When $n = 2$, the state is EPR pair, and when $n \geq 3$, they are called the GHZ state.

According to Eqs. 4.63, 4.64, and 4.65, the n-qubit generalized GHZ state can be expressed by the X basis and Z basis as follows:

$$\begin{aligned}
|\Phi_0\rangle_{1,2,\ldots,n} &= \frac{1}{\sqrt{2}}\left(|0\rangle^{\otimes n} + |1\rangle^{\otimes n}\right) = \frac{1}{\sqrt{2}}\sum_{j=0}^{1}|Z_j\rangle_1 \ldots |Z_j\rangle_n \\
&= \frac{1}{\sqrt{2}}\sum_{j=0}^{1}\left\{\left(\frac{1}{\sqrt{2}}\sum_{k_1=0}^{1} e^{-\pi i j k_1}|X_{k_1}\rangle\right)\ldots\left(\frac{1}{\sqrt{2}}\sum_{k_n=0}^{1} e^{-\pi i j k_n}|X_{k_n}\rangle\right)\right\} \\
&= 2^{-\frac{n+1}{2}}\sum_{j=0}^{1}\sum_{k_1,\ldots,k_n=0}^{1} e^{-\pi i j(k_1+\ldots+k_n)}|X_{k_1}\rangle\ldots|X_{k_n}\rangle \\
&= 2^{-\frac{n+1}{2}}\sum_{k_1,\ldots,k_n=0}^{1}\left(\sum_{j=0}^{1} e^{-\pi i j(k_1+\ldots+k_n)}\right)|X_{k_1}\rangle\ldots|X_{k_n}\rangle
\end{aligned} \tag{4.68}$$

Herein, $k_1 + k_2 + \ldots + k_n$ denotes mod 2 plus. For $K = k_1 + k_2 + \ldots + k_n (\text{mod } 2)$, for k_i ($i = 1, 2, \ldots, n$), all the possible values of K are 0 or 1. According to Euler's formula, when $K = 0$, $\sum_{j=0}^{1} e^{-\pi i j(k_1+\ldots+k_n)} = 2$; When $K = 1$, $\sum_{j=0}^{1} e^{-\pi i j(k_1+\ldots+k_n)} = 0$. Therefore, the Eq. 4.68 simplifies to

$$\begin{aligned}
|\Phi_0\rangle_{12\ldots n} &= \frac{1}{\sqrt{2}}\left(|0\rangle^{\otimes n} + |1\rangle^{\otimes n}\right) = \frac{1}{\sqrt{2}}\sum_{j=0}^{1}|Z_j\rangle_1 \ldots |Z_j\rangle_n \\
&= 2^{-\frac{n+1}{2}}\sum_{k_1,\ldots,k_n=0}^{1}\left(\sum_{j=0}^{1} e^{-\pi i j(k_1+\ldots+k_n)}\right)|X_{k_1}\rangle\ldots|X_{k_n}\rangle \\
&= 2^{-\frac{n-1}{2}}\sum_{\substack{k_1,\ldots,k_n=0 \\ k_1+\ldots+k_n=0(\text{mod}2)}}^{1}|X_{k_1}\rangle\ldots|X_{k_n}\rangle
\end{aligned} \tag{4.69}$$

Similarly, $|\Phi_1\rangle$ can be represented as follows:

$$
\begin{aligned}
|\Phi_1\rangle_{12\ldots n} &= \frac{1}{\sqrt{2}}\left(|0\rangle^{\otimes n} - |1\rangle^{\otimes n}\right) = \frac{1}{\sqrt{2}}\left(|Z_0\rangle_1|Z_0\rangle_2\ldots|Z_0\rangle_n - |Z_1\rangle_1|Z_1\rangle_2\ldots|Z_1\rangle_n\right) \\
&= 2^{-\frac{n+1}{2}}\left[\left(\sum_{\substack{k_1,\ldots,k_n=0\\ k_1+\ldots+k_n=0(\mathrm{mod}\ 2)}}^{1}|X_{k_1}\rangle|X_{k_2}\rangle\ldots|X_{k_n}\rangle + \sum_{\substack{k_1,\ldots,k_n=0\\ k_1+\ldots+k_n=1(\mathrm{mod}\ 2)}}^{1}|X_{k_1}\rangle|X_{k_2}\rangle\ldots|X_{k_n}\rangle\right)\right. \\
&\qquad \left. - \left(\sum_{\substack{k_1,\ldots,k_n=0\\ k_1+\ldots+k_n=0(\mathrm{mod}2)}}^{1}|X_{k_1}\rangle|X_{k_2}\rangle\ldots|X_{k_n}\rangle + \sum_{\substack{k_1,\ldots,k_n=0\\ k_1+\ldots+k_n=1(\mathrm{mod}2)}}^{1}e^{-\pi i}|X_{k_1}\rangle|X_{k_2}\rangle\ldots|X_{k_n}\rangle\right)\right] \\
&= 2^{-\frac{n-1}{2}}\sum_{\substack{k_1,\ldots,k_n=0\\ k_1+\ldots+k_n=1(\mathrm{mod}2)}}^{1}|X_{k_1}\rangle|X_{k_2}\rangle\ldots|X_{k_n}\rangle
\end{aligned}
\tag{4.70}
$$

If each particle in $|\Phi_0\rangle_{1,\ 2,\ \ldots,\ n}$ and $|\Phi_1\rangle_{1,\ 2,\ \ldots,\ n}$ performed the X basis measurement with outcome $|X_{k_i}\rangle$ as $|X_{k_i}\rangle$ $(i = 1, 2, \ldots, n)$. Furthermore, the measurement result of $|X_0\rangle$ represents the classical bit '*0*' while $|X_1\rangle$ represent the classical bit '*1*'. Thus, the measurement results will satisfy $X_{k_1} \oplus X_{k_2} \oplus \ldots \oplus X_{k_n} = 0$ for $|\Phi_0\rangle_{1,\ 2,\ \ldots,\ n}$ and $X_{k_1} \oplus X_{k_2} \oplus \ldots \oplus X_{k_n} = 1$ for $|\Phi_1\rangle_{1,\ 2,\ \ldots,\ n}$, where $\oplus$ is the bitwise exclusive-OR. For example, suppose that $n = 2$. The state is $|\varphi_0\rangle_{12} = \frac{1}{\sqrt{2}}(|00\rangle + |11\rangle)_{12} = \frac{1}{\sqrt{2}}(|{+}{+}\rangle + |{-}{-}\rangle)_{12}$, and the value of $X_{k_1} \oplus X_{k_2} = 0$. While the other state is $|\varphi_1\rangle_{12} = \frac{1}{\sqrt{2}}(|00\rangle - |11\rangle)_{12} = \frac{1}{\sqrt{2}}(|{+}{-}\rangle + |{-}{+}\rangle)_{12}$ and $X_{k_1} \oplus X_{k_2} = 1$.

4.4.3.2 MDQSS Scheme

Assume that n participants {Alice$_1$, Alice$_2$, ..., Alice$_{n\text{-}1}$, Alice$_n$} have shared a secret S, which is made up of m classical binary bits. The share of Alice$_i$ is $A_i = (a_{i1}, a_{i2}, \ldots, a_{im})$, $i \in \{1, 2, \ldots, n\}$, and all the shares satisfy $A_1 \oplus A_2 \oplus \ldots \oplus A_n = S$, where $\oplus$ stands for the bit-XOR. Now, {Alice$_1$, Alice$_2$, ..., Alice$_{n\text{-}1}$, Alice$_n$} want to share the secret with another group of participants {Bob$_1$, Bob$_2$, ..., Bob$_{n-1}$, Bob$_n$}. Of course, no Bob can restore the secret individually unless all Bobs cooperate together. In the MDQSS scheme, it is assumed that there is a trusted party Trent. All quantum resources are prepared by Trent, and other participants do not need to prepare any quantum resources. The detailed MDQSS scheme Steps are as follows and the schematic diagram of the MDQSS scheme is shown in Fig. 4.8.

Step 1: Trent generates n groups of quantum sequences containing m particles, denoted as $S_i = \{p_{i1}, p_{i2}, \ldots, p_{im}\}$, where $p_{ij} \in \{|0\rangle\}, i \in \{1, 2, \ldots, n\}$, $j \in \{1, 2, \ldots, m\}$.

Step 2: Trent performs Hadamard gate on each particle in S_1 sequence, and the particle in S_1 changes from $|0\rangle$ state to the superposition state of $|0\rangle$ state and $|1\rangle$ state.

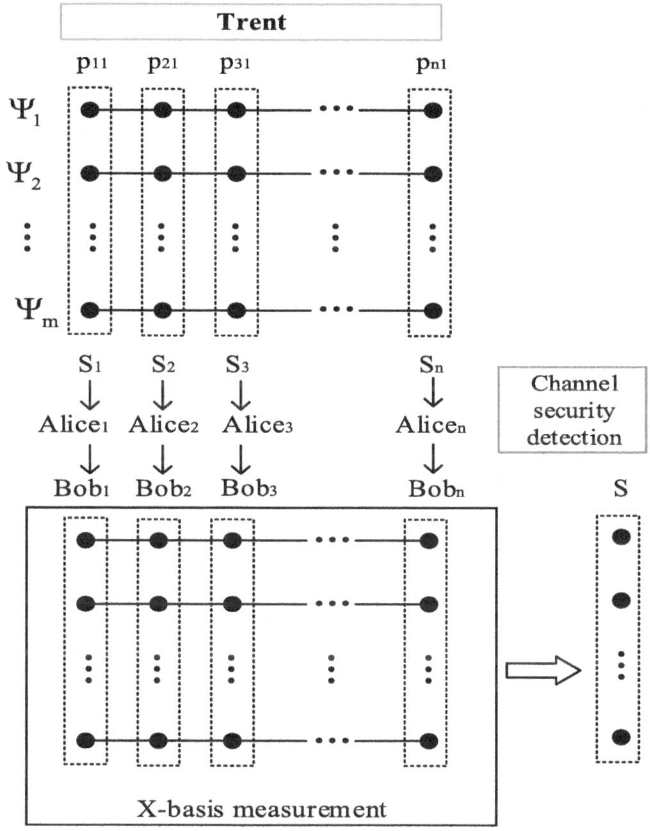

Fig. 4.8 The brief flowchart of DMQSS scheme

Step 3: Trent performs $m*(n-1)$ CNOT operations, denoted as CNOT_{ij}, $i \in \{2, 3, \ldots, n\}$, $j \in \{1, 2, \ldots, m\}$. In a CNOT_{ij} operation, the particle in S_1 is the control particle and the other sequence particles are the target particle, that is, the control particle is p_{1j} and the target particle is p_{ij}.

Step 4: After the above operation, all particle sequences share m generalized GHZ states, $\Psi_j = \frac{1}{\sqrt{2}}\left(|0\rangle^{\otimes n} + |1\rangle^{\otimes n}\right), j \in \{1, 2, \ldots, m\}$.

Step 5: Trent randomly prepares some decoy particles, which are random in the state $\{|0\rangle, |1\rangle, |+\rangle, |-\rangle\}$. Trent randomly inserts these decoy particles into S_i sequence, denoted as S_{i*}, and records the insertion position and initial state of the decoy particles. Trent then sends S_{i*} to Alice_i respectively, $i=1, 2, \ldots, n$.

Step 6: After determining that Alice_i ($i=1, 2, \ldots, n$) has received the sequence S_{i*}, Trent publishes the locations of the decoy particles and asks Alice_i to measure the particles randomly in two bases Z basis or X basis. Alice_i publishes the results of the measurements. Trent calculates the error rate by comparing the measured results with the initial state. If the error rate exceeds the threshold, Trent terminates the distribution and starts a new share. Otherwise, Alice$_i$ receives the sequence S_i sent by Trent.

Step 7: According to her own share $A_i = (a_{i1}, a_{i2}, \ldots, a_{im})$, Alice$_i$ performs unitary operations on each particle of her sequence S_i. If a_{ij}=0, Alice$_i$ executes the unitary operation $I = |0\rangle\langle 0| + |1\rangle\langle 1|$ and performs the unitary operation $\sigma_Z = |0\rangle\langle 0| - |1\rangle\langle 0|$ when a_{ij}=1, where i=1, 2, ..., n, j=1, 2, ..., m.

Step 8: Then Alice$_i$ (i=1, 2, ..., n) sends S_i to Bob$_i$ after confirming that the eavesdropping check has passed through operations similar to Steps 5 and 6.

Step 9: Bob$_i$ (i=1,2,...,n) uses X *basis* to measure each particle in sequence $S_i = \{p_{i1}, p_{i2}, \ldots, p_{im}\}$, and gets the measurement result $B_i = (b_{i1}, b_{i2}, \ldots b_{im})$ where $b_{ij} = 0$ if the measurement result is $|+\rangle$ and $b_{ij} = 1$ if the measurement result is $|-\rangle$. Obviously, $B_i = (b_{i1}, b_{i2}, \ldots b_{im})$ is the share obtained by Bob$_i$.

Step 10: After the above operation, {Alice$_1$, Alice$_2$, ..., Alice$_{n-1}$, Alice$_n$} has shared the key S to {Bob$_1$, Bob$_2$, ..., Bob$_{n-1}$, Bob$_n$}. Then we will prove that $B_1 \oplus B_2 \oplus \ldots \oplus B_n = S$.

4.4.3.3 Proof

To simplify the proof, we assume that the secret M shared by {Alice$_1$, Alice$_2$, ..., Alice$_{n-1}$, Alice$_n$} has only one bit, that is, m=1. After Step 3 of Sect. 4.4.3.2, Trent prepares a sequence S (i=1, 2, ..., n) will share a generalized GHZ state

$$\Psi = \frac{1}{\sqrt{2}}\left(|0\rangle^{\otimes n} + |1\rangle^{\otimes n}\right) = 2^{-\frac{n-1}{2}} \sum_{\substack{k_1, k_2, \ldots k_n = 0 \\ k_1 \oplus k_2 \oplus \ldots \oplus k_n = 0}}^{1} |X_{k_1}\rangle |X_{k_2}\rangle \ldots |X_{k_n}\rangle \tag{4.71}$$

in which $|X_0\rangle = |+\rangle = \frac{1}{\sqrt{2}}(|0\rangle + |1\rangle)$, $|X_1\rangle = |-\rangle = \frac{1}{\sqrt{2}}(|0\rangle - |1\rangle)$.

When Alice$_i$ (i=1, 2, ..., n) receives the eavesdropping detected sequence S, Alice$_i$ performs unitary operations on the particles in the sequence S according to its share $A_i = (a_i)$. If $a_i = 0$, Alice$_i$ performs the operation $I = |0\rangle\langle 0| + |1\rangle\langle 1|$, and if $a_i = 1$, Alice$_i$ performs the operation $\sigma_Z = |0\rangle\langle 0| - |1\rangle\langle 0|$. After these operations, {Alice$_1$, Alice$_2$, ..., Alice$_{n-1}$, Alice$_n$} shared quantum state will become

$$\Psi' = \frac{1}{\sqrt{2}}\left(|0\rangle^{\otimes n} + |1\rangle^{\otimes n}\right) = 2^{\frac{-(n-1)}{2}} \sum_{\substack{r_1, r_2, \ldots r_n = 0 \\ r_1 \oplus r_2 \oplus \ldots \oplus r_n = 0}}^{1} |X_{r_1}\rangle |X_{r_2}\rangle \ldots |X_{r_n}\rangle \tag{4.72}$$

or

$$\Psi'' = \frac{1}{\sqrt{2}}\left(|0\rangle^{\otimes n} - |1\rangle^{\otimes n}\right) = 2^{\frac{-(n-1)}{2}} \sum_{\substack{r_1, r_2, \ldots r_n = 0 \\ r_1 \oplus r_2 \oplus \ldots \oplus r_n = 1}}^{1} |X_{r_1}\rangle |X_{r_2}\rangle \ldots |X_{r_n}\rangle \tag{4.73}$$

where $r_j = k_j \oplus a_j, j \in \{1, 2, \ldots, n\}$.

After Step 8 of Sect. 4.4.3.2, after Bob$_i$ confirms that there is no eavesdropping, the particles in the received sequence S are measured by X basis $\{|+\rangle, |-\rangle\}$. If the measurement result is $|+\rangle$, then $b_i = 0$; if the measurement result is$|-\rangle$, $b_i = 1$. Then we can get that

$$\begin{aligned} b_1 \oplus b_2 \oplus \ldots \oplus b_n &= r_1 \oplus r_2 \oplus \ldots \oplus r_n \\ &= (k_1 \oplus a_1) \oplus (k_2 \oplus a_2) \oplus \ldots \oplus (k_n \oplus a_n) \\ &= (k_1 \oplus k_2 \oplus \ldots \oplus k_n) \oplus (a_1 \oplus a_2 \oplus \ldots \oplus a_n) = M \end{aligned} \tag{4.74}$$

Obviously, the scheme will work well and {Alice$_1$, Alice$_2$, ..., Alice$_{n-1}$, Alice$_n$} can share the secret M with {Bob$_1$, Bob$_2$, ..., Bob$_{n-1}$, Bob$_n$}. No one Bob or less than n Bobs can restore the secret M unless all Bobs cooperate.

4.4.3.4 Add an Agent

Suppose a new agent, named Bob$_X$, wants to add this MDQSS scheme before the distributed quantum is measured. Sharing can be achieved with the help of older members and the trusted party according to the following steps:

Step 1: The trusted party Trent randomly selects any of the Bob groups. Let's say that Bob$_k$ is selected. Bob$_k$ randomly selects some decoys randomly in $\{|0\rangle, |1\rangle, |+\rangle, |-\rangle\}$ and inserts them into its quantum sequence S_k, which is denoted as S_{k*}, and records the insertion position and initial state of decoy particles. Then S_{k*} is sent to the trusted party Trent.

Step 2: After determining that Trent has received the S_{k*} sequence, Bob$_k$ announces the position, initial state and corresponding measurement basis of the decoy states. Trent measures the decoy states based on Bob$_k$'s announcement and then tells Bob$_k$ the measurement results. Bob$_k$ calculates the error rate by comparing the measured results with the initial state. If the error rate is higher than the threshold value, Bob$_k$ will tell Trent to cancel the process of adding a new member. Otherwise, proceed to the next step.

Step 3: Now Trent has the sequence S_k. In order to verify that S_k sent by Bob$_k$ is correct instead of Bob$_k$'s bogus sequence, Trent randomly selects enough C particles in S_k, which is recorded as S_{kc}, and the remaining particles of S_k are recorded as sequence S_{kr}. For convenience of marking, the first C particles in the sequence S_k are selected here. Trent reveals to the rest agents of the Bob group the location of the particles in the selected sequence S_{kc}. The Bobs select the particles at these locations, denoted as $S_{1c}, S_{2c}, \ldots, S_{k-1c}, S_{k+1c}, \ldots, S_{nc}$. The remaining particles are labeled as S_{1r}, S_{2r}, ..., S_{k-1r}, S_{k+1r}, ..., S_{nr}. Bobs carried out eavesdropping check on the transmitted particle sequences S_{1c}, S_{2c}, ..., S_{k-1c}, S_{k+1c}, ...S_{nc} through the steps similar to Step 1–2 and send them to Trent. If the eavesdropping check is passed, then Trent will own quantum sequences S_{1c}, S_{2c}, ..., S_{nc}.

Step 4: Trent measure the particles of $S_{ic} = (p_{i1}, p_{i2}, \ldots, p_{ic})(i=1,2,\ldots,n)$ in the base $\{|+\rangle, |-\rangle\}$, and gets the measurement results $B_{ic} = (b_{ic1}, b_{ic2}, \ldots, b_{icc})$, where $b_{icj} = 0$ if the measurement result of the p_{ij} is $|+\rangle$ and $b_{icj} = 1$ if the measurement result of the p_{ij} is $|-\rangle$, $j=1,2,\ldots,c$. Then Trent can obtain those shares $B_{ic} = (b_{ic1}, b_{ic2}, \ldots, b_{icc})$. Now, {Bob$_1$, Bob$_2$, ..., Bob$_k$, ..., Bob$_n$} share *m-c* GHZ state entanglement of *n* particles.

$$\Psi_j = \frac{1}{\sqrt{2}}\left(|0\rangle^{\otimes n} + |1\rangle^{\otimes n}\right) \text{ or } \Psi_j = \frac{1}{\sqrt{2}}\left(|0\rangle^{\otimes n} - |1\rangle^{\otimes n}\right), j = 1, 2, \ldots, m\text{-}c \tag{4.75}$$

Step 5: Trent also asks {Alice$_1$, Alice$_2$, ..., Alice$_{n-1}$, Alice$_n$} to send $A_{ic} = (a_{i1}, a_{i2}, \ldots, a_{ic})$ to him. Trent judges whether Eq. 4.76 is true or not:

$$A_{1c} \oplus A_{2c} \oplus \ldots \oplus A_{nc} = M_C = B_{1c} \oplus B_{2c} \oplus \ldots \oplus B_{nc} \tag{4.76}$$

If it is not true, it means that the S_k sent by Bob$_k$ is not true. If it is true, then proceed to the next Step.

Step 6: Trent prepared a sequence of *m-c* particles, denoted as $S_{xr} = (p_{x1}, p_{x2}, \ldots, p_{xm-c})$, $p_{xj} = |0\rangle$. According to the particle in S_{kr}, CNOT$_{ij}$ operation is done on the particle in S_{xr}, the control particle is p_{kj}, and the target particle is p_{xj}. Then *m-c* GHZ state entanglement of *n* particles that {Bob$_1$, Bob$_2$, ..., Bob$_k$, ..., Bob$_n$} shared become *m-c*+1 GHZ state entanglement of *n*+1 particles that {Bob$_1$, Bob$_2$, ..., Bob$_k$, ..., Bob$_n$, Bob$_x$} share Eq. 4.77.

$$\Psi_j = \frac{1}{\sqrt{2}}\left(|0\rangle^{\otimes n+1} + |1\rangle^{\otimes n+1}\right) \text{ or } \Psi_j = \frac{1}{\sqrt{2}}\left(|0\rangle^{\otimes n+1} - |1\rangle^{\otimes n+1}\right), \mathrm{j} = 1, 2, \ldots, \mathrm{m\text{-}c} + 1 \tag{4.77}$$

Step 7: Trent sends S_{kr} and S_{xr} to Bob$_k$ and Bob$_x$ through a process similar to Step1–2 in order to check whether exist eavesdropping.

Step 8: {Bob$_1$, Bob$_2$, ..., Bob$_k$, ..., Bob$_n$, Bob$_x$} measure the particles of $S_{ir} = (p_{i1}, p_{i2}, \ldots, p_{ir})$ $(i=1, 2, 3, \ldots, n)$ in the base $\{|+\rangle, |-\rangle\}$ and gets the measurement results $B_{ir} = (b_{ir1}, b_{ir2}, \ldots b_{irm-c+1})$, where $b_{irj} = 0$ if the measurement result of the p_{ij} is $|+\rangle$ and $b_{irj} = 1$ if the measurement result of the p_{ij} is $|-\rangle$, $j=1, 2, \ldots, m\text{-}c+1$. Bob$_i$ can obtain the share $B_{ir} = (b_{ir1}, b_{ir2}, \ldots b_{irm-c+1})$.

Step 9: The rest of the secret noted as $M_r = B_{1r} \oplus B_{2r} \oplus \ldots \oplus B_{kr} \ldots \oplus B_{nr} \oplus B_{xr}$. At this time, Trent announces the information M_c and position. {Bob$_1$, Bob$_2$, ..., Bob$_k$, ..., Bob$_n$, Bob$_x$} insert bits from M_c into M_r and then the complete secret *S* can be recovered.

4.4.3.5 Delete an Agent

Assuming that Bob$_k$, a member of Bob's group, wants to exit the MDQSS scheme before the quantum is measured, Trent and the rest of Bobs need to do the following:

Step 1: Bob$_k$ randomly selects decoy particles and inserts them into the quantum sequence S_k randomly, denoted as S_{k*}, and sends S_{k*} to Trent. After going through similar Step 5–6 of Sect. 4.4.3.2, Trent receives the sequence S_k sent by Bob$_k$.

Step 2: To check the correctness of sequence S_k, Trent do the same Step 3–5 of Sect. 4.4.3.2 and Trent will get $M_c = B_{1c} \oplus B_{2c} \oplus \ldots \oplus B_{nc}$.

Step 3: After determining the correctness of sequence S_k, Trent adds decoy particles randomly to the quantum sequence S_{kr}, which are the remaining particles of the deleted Bob$_k$. Then Trent sends the sequence to any Bob (except Bob$_k$), assuming that Bob$_g$ is selected. After an eavesdropping check similar to Step 5–6 of Sect. 4.4.3.2, Bob$_g$ receives the quantum sequence S_{kr}.

Step 4: {Bob$_1$, Bob$_2$, ..., Bob$_{k-1}$, Bob$_{k+1}$, ..., Bob$_g$, ..., Bob$_n$} measure the particles of $S_{ir} = (p_{i1}, p_{i2}, \ldots, p_{ir})$ (i=1, 2, ..., k-1, k, k+1, ..., n) in the base $\{|+\rangle, |-\rangle\}$, and gets the measurement results $B_{ir} = (b_{ir1}, b_{ir2}, \ldots b_{irm-c})$, where $b_{irj} = 0$ if the measurement result of the p_{ij} is $|+\rangle$ and $b_{irj} = 1$ if the measurement result of the p_{ij} is $|-\rangle$, j=1, 2, ..., m-c. Bob$_i$ can obtain the share $B_{ir} = (b_{ir1}, b_{ir2}, \ldots b_{irm-c})$ except Bob$_k$ and Bob$_g$ will get $B_{gr} = (b_{gr1}, b_{gr2}, \ldots b_{grm-c})$ and $B_{kr} = (b_{kr1}, b_{kr2}, \ldots b_{krm-c})$.

Step 5: Bob$g$ updates his share B_{gr} toB'_{gr} and the new key is $B'_{gr} = \{(b_{gr1} \oplus b_{kr1}), (b_{gr2} \oplus b_{kr2}), \ldots, (b_{grm-c} \oplus b_{krm-c})\}$.

Step 6: Now the rest of the secret noted as $M_r = B_{1r} \oplus B_{2r} \oplus \ldots \oplus B'_{gr} \ldots \oplus B_{nr}$. At this time, Trent announces the information M_c and position. {Bob$_1$, Bob$_2$, ..., Bob$_{k-1}$, Bob$_{k+1}$, ..., Bob$_g$, ..., Bob$_n$} insert bits from M_c into M$_r$ and then the complete secret S can be recovered.

4.4.3.6 Security Analyses

This subsection will prove the MDQSS scheme is secure. We analyze the security under several common attacks, such as intercept-resend attack, Trojan horse attack, collusion attack and entangle-measure attack. Then the efficiency of the scheme is also analyzed.

A. **Intercept-Resend Attack**

In the proposed MDQSS schemes, suppose that an eavesdropper (Eve) can perform the intercept-resend attack as follows. Suppose that Eve wants to obtain Bob's shadow key. Eve can intercept the particle sequences sent by all Alice and resend other forged particles prepared by themselves in hope of passing the checking process. Therefore, they can obtain the initial particle sequences S$_i$ (i=1, 2, ..., n) and then they intend to get the information of Alice. However, each quantum transmission is protected by the decoy particles to prevent the eavesdropping attack. Since decoy particles $((|0\rangle, |1\rangle, |+\rangle, |-\rangle))$ are randomly inserted in the transmitted sequence (e.g., S_i), without knowing the positions, basis, and value of each decoy particles, Eve may be detected with the probability $1-(3/4)^k$, where k is the number of the decoy particles, and 3/4 is the probability that Eve can pass the eavesdropping check if one decoy particle is disturbed. Thus, if k is large enough, the probability $1-(3/4)^k$ is approximately 1.

B. **Trojan Horse Attack**

Generally speaking, Trojan horse attacks include the invisible photon attack and the delayed photon (multi-photon) attack [59, 60].

In order to resist Eve's invisible photon attack, Bob_i ($i=1, 2, ,.., n$) installs a wavelength filter in front of the receiving device, which only allows photons of legal wavelength to enter Bob_i's device, so that invisible photons cannot enter Bob_i's device. Therefore, Eve cannot steal Bob_i's encoded information with invisible photons.

In order to resist Eve's delayed photon attack, Bob_i sets up a photon number separator (PNS) in front of the receiving device. The PNS performs quantum measurement on the received signal. When two measurement results are obtained, Bob_i believes that there is a multi-photon signal on the quantum channel and rejects the signal. When only one measurement is obtained, Bob_i recognizes that the received signal contains only one photon and accepts the signal. Therefore, Eve cannot steal the encoded information received from Bob through multi-photon signals.

With these two devices, Bob_i can effectively overcome the impact of Trojan attacks.

C. **Collusion Attack**

In proposed MDQSS scheme, Alice_i will send S_i with enough decoy particles to Bob_i and their corresponding to measurement results B_i would satisfy the equation of $B_1 \oplus B_2 \oplus \ldots \oplus B_n = S$, where S is the secret that the group of Alices have shared. Suppose that only Bob_k is honest and other $n-1$ agents are dishonest. Bob_1, Bob_2, ..., Bob_{k-1}, Bob_{k+1}, ..., Bob_n want to recover Alices's secret without the help of Bob_k. Then they try to intercept the sequence S_k to get the shadow B_k of Bob_k. However, since Alice_i distributes all photons to the agents without preserving anything and randomly insert sufficient number of decoy states in each quantum transmission, Bob_i ($i=1, 2, ..., k-1, k+1, ..., n$) cannot pass the eavesdropping check in Step 8 of Sect. 4.4.3.2 and cannot obtain any information about the secret. That is, MDQSS scheme is secure against the collusion attack as long as there is an honest agent.

D. **Entangle-Measure Attack**

Suppose Eve has intercepted any particle sequence S_{i*} sent by Alice_i ($i=1, 2, ..., n$), she then uses a unitary operation U_E to entangle some ancillas $E = (|E_1\rangle, |E_2\rangle \ldots |E_{m+q}\rangle)$ (assume S_{i*} contains m particles and q decoy particles) and measure them to steal the secret. However, the effect of Eve's operation on the decoy particles will produce the following results:

$$
\begin{aligned}
&U_E \bullet |0\rangle|E_i\rangle = a|0\rangle|e_{00}\rangle + b|0\rangle|e_{01}\rangle \\
&U_E \bullet |1\rangle|E_i\rangle = c|0\rangle|e_{10}\rangle + d|0\rangle|e_{11}\rangle \\
&U_E \bullet |+\rangle|E_i\rangle = \frac{1}{2}\begin{bmatrix} |+\rangle(a|e_{00}\rangle + b|e_{01}\rangle + c|e_{10}\rangle + d|e_{11}\rangle) \\ +|-\rangle(a|e_{00}\rangle - b|e_{01}\rangle + c|e_{10}\rangle - d|e_{11}\rangle) \end{bmatrix} \\
&U_E \bullet |-\rangle|E_i\rangle = \frac{1}{2}\begin{bmatrix} |+\rangle(a|e_{00}\rangle + b|e_{01}\rangle - c|e_{10}\rangle - d|e_{11}\rangle) \\ +|-\rangle(a|e_{00}\rangle - b|e_{01}\rangle - c|e_{10}\rangle + d|e_{11}\rangle) \end{bmatrix},
\end{aligned}
\tag{4.78}
$$

where $U_E \cdot U_E^* = U_E^* \cdot U_E = I$; $|E_i\rangle$ is the initial state of Bob's ancilla; $|e_{00}\rangle$, $|e_{01}\rangle$, $|e_{10}\rangle$, $|e_{11}\rangle$ are the four states which can be distinguished by Eve and $|a^2| + |b^2| = |c^2| + |d^2| = 1$. In order to pass the eavesdropping check, Eve will set $b=c=0$ when Alice's decoy particle is $|0\rangle$ or $|1\rangle$. While if the decoy particle is $|+\rangle$ or $|-\rangle$, Eve has to set $a|e_{00}\rangle - b|e_{01}\rangle + c|e_{10}\rangle - d|e_{11}\rangle = a|e_{00}\rangle + b|e_{01}\rangle - c|e_{10}\rangle - d|e_{11}\rangle = \vec{0}$, where $\vec{0}$ denotes a zero vector. In this way, Eve can pass the eavesdropping check. However, if $b=c=0$, then $a|e_{00}\rangle - d|e_{11}\rangle = \vec{0}$, that is $a|e_{00}\rangle = d|e_{11}\rangle$. Obviously Eve cannot distinguish $|e_{00}\rangle$ from $|e_{11}\rangle$. As a result, Eve cannot get any valuable information from the auxiliary particles. On the other hand, if Eve makes $a|e_{00}\rangle \neq d|e_{11}\rangle$, then she can get the shadow by the ancillas. However, Eve's attack will be detected because the state of the decoy particle is disturbed.

E. **Security Under the Noisy Quantum Channel**

In proposed MDQSS scheme, agents can be added or deleted from the existing DQSS scheme, and at the meantime, the shared secret still remains same. When adding a new agent, the trusted party, Trent, will joint an existing agent and use a subset of the particles to verify that the selected old agent is secure, which can also obtain a part of the secret. Then Trent generates a generalized GHZ state and sends the particles to the new and joint agents, and finally all the agents measure the remaining particles with the X basis to get the remaining secrets. The old agents cannot know the particles of the new agent so they cannot get the shadow of the new agent. In addition, because Trent has verified the security of the selected agent, the old agents are prevented from sending spurious particles, and they cannot collude to gain secrets without the participation of a new agent. When deleting a current agent, Trent asks the deleted agent to send him its own particles and randomly select a subset of the particles to verify the correctness. If the deleted agent submits a false particle, it will be detected. Trent then notifies the other agents to terminate the process and start a new one.

4.4.3.7 Efficiency Analysis

The proposed MDQSS scheme also has high qubit efficiency. According to literature [19], the quantum bit efficiency η_1 is defined as $\eta_1 = \frac{q_s}{q_t}$, where q_t represents the total number of bits transmitted through the quantum channel, and q_s represents the number of classical bits shared. In addition to the qubits consumed by the eavesdropping check, when Alice$_i$ sends m qubits to Bob$_i$ via the quantum channel, Bob$_i$ can get m classical bits, that is, $\eta_1 = \frac{m}{m} \times 100\% = 100\%$. According to literature [61], the quantum bit efficiency $_{n2}$ can also be defined as $\eta_2 = \frac{q_u}{q_t}$, where q_u is the final valid quantum bit number and q_t is the total number of bits transmitted through quantum channel. Apart from the quantum bits used for eavesdropping check, the rest are used for effective transmission key. Therefore, the theoretical value of the quantum bit efficiency n_2 (excluding the eavesdropping consumption) of the scheme also reaches 100%. Therefore, the efficiency of the MDQSS scheme reaches the maximum theoretically.

4.4.3.8 Summary

This section proposes a MDQSS scheme containing a trusted center, which can prepare all quantum resources, and other participants do not need to prepare any quantum entangled resources, effectively reducing the number of qubit generators. Participants only need perform unitary operation and single particle measurements to realize secret sharing, and the operation is easy to realize. The flexibility of the MDQSS scheme is reflected in the fact that agents can be added and removed dynamically. The security analysis shows that MDQSS scheme can effectively resist intercept-resend attack, Trojan horse attack, collusion attack and entangle-measure attack. When this scheme is applied to quantum network, a trusted center can provide quantum resource supply service for multiple groups of users, so that expensive quantum devices can be shared by many users and the cost borne by each user is reduced, which will provide a referable method for the practical application of quantum secret sharing.

4.5 Brief Summary

In this chapter, we mainly study the scheme of quantum secret sharing. By studying and analyzing the related theories, based on two points of view, a improved two-qubit quantum state sharing scheme based on entanglement swapping of Bell states, a dynamic multiparty quantum secret sharing scheme with a trusted party based on generalized GHZ State and a novel dynamic quantum secret sharing scheme in high-dimensional quantum system are proposed, and all three schemes have high security.

References

1. A. Shamir, How to share a secret. Commun. ACM. **22**, 612–613 (1979)
2. G.R. Blakley, *Safeguarding cryptographic key*, in 1979 International Workshop on Managing Requirements Knowledge (MARK), New York (1979), pp. 313–318. https://doi.org/10.1109/MARK.1979.8817296.
3. F.G. Deng, X.H. Li, C.Y. Li, et al., Multiparty quantum-state sharing of an arbitrary two-particle state with Einstein-Podolsky-Rosen pairs. Phys. Rev. A. **72**(4), 044301 (2005)
4. X.H. Li, P. Zhou, C.Y. Li, et al., Efficient symmetric multiparty quantum state sharing of an arbitrary m-qubit state. J. Phys. B **39**(8), 1975–1983 (2006)
5. G. Gordon, G. Rigolin, Generalized quantum-state sharing. Phys. Rev. A **73**(6), 062316 (2006)
6. Z.Y. Wang, Y.M. Liu, D. Wang, et al., Generalized quantum state sharing of arbitrary unknown two-qubit state. Opt. Commun. **276**(2), 322–326 (2007)
7. Z.X. Man, Y.J. Xia, N.B. An, Quantum state sharing of an arbitrary multiqubit state using nonmaximally entangled GHZ states. Eur. Phys. J. D **42**, 333–340 (2007)
8. Z.Y. Wang, H. Yuan, S.H. Shi, et al., Three-party qutrit-state sharing. Eur. Phys. J. D **41**(2), 371–375 (2007)
9. T.J. Wang, H.Y. Zhou, F.G. Deng, Quantum state sharing of an arbitrary m-qudit state with two-qudit entanglements and generalized Bell-state measurements. Phys. A **387**(18), 4716–4722 (2008)
10. D. Li, R. Wang, F. Zhang, et al., Quantum information splitting of arbitrary two-qubit state by using four-qubit cluster state and Bell-state. Quantum Inf. Process. **14**, 1103–1116 (2015)
11. H. Qin, R. Tso, Y.W. Dai, Multi-dimensional quantum state sharing based on quantum Fourier transform. Quantum Inf. Process. **17**, 48 (2018)
12. H. Cao, W. Ma, Verifiable threshold quantum state sharing scheme. IEEE Access **6**(1), 10453–10457 (2018)
13. S.M. Lee, S.W. Lee, H. Jeong, H.S. Park, Quantum teleportation of shared quantum secret. Phys. Rev. Lett. **124**(6), 060501 (2020)
14. F. Gao, S.J. Qin, F.Z. Guo, et al., Cryptanalysis of the arbitrated quantum signature schemes. Phys. Rev. A **84**(2), 022344 (2011)
15. T.Y. Wang, Z.L. Wen, One-time proxy signature based on quantum cryptography. Quantum Inf. Process. **11**(2), 455–463 (2012)
16. M. Hillery, V. Bužek, A. Berthiaume, Quantum secret sharing. Phys. Rev. A **59**, 1829 (1999)
17. R. Cleve, D. Gottesman, H.K. Lo, How to share a quantum secret. Phys. Rev. Lett. **83**, 648 (1999)
18. T. Gao, F.L. Yan, Y.C. Li, Quantum secret sharing between m-party and n-party with six states. Sci. China Ser. G Phys. Mech. Astron. **52**, 1191–1202 (2009)
19. R.H. Shi, L.S. Huang, W. Yang, H. Zhong, Quantum secret sharing between multiparty and multiparty with Bell states and Bell measurements. Sci. China Phys. Mech. Astron. **53**, 2238–2244 (2010)
20. C.W. Tsai, T. Hwang, Multi-party quantum secret sharing based on two special entangled states. Sci. China Phys. Mech. Astron. **55**, 460–464 (2012)
21. H.D. Massoud, F. Elham, A novel and efficient multiparty quantum secret sharing scheme using entangled states. Sci. China Phys. Mech. Astron. **55**, 1828–1831 (2012)
22. Y.G. Yang, Q.Y. Wen, Sci. China Ser. G Phys. Mech. Astron. **51**(9), 1308–1315 (2008)
23. X.L. Song, Y.B. Liu, H.Y. Deng, Y.G. Xiao, Threshold quantum secret sharing between multi-party and multi-party. Sci. Rep. **7**, 6366 (2017)
24. S. Kartick, O. Hari, Efficient quantum secret sharing without a trusted player. Quantum Inf. Process. **19**, 73 (2020)
25. M.M. Wang, X.B. Chen, Y.X. Yang, Quantum secret sharing for general access structures based on multiparticle entanglements. Quantum Inf. Process. **13**, 429–443 (2014)
26. S. Mashhadi, General secret sharing based on quantum Fourier transform. Quantum Inf. Process. **18**, 114 (2019)

27. H.Y. Jia, Q.Y. Wen, F. Gao, S.J. Qin, F.Z. Guo, Dynamic quantum secret sharing. Phys. Letts. A **376**, 1035–1041 (2012)
28. J.L. Hsu, S.K. Chong, T. Hwang, C.W. Tsai, Dynamic quantum secret sharing. Quantum Inf. Process. **12**, 331–344 (2013)
29. T.Y. Wang, Y.P. Li, Cryptanalysis of dynamic quantum secret sharing. Quantum Inf. Process. **12**, 1991–1997 (2013)
30. C.H. Liao, C.W. Yang, T. Hwang, Comment on "dynamic quantum secret sharing". Quantum Inf. Process. **12**, 3143–3147 (2013)
31. C.H. Liao, C.W. Yang, T. Hwang, Dynamic quantum secret sharing scheme based on GHZ state. Quantum Inf. Process. **13**, 1907–1916 (2014)
32. S. Mishra, C. Shukla, A. Pathak, R. Srikanth, A. Venugopalan, An integrated hierarchical dynamic quantum secret sharing scheme. Int. J. Theor. Phys. **54**, 3143–3154 (2015)
33. H. Liu, H. Ma, K. Wei, X. Yang, W. Qu, T. Dou, Y. Chen, R. Li, W. Zhu, Multi-group dynamic quantum secret sharing with single photons. Phys. Lett. A **380**, 2349–2353 (2016)
34. X. Li, G.L. Long, F.G. Deng, J.W. Pan, Efficient multiparty quantum-secret-sharing schemes. Phys. Rev. A **69**, 052307 (2004)
35. H.W. Qin, Y.W. Dai, Dynamic quantum secret sharing by using d-dimensional GHZ state. Quantum Inf. Process. **16**, 64 (2017)
36. Y. Song, Z.H. Li, Y.M. Li, A dynamic multiparty quantum direct secret sharing based on generalized GHZ states. Quantum Inf. Process. **17**, 244 (2018)
37. C.W. Yang, C.W. Tsai, Efficient and secure dynamic quantum secret sharing scheme based on bell states. Quantum Inf. Process. **19**, 162 (2020)
38. L.Y. MA, Two-qubit quantum state sharing scheme based on Bell state. Int. J. Theor. Phys. **59**, 1844–1853 (2020)
39. X.J. Ma, Z.R.-G. Hu, WenWen, Improved two-qubit quantum state sharing scheme based on entanglement swapping of bell states. Quantum Inf. Process. **21**(3), 44 (2022)
40. P.G. Kwiat, K. Mattle, H. Weinfurther, A. Zeilinger, New high-intensity source of polarization-entangled Photon Pairs. Phys. Rev. Lett. **75**, 4337 (1995)
41. T. Li, A. Miranowicz, K.Y. Xia, F. Nori, Resource-efficient analyzer of Bell and Greenberger-Horne-Zeilinger states of multiphoton systems. Phys. Rev. A **100**, 052302 (2019)
42. Y. Song, Y. Li, W. Wang, Multiparty quantum direct secret sharing of classical information with bell states and bell measurements. Int. J. Theor. Phys. **57**(5), 1559–1571 (2018)
43. M.M. Wang, L.T. Tian, Z.G. Qu, *Efficient Multiparty Quantum Secret Sharing Scheme in High-dimensional System*, in International Conference on Cloud Computing and Security, Lecture Notes in Computer Science, pp. 23–31.
44. K. Vahid, B. Alireza, B. Saber, Entanglement swapping of generalized cat states and secret sharing. Phys. Rev. A **65**, 042320 (2002)
45. X. Wang, Y.M. Liu, L.F. Han, Z.J. Zhang, Multiparty quantum secret sharing of secure direct communication with high-dimensional quantum superdense coding. Int. J. Quantum Inf. **6**, 1155–1163 (2008)
46. I.C. Yu, F.L. Lin, C.Y. Huang, Quantum secret sharing with multi-level mutually (un-)biased bases. Phys. Rev. A **78**, 012344 (2008)
47. Z.H. Liu, H.W. Chen, J. Xu, High-dimensional deterministic multiparty quantum secret sharing without unitary operations. Quantum Inf. Process. **11**, 1785–1795 (2012)
48. H.W. Qin, W. Tang, R.L. Tso, Hierarchical quantum secret sharing based on special high-dimensional entangled state. IEEE J. Sel. Top. Quant. **26**, 6600106 (2020)
49. W.W. Hu, R.-G. Zhou, X. Li, P. Fan, C.Y. Tan, A novel dynamic quantum secret sharing in high-dimensional quantum system. Quantum Inf. Process. **20**(5), 159 (2021)
50. K. Vahid, B. Alireza, B. Saber, Quantum key distribution for d-level systems with generalized Bell states. Phys. Rev. A **65**, 052331 (2002)
51. W.K. Wootters, B.D. Fields, Optimal state-determination by mutually unbiased measurements. Ann. Phys. **191**, 363–381 (1989)

52. A.O. Pittenge, M.H. Rubin, Mutually unbiased bases, generalized spin matrices and separability. Linear Algebra Appl. **390**, 255 (2004)
53. M.A. Nielsen, I.L. Chuang, *Quantum Computation and Quantum Information* (Cambridge University, Cambridge, 2010)
54. A. Fonseca, High-dimensional quantum teleportation under noisy environments. Phys. Rev. A **100**, 062311 (2019)
55. R.A. Bertlmann, P.J. Krammer, Bloch vectors for qudits. Phys. A **41**, 235303 (2008)
56. R.-G. Zhou, M. Huo, W. Hu, Y. Zhao, Dynamic multiparty quantum secret sharing with a trusted party based on generalized GHZ State. IEEE Access **9**, 22986–22995 (2021)
57. M.A. Nielsen, I.L. Chuang, Quantum computation and quantum information[J]. Math. Struct. Comput. Sci. **17**(6), 1115–1115 (2002)
58. W. Yang, L. Huang, R. Shi, L. He, Secret sharing based on quantum Fourier transform [J]. Quantum Inf. Process. **12**(7) (2013)
59. N. Gisin, S. Fasel, B. Kraus, et al., Trojan-horse attacks on quantum-key-distribution systems [J]. Phys. Rev. A **73**(2) (2006)
60. Z. Zhu, *Research on Secret Sharing Scheme and Dialogue Scheme Based On Quantum Theory [D]* (Xidian University, 2011)
61. Z.-h. FENG, X.-q. TAN, C. LIANG, Quantum key distribution scheme based on Bell states and its entanglement[J]. Appl. Res. Comput. **32**(3), 873–876 (2015)

Chapter 5
Quantum Direct Communication

Abstract Quantum direct communication is an important branch of quantum communication, which is a kind of communication mode that transmits secret information directly without establishing a key in advance. Quantum direct communication plays an irreplaceable role in classical communication and quantum communication because of its unique advantages. Quantum direct communication can be divided into two kinds: deterministic secure quantum communication and quantum secure direct communication, both of which have high security and can guarantee the unconditional secure transmission of information in theory.

5.1 Introduction

Quantum direct communication (QDC), as an important technology parallel to quantum cryptography, is a kind of reliable communication that is different from the QKD and is the security of communications. Since 2016, it has shown accelerated development trend, along with broad application prospects. The QDC scheme loads information in quantum states and transmits it directly in quantum channels, independent of encryption and decryption algorithms and key distribution. The scheme relies on quantum noncloning, quantum measurement collapse, and other quantum principles to detect and prevent eavesdropping and to ensure the security of information transmission, that is, when someone eavesdrops, the quantum state will be destroyed so that the eavesdropper cannot get any information, even if it has a strong computing power, and also cannot be deciphered. QDC has changed the two-channel structure of classical secret communication and uses the single-channel structure, which only contains quantum communication and which simplifies the steps that may lead to information leakage and improves the security level of data transmission.

QDC technology can be compatible with the existing network, promote the research results of practical [1–8], and be applied to finance, government affairs, defense, and other fields. From the perspective of industrial development, QDC promotes the deep integration of quantum technology and the field of social economy and increases the development of industrial economy. QDC can provide

R.-G. Zhou et al., *Design and Analysis of Secure Quantum Communication Schemes*,
https://doi.org/10.1007/978-3-031-78428-6_5

powerful technical support for network security, national defense, and military affairs and promote the establishment of QDC standardization and strategic development of national quantum innovation technology industry. It can be seen that QDC will become one of the important technical means for the development of information security in the future.

5.2 Principle and Research Progress of Quantum Direct Communication

QKD uses quantum key to achieve the purpose of secure communication, and its quantum key is the bridge and guarantee of classical information security. QDC is a kind of communication mode that can directly transmit secret information without establishing key in advance. When it is used to transmit information of key, it can be regarded as a form of QKD. The difference between QKD and QDC is as follows: First, QKD transmits the key, while QDC directly transmits the secret information; second, when it is found that there is eavesdropping, QKD can directly give up the key, and QDC because the transmission is the information itself, so it cannot choose to give up the information, which requires some processing methods, such as block transmission and the two-step method, to make Eve even if it can get some data, but only some random data, to prevent the disclosure of secret information. It can be seen that the security requirements of QDC are higher than that of QKD.

The QDC scheme includes quantum secure direct communication (QSDC) and deterministic secure quantum communication (DSQC).

The QSDC scheme realizes the direct transmission of secret messages using quantum channels. The scheme has four main requirements:

(a) Secret messages can be transmitted and read directly without the need for additional classical information.
(b) Eavesdroppers cannot obtain useful information about ciphertext.
(c) Secret messages cannot be leaked before the message coding process.
(d) Use bulk transfer to send and receive particles.

Since 2000, Long and Liu have proposed the first QSDC scheme [9], in which they transmitted information through Bell states and used quantum data block and distributed transmission techniques, which led to the development of QSDC. With the proposal of a two-step QSDC based on dense coding by Deng [10], the construction principle and safety criterion of QSDC have been clarified gradually. In 2004, Deng et al. put forward the "quantum one-time-pad" scheme by means of the "one-time-pad" communication scheme using a single photon [11]. These initial schemes explain the concept of QSDC in detail and give the construction principle and security conditions, which promote the rapid development of QSDC in theory. Subsequently, QSDC research gradually involved many directions, and many communication schemes based on single photon [12–15] and entanglement pair [16–21] emerged. At present, most research on QSDC is still limited to one-way or two-way transmission between two parties, and the research on communication between three or more parties is still in the exploratory stage. In 2006, Jin et al. [22] first proposed a three-party QSDC scheme with GHZ state as the medium and proposed a method to extend it to N-party communication. In the same year, Man et al. improved the coding rule [23] proposed by Jin and proposed a more secure QSDC scheme. In 2007, Wang et al. proposed a QSDC scheme using EPR pairs for simultaneous three-party communication [24]. Subsequently, in 2011, Wang et al. analyzed the information leakage problem in this scheme [25]. In the same year, Chong et al. proposed a parallel EPR-based three-party simultaneous communication scheme [26] to improve communication efficiency. In 2013, Yin et al. improved the efficiency of a three-party QSDC scheme based on EPR pairs to 60% [27]. In 2016, Wang et al. proposed a three-way QSDC scheme using two degrees of freedom of polarization and path of a single photon [28], which greatly improves the utilization ratio of a single photon. In 2017, He et al. introduced a new three-party QSDC scheme that resists collective environmental noise [29].

The principle of DSQC is that the sender generates a key before communicating, and according to the key, the different quantum states used to encode the message are selected. The sender then encodes the message on a quantum state, and the receiver measures it on a preagreed measurement basis. The quantum state and measurement basis are designed so that the receiver can obtain part of the information of the quantum state. After judging that there is no eavesdropping threat, the sender then announces the key so that the receiver can obtain all the information of the quantum state and decode it. In this process, the message decoding process must rely on the key, that is, the classical information, while the message decoding process of QSDC does not need the assistance of classical information. Another QSDC-like scheme is the deterministic QKD scheme. This kind of scheme does not require classical information assistance in the decoding process. However, the security detection process is carried out after the message transmission. If eavesdropping is detected, it will lead to the leakage of previously transmitted messages, so it can only be used to transmit random number sequences.

Himizu and Imoto proposed a pioneering DSQC scheme [30]. In their scheme, Alice encodes ciphertext into an entangled state and transfers it to Bob. Bob can read the secret information by performing the Bell base measurement and getting the encoded base information from the Alice side through the classical channel. In 2002, Beige et al. proposed another DSQC scheme based on single-photon two-qubit states [31]. In this scheme, each qubit needs to transmit an additional classical bit before Bob can read the message. However, the scheme is somewhat insecure because eavesdroppers can easily steal messages via quantum teleportation. Man proposed several DSQC schemes based on quantum teleportation and entanglement swapping [32]. Next, Li designed two DSQC schemes, one of which is based on pure entangled states, and the other is based on d-dimensional single photon states [33]. Long presented a DSQC scheme, in which the receiver can read out the secret message only after each qubit exchanges at least one bit of classical information [34]. Shukla extended a DSQC scheme based on orthogonal states so that it does not require the actual transmission of message qubits [35]. Joy proposed two DSQC schemes that use three-qubit GHZ-like states and five-qubit Brown states as quantum channels [36], respectively.

While many QSDC schemes had been proposed one after another, people proposed a bidirectional QSDC scheme based on the EPR states, which is also known as the quantum dialogue scheme (QD). QD allows two communicants to exchange their secret messages simultaneously without any information leakage. Nguyen [37] first proposed a bidirectional QSDC scheme based on the EPR states. In this scheme, two communicants could exchange secret message with each other at the same time. Therefore, it is also known as the quantum dialogue scheme (QD). Since then, more and more scholars have focused their research on the design and analysis of quantum dialogue. Ji [38] proposed a QD scheme based on N groups of single-photon sequences. Yang [39] also presented a quasi-safe QD scheme using single photons. With many QD schemes [40–42] based on GHZ states, W states and four-particle entangled states had been proposed one after another. Gao [43] found that most of these schemes have an information leakage problem. In order to solve this problem, researchers have successively proposed a series of QD schemes with higher security [44–46].

Although QDC has been developing rapidly in recent years, it still faces many problems, such as limited application scenarios, unsolved technical bottlenecks, and security risks. Based on this, the semi-quantum environment can solve these problems properly, which can further reduce the cost of QDC, further expand the application scenarios, and improve security. It plays an important role in the application of communication and the study of scientific theory, which can link up quantum communication and classical communication and promote their balanced development.

5.3 Deterministic Secure Quantum Communication

5.3.1 DSQC Scheme

Tsai et al. designed an effective DSQC scheme based on dense coding [47]. Let four unitary operations I, σ_x, $i\sigma_y$, σ_z represent 00, 01, 10, and 11, respectively. Alice performs two unitary operations on the first and third particles of the cluster state and passes them to Bob. Bob can distinguish the two unitary operations chosen by Alice by von Neumann measurements of the four-particle cluster states.

$$I = |0\rangle\langle 0| + |1\rangle\langle 1| \tag{5.1}$$

$$\sigma_x = |0\rangle\langle 1| + |1\rangle\langle 0| \tag{5.2}$$

Table 5.1 Dense decoding for $|C_1\rangle$

U_{A1}/U_{A2}	I	σ_x	$i\sigma_y$	σ_z
I	$\|C_1\rangle$	$\|C_5\rangle$	$-\|C_6\rangle$	$\|C_2\rangle$
σ_x	$\|C_{10}\rangle$	$\|C_{14}\rangle$	$-\|C_{13}\rangle$	$\|C_9\rangle$
$i\sigma_y$	$\|C_{16}\rangle$	$\|C_{12}\rangle$	$-\|C_{11}\rangle$	$\|C_{15}\rangle$
σ_z	$\|C_7\rangle$	$\|C_{13}\rangle$	$-\|C_4\rangle$	$\|C_8\rangle$

$$\sigma_{iy} = |0\rangle\langle 1| - |1\rangle\langle 0| \tag{5.3}$$

$$\sigma_z = |0\rangle\langle 0| - |1\rangle\langle 1| \tag{5.4}$$

The specific steps of the scheme are as follows:

Step 1: Alice prepares N cluster states $|C_1\rangle$ sequence S, $S = \{s_1, s_2, \cdots, s_n\}$, where $s_i = \{q_1^i, q_2^i, q_3^i, q_4^i\}$ and $i = 1, 2, \cdots, n$. She performs two unitary operations U_{A1} and U_{A2} on q_1^i and q_3^i, respectively, based on four bits message, where U_{A1}, $U_{A2} \in \{I, \sigma_x, i\sigma_y, \sigma_z\}$. For example, if her message is 0010, she performs $U_{A1} = I$ and $U_{A2} = i\sigma_y$ on q_1 and q_3, respectively. Next, she prepares sufficient decoy states $Q_C = \{q_c^1, q_c^2, \cdots, q_c^{4n}\}$, where $q_c^i \in \{|0\rangle, |1\rangle, |+\rangle, |-\rangle\}$ and $i = 1, 2, \cdots, 4n$ for eavesdropper check. Finally, she shuffles these particles in S and Q_C to get S' and sends it to Bob.

$$|C_1\rangle = \frac{1}{2}(|0000\rangle + |0011\rangle + |1100\rangle - |1111\rangle)_{1234} \tag{5.5}$$

Step 2: After confirming that Bob has received the particles, Alice announces the decoy state's position and measurement basis in sequence S'.

Step 3: Bob measures the corresponding particles q_c^i in the sequence S'.

Step 4: Bob sent the measuring results to Alice, who compared Bob's results to estimate the error rate. If the error rate exceeds the default threshold, Alice will abort this communication and restart a new one.

Step 5: Alice announces the order of the remaining particles in the sequence S'.

Step 6: Bob restores the original order of S' and then measures the remaining particles by premeasuring the cluster states of the corresponding particles. According to Table 5.1, Alice's unitary operations and thus the message can be deduced by Bob based on his measuring results.

5.3.2 *DSSQC Scheme*

Analysis shows that most of the current semi-quantum direct communication schemes are proposed based on single photons with one polarization degree of freedom in the field of QSDC, such as Bell states and GHZ states. Comprehensive consideration to improve communication security and transmission efficiency, combining with some advantages of QSDC schemes based on two degrees of freedom, such as increasing channel capacity, improving transmission efficiency, simplifying preparation and the measurement method of quantum state, enhancing the enforceability of the scheme, and etc. [28, 48, 49], this section propose an efficient and novel deterministic secure semi-quantum communication (DSSQC) scheme based on single photons in both polarization and spatial-mode degrees of freedom [50]. Compared with single photons in only one degree of freedom mentioned in the existing schemes, the DSSQC scheme has the following advantages:

1. The DSSQC scheme only adopts single-photon states as the initial quantum resource, which saves the consumption of additional resources and improves the efficiency of quantum state preparation.
2. The measurement of single-photon states involved in the DSSQC scheme is more enforceable.
3. Each photon in the DSSQC scheme can carry two bits of secret message, which increases the amount of information to be transmitted and improves the communication efficiency.
4. The DSSQC scheme has high security because it is resistant to the Trojan horse attack, the intercept-resend attack, the measure-resend attack, the entangle-measure attack, the DoS attack, etc.

5.3.2.1 A Single-Photon State in Both Polarization and Spatial-Mode Degrees of Freedom

A single-photon state in both the polarization and the spatial-mode degrees of freedom can be expressed as follows [48]:

$$|\phi\rangle = |\phi\rangle_P \otimes |\phi\rangle_S, \tag{5.6}$$

where $|\phi\rangle_P$ and $|\phi\rangle_S$ are the single-photon states in the polarization and the spatial-mode degrees of freedom, respectively.

For a single-photon state being in the polarization degree of freedom, there are four possible forms of $|\phi\rangle_P$, namely, $|H\rangle_P$, $|V\rangle_P$, $|S\rangle_P$, and $|A\rangle_P$. Among them, $|H\rangle_P$ and $|V\rangle_P$ denote the horizontal-vertical polarization states, and the corresponding measurement basis is $Z_P = \{|H\rangle_P, |V\rangle_P\}$; $|S\rangle_P$ and $|A\rangle_P$ represent the diagonal polarization states, and the corresponding measurement basis is $X_P = \{|S\rangle_P, |A\rangle_P\}$. Here,

$$|S\rangle_P = \frac{1}{\sqrt{2}}\left(|H\rangle_P + |V\rangle_P\right), |A\rangle_P = \frac{1}{\sqrt{2}}\left(|H\rangle_P - |V\rangle_P\right) \tag{5.7}$$

For a single-photon state, which is in the spatial-mode degree of freedom, there are also four possible forms of $|\phi\rangle_S$, namely, $|b_1\rangle_S$, $|b_2\rangle_S$, $|s\rangle_S$, and $|a\rangle_S$. Among them, $|b_1\rangle_S$ and $|b_2\rangle_S$ denotes the single-photon states in two spatial path modes and

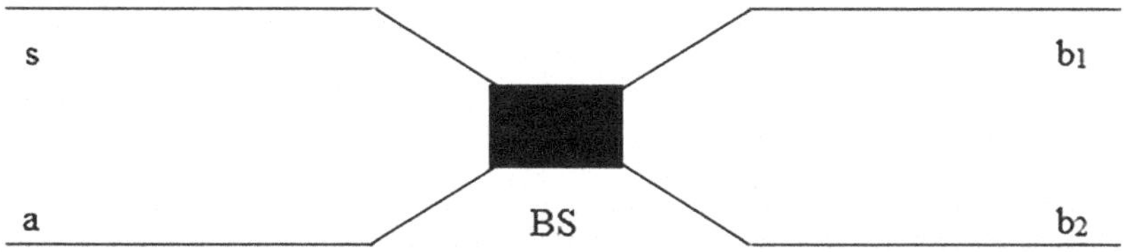

Fig. 5.1 Schematic diagram of a Hadamard operation in the spatial degrees of freedom

measurement basis is $Z_S = \{|b_1\rangle_S, |b_2\rangle_S\}$; $|s\rangle_S$ and $|a\rangle_S$ can be converted from $|b_1\rangle_S$ and $|b_2\rangle_S$, respectively, and the measurement basis is $X_S = \{|s\rangle_S, |a\rangle_S\}$. Here,

$$|s\rangle_S = \frac{1}{\sqrt{2}}(|b_1\rangle_S + |b_2\rangle_S), |a\rangle_S = \frac{1}{\sqrt{2}}(|b_1\rangle_S - |b_2\rangle_S) \tag{5.8}$$

The single-photon state $|\phi\rangle = |\phi\rangle_P \otimes |\phi\rangle_S$ in principle can be produced with a beam splitter (BS: 50/50). In fact, the quantum state in the polarization degree of freedom and those in the spatial-mode degree of freedom are commutative, which means users can operate them independently. In detail, the sender first prepares a sequence of single-photon polarization state $|\phi\rangle_P$ and then use a BS or not on each photon to produce the spatial-mode state $|\phi\rangle_S$, as shown in Fig. 5.1, where the BS can realize the Hadamard operation in the spatial-mode quantum state of a single photon, and the transformation relationship is as follows [48]:

$$|s\rangle \rightarrow \frac{1}{\sqrt{2}}(|b_1\rangle + |b_2\rangle), |a\rangle \rightarrow \frac{1}{\sqrt{2}}(|b_1\rangle - |b_2\rangle) \tag{5.9}$$

Next, two unitary operations are introduced into each degree of freedom of single photons. Similar to Refs. [28, 51], the basis of each photon cannot be changed. That is,

$$\begin{aligned} I_P &= |H\rangle\langle H| + |V\rangle\langle V| \\ I_S &= |b_1\rangle\langle b_1| + |b_2\rangle\langle b_2| \\ U_P &= |V\rangle\langle H| - |H\rangle\langle V| \\ U_S &= |b_2\rangle\langle b_1| - |b_1\rangle\langle b_2| \end{aligned} \tag{5.10}$$

According to the above unitary operations, the single-photon states in two degrees of freedom can be transformed into the following:

$$\begin{aligned} I_P|H\rangle_P &= |H\rangle_P \quad & I_S|b_1\rangle_S &= |b_1\rangle_S \\ I_P|V\rangle_P &= |V\rangle_P \quad & I_S|b_2\rangle_S &= |b_2\rangle_S \\ I_P|S\rangle_P &= |S\rangle_P \quad & I_S|s\rangle_S &= |s\rangle_S \\ I_P|A\rangle_P &= |A\rangle_P \quad & I_S|a\rangle_S &= |a\rangle_S \end{aligned}$$

$$\begin{array}{ll} U_P|H\rangle_P = |V\rangle_P & U_S|b_1\rangle_S = |b_2\rangle_S \\ U_P|V\rangle_P = -|H\rangle_P & U_S|b_2\rangle_S = -|b_1\rangle_S \\ U_P|S\rangle_P = -|A\rangle_P & U_S|s\rangle_S = -|a\rangle_S \\ U_P|A\rangle_P = |S\rangle_P & U_S|a\rangle_S = |s\rangle_S \end{array} \tag{5.11}$$

5.3.2.2 DSSQC Scheme

In the proposed DSSQC scheme, quantum sender Alice wants to use single-photon state $|\phi\rangle = |H\rangle_P \otimes |b_1\rangle_S$ to transmit a secret message M with the length of n bits to classical receiver Bob. As stated in Ref. [52], Bob's operations are restricted. In addition, in order to prevent Trojan horse attack [53], Bob places a wavelength filter and a photon number splitter (PNS) [54] in front of the receiving device. Before the scheme starts, Alice and Bob agree on encryption rule and use unitary operations to independently encode secret message, see Sect. 5.3.2.1. For example, when Alice performs unitary operation $I_P \otimes I_S$ on $|\phi\rangle$, the corresponding secret message bits to be encrypted are *00*. The results obtained and several other situations are shown in Table 5.2.

With reference to a flow chart shown in Fig. 5.2, the detailed steps of the DSSQC scheme are described as follows:

Step 1: Alice prepares a sequence of ordered n single photons S in both polarization and spatial-mode degrees of freedom as detection particles. All of them are from the state of $|S\rangle_P \otimes |s\rangle_S$. Then, she selects half of S as S_1 and sends it to Bob, while the remaining half is called sequence S_2.

Step 2: When sequence S_1 arrives, Bob randomly chooses to perform reflection or $Z_P \otimes Z_S$ basis measurement on each photon in it. In reflection operation, Bob reflects single photons back to Alice without disturbance; in measurement operation, Bob generates the same states as the measurement results and sends them to Alice.

Step 3: After confirming that Alice has received, Bob announces the positions, operations, and measurement results of the corresponding single photons. Next, Alice performs the first eavesdropping check to determine whether the quantum channel is safe.

Table 5.2 Encryption rule

Single photons	Secret messages	Unitary operations	Corresponding results
$\|H\rangle_P \otimes \|b_1\rangle_S$	00	$I_P \otimes I_S$	$\|H\rangle_P \otimes \|b_1\rangle_S$
	01	$I_P \otimes U_S$	$\|H\rangle_P \otimes \|b_2\rangle_S$
	10	$U_P \otimes I_S$	$\|V\rangle_P \otimes \|b_1\rangle_S$
	11	$U_P \otimes U_S$	$\|V\rangle_P \otimes \|b_2\rangle_S$

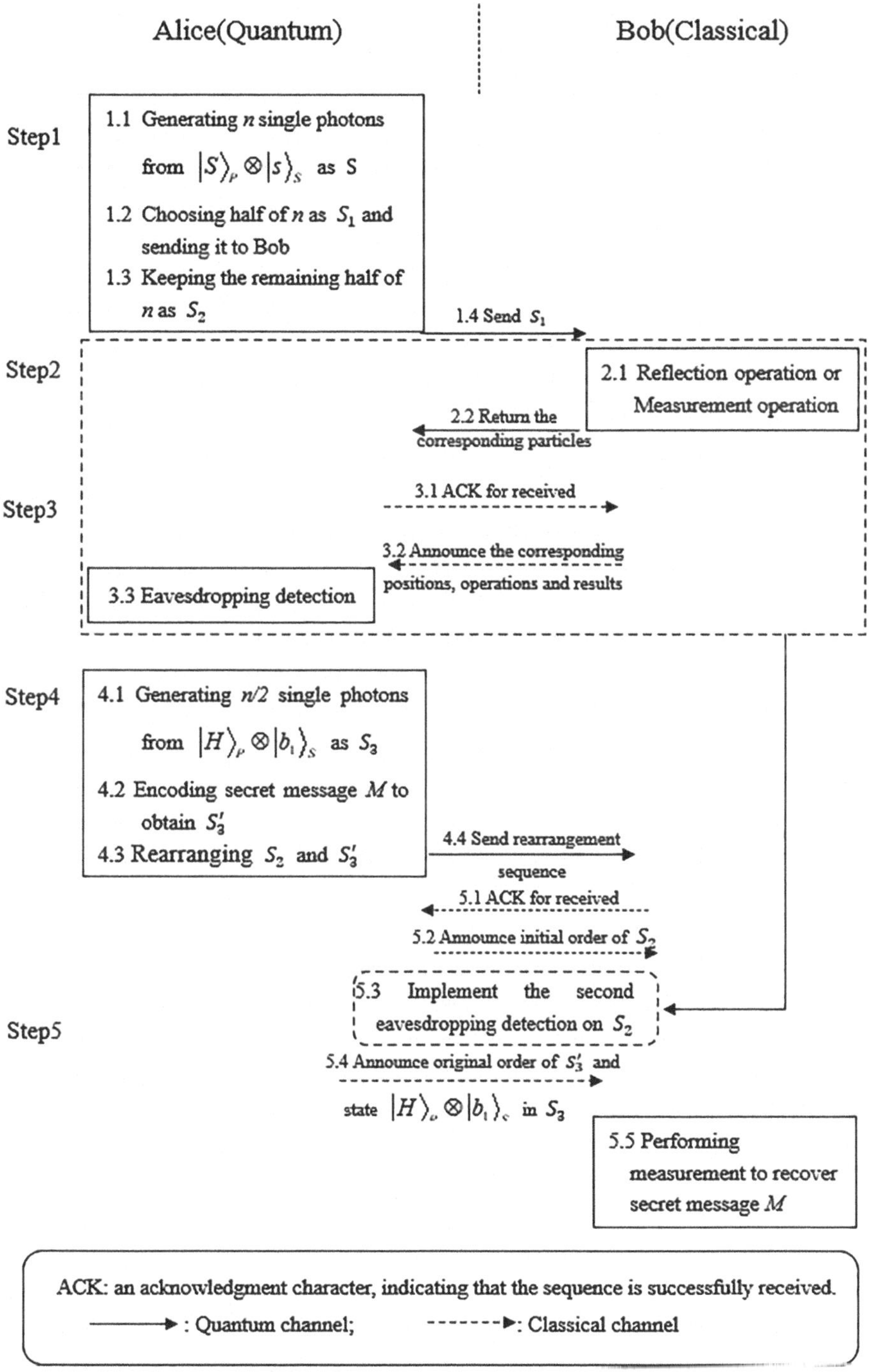

Fig. 5.2 The flow chart of the DSSQC scheme

1. For measurement operation, Alice also performs $Z_P \otimes Z_S$ basis measurement on the received single photons and then observes whether the results are the same as that announced by Bob.
2. For reflection operation, according to the initial single-photon state $|S\rangle_P \otimes |s\rangle_S$, Alice measures each of the reflected single photons in the $X_P \otimes X_S$ basis and compares whether the results are the same as the initial quantum states.

After the detection is completed, Alice will get an error rate. If the error rate exceeds the predetermined security threshold, the scheme would be abandoned; otherwise, continue to the next step.

Step 4: Alice prepares a sequence of ordered *n/2* single-photons S_3, all of which come from the single photon state of $|H\rangle_P \otimes |b_1\rangle_S$. Then, according to the state of n bits secret message M to be transmitted, Alice uses the encryption rule in Table 5.2 to encode it to obtain a sequence S_3', and rearranges the sequence of S_2 and S_3', then sends the rearranged sequence to Bob.

Step 5: After confirming that Bob has received the rearranged sequence, Alice first announces the original order of S_2 in it. At this time, Alice and Bob perform the same operations as Steps 2–3 on S_2 to implement the second eavesdropping check to determine whether the eavesdropper has destroyed the quantum state coherence during the information transmission process. Then Alice announces the original order of S_3' and the initial degrees of freedom corresponding to the quantum state in the sequence S_3, that is $|H\rangle_P \otimes |b_1\rangle_S$. Next, Bob performs $Z_P \otimes Z_S$ basis measurement on each photon state in S_3'. Finally, Bob can recover the secret message M, which is sent by Alice according to the initial degree of freedom $|H\rangle_P \otimes |b_1\rangle_S$, the original sequence of S_3', and the measurement results. For example, corresponding to the relationship in Table 5.2, if the result measured by Bob is $|V\rangle_P \otimes |b_1\rangle_S$, it indicates that the corresponding secret message bits are *10*.

5.3.2.3 Security Analysis

So far, many effective attack strategies have been proposed in the research of quantum cryptography, such as the intercept-resend attack [55], Trojan horse attack [53], the entangle-menaure attack [56], and other attacks [57–59]. For the proposed DSSQC scheme, we mainly analyze its security against several common attacks, such as the intercept-resend attack, the Trojan horse attack, the measure-resend attack, the entangle-measure attack, the DoS attack, etc.

A. Intercept-Resend Attack

In the intercept-resend attack, attacker Eve prepares fake single photons in advance and then intercepts the real single photons propagating in the quantum channel and replaces them with the fake single photons. For this attack, Eve will inevitably be discovered from two aspects: on the one hand, the states of fake single photons prepared by Eve are random; on the other hand, Bob's operations also are random to Eve. Suppose that Eve's fake single photon is $|S\rangle_P \otimes |b_1\rangle_S$. After intercepting the real single-photon state from Alice, she sends the state of $|S\rangle_P \otimes |b_1\rangle_S$ to Bob. If Bob chooses reflection operation, Alice would receive $|S\rangle_P \otimes |b_1\rangle_S$ and use the $X_P \otimes X_S$ basis to measure it. Then Alice can get $|S\rangle_P \otimes |s\rangle_S$ or $|S\rangle_P \otimes |a\rangle_S$ with a probability of 50%. At this time, when Alice performs eavesdropping check process, Eve can be detected with a probability of 50%. If Bob chooses measurement operation, the state of $|S\rangle_P \otimes |b_1\rangle_S$ would be collapsed into $|H\rangle_P \otimes |b_1\rangle_S$ or $|V\rangle_P \otimes |b_1\rangle_S$ with a probability of 50%. Then Alice uses the $Z_P \otimes Z_S$ basis to measure the quantum states sent back from Bob and obtains $|H\rangle_P \otimes |b_1\rangle_S$ or $|V\rangle_P \otimes |b_1\rangle_S$ with a probability of 50%. However, if Eve's attack does not occur, when Bob is measuring the genuine single-photon state of $|S\rangle_P \otimes |s\rangle_S$ under the measurement operation, Alice then performs subsequent measurement and can get $|H\rangle_P \otimes |b_1\rangle_S$ or $|H\rangle_P \otimes |b_2\rangle_S$ or $|V\rangle_P \otimes |b_1\rangle_S$ or $|V\rangle_P \otimes |b_2\rangle_S$, each with a probability of 25%. Therefore, in this case, when Alice implements eavesdropping check process, Eve can also be detected with a probability of 50%. In addition, only when the single photons prepared by Eve is the same as that state sent by Alice, this attack would not be discovered and the corresponding probability is $(\frac{1}{16})^n$, so when n is large enough, the probability will converge to *0*. That is to say, this attack is unable to proceed successfully.

B. Measure-Resend Attack

In the measure-resend attack, Eve intercepts single photon sent from Alice to Bob, then randomly selects a measurement basis to measure it and sends the result to Bob. In this condition, the measurement basis comes from a random one in the set $\{Z_P \otimes Z_S, Z_P \otimes X_S, X_P \otimes Z_S, X_P \otimes X_S\}$. For this attack, Eve will also be discovered from two aspects: For one thing, Eve's measurement may destroy the state of a single photon; for another, Bob's operations are random to Eve. The specific analysis is similar to the intercept-resend attack. Since Eve does not know the two initial degrees of freedom of detection particles and the corresponding measurement basis to be executed, she is impossible to accurately obtain the secret messages.

C. Trojan Horse Attack

Trojan horse attack includes a multi-photon-signal attack [60], an invisible-photon attack [61] or a delay-photon attack [54], in which the receiver not only estimates the number of photons in each legitimate single-photon signal in received sequence but also filters out illegal photon signals. In the DSSQC scheme, Bob places a wavelength filter and photon number splitter (PNS) in front of the receiving device. The wavelength filter is mainly used to filter out single photons with illegitimate wavelengths and throw it away before reaching the receiving device. In addition, a photon number splitter (PNS) is used in each path to divide each signal into two pieces, each of which is performed single-photon measurement. Then Bob can distinguish whether there is a multi-photon signal based on the results. Hence, the DSSQC scheme is safe against Trojan horse attack in principle.

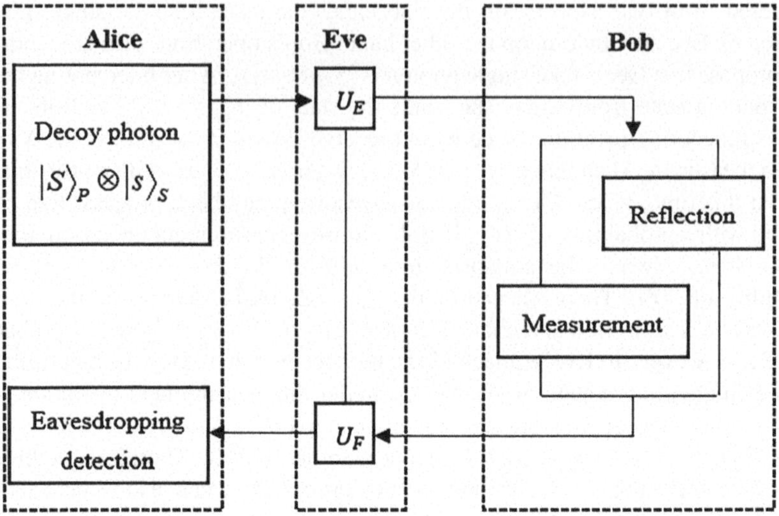

Fig. 5.3 The flow chart of the entangle-measure attack

D. Entangle-Measure Attack

In general, Eve's entangle-measure attack is composed of two operations: U_E and U_F. U_E attacks qubits as they go from Alice to Bob and U_F as they go back from Bob to Alice, where U_E and U_F share a common probe space [60]. The flow chart is shown in Fig. 5.3 below.

Next, we suppose Eve performs operation U_E on the system composed of the detection sequence S_1 and the auxiliary particle, which is prepared by Eve as $|\alpha\rangle$. Then the state of composite system can be described as

$$\begin{aligned} U_E|\phi\rangle \otimes |\alpha\rangle &= U_E(|S\rangle \otimes |s\rangle) \otimes |\alpha\rangle \\ &= |Hb_1\rangle|\alpha_{00}\rangle + |Hb_2\rangle|\alpha_{01}\rangle + |Vb_1\rangle|\alpha_{10}\rangle + |Vb_2\rangle|\alpha_{11}\rangle \end{aligned} \tag{5.12}$$

where $|\alpha_{00}\rangle$, $|\alpha_{01}\rangle$, $|\alpha_{10}\rangle$ and $|\alpha_{11}\rangle$ are un-normalized states of Eve's probe.

According to Step 2, when Bob receives the detection sequence, he randomly selects the measurement or reflection operation. Then, Eve implements the operation U_F on the particles sent back to Alice by Bob and the auxiliary particle $|\beta\rangle$. For the specific situation of measurement and reflection operation, see the following analysis.

1. On the one hand, when Bob chooses the measurement operation, the system state will collapse into either $|Hb_1\rangle|\alpha_{00}\rangle$, $|Hb_2\rangle|\alpha_{01}\rangle$, $|Vb_1\rangle|\alpha_{10}\rangle$, or $|Vb_2\rangle|\alpha_{11}\rangle$. If Eve wants to be undetected during the eavesdropping check, for operation U_F, there should satisfy the following relationship:

$$U_F(|Hb_1\rangle|\alpha_{00}\rangle) = |Hb_1\rangle|\beta_{00}\rangle \tag{5.13}$$

$$U_F(|Hb_2\rangle|\alpha_{01}\rangle) = |Hb_2\rangle|\beta_{01}\rangle \tag{5.14}$$

$$U_F(|Vb_1\rangle|\alpha_{10}\rangle) = |Vb_1\rangle|\beta_{10}\rangle \tag{5.15}$$

$$U_F(|Vb_2\rangle|\alpha_{11}\rangle) = |Vb_2\rangle|\beta_{11}\rangle \tag{5.16}$$

where $|\beta_{00}\rangle$, $|\beta_{01}\rangle$, $|\beta_{10}\rangle$, and $|\beta_{11}\rangle$ also are unnormalized states of Eve's probe. It can be seen from the above analysis that after Bob performs the measurement operation, if U_F cannot change the state of the single photon, no error will be introduced; otherwise, Eve will eventually be detected with a nonzero probability.

2. On the other hand, when Bob chooses the reflection operation, the system state will be kept unchanged. At this time, after Eve performs the operation U_F, the following relationship exists:

$$\begin{aligned} &U_F(|Hb_1\rangle|\alpha_{00}\rangle + |Hb_2\rangle|\alpha_{01}\rangle + |Vb_1\rangle|\alpha_{10}\rangle + |Vb_2\rangle|\alpha_{11}\rangle) \\ &= |Hb_1\rangle|\beta_{00}\rangle + |Hb_2\rangle|\beta_{01}\rangle + |Vb_1\rangle|\beta_{10}\rangle + |Vb_2\rangle|\beta_{11}\rangle \end{aligned} \tag{5.17}$$

Similarly, in order to pass the eavesdropping check, after Eve performs the operation U_F, she must ensure that Alice finally measures the system state in the result $|S\rangle \otimes |s\rangle$. So it can be deduced [62],

$$|\beta_{00}\rangle = |\beta_{01}\rangle = |\beta_{10}\rangle = |\beta_{11}\rangle = |\beta\rangle \tag{5.18}$$

That is,

$$U_F(|Hb_1\rangle|\alpha_{00}\rangle) = |Hb_1\rangle|\beta\rangle \tag{5.19}$$

$$U_F(|Hb_2\rangle|\alpha_{01}\rangle) = |Hb_2\rangle|\beta\rangle \tag{5.20}$$

$$U_F(|Vb_1\rangle|\alpha_{10}\rangle) = |Vb_1\rangle|\beta\rangle \tag{5.21}$$

$$U_F(|Vb_2\rangle|\alpha_{11}\rangle) = |Vb_2\rangle|\beta\rangle \tag{5.22}$$

It can be concluded that the final state of Eve's probe should be independent of Bob's measurement result.

Finally, combining all the above analysis, it can be seen that Eve cannot meet the requirements, so it can be judged that the DSSQC scheme is safe against Eve's entangle-measure attack.

E. Man-in-the-Middle Attack

Man-in-the-middle attack means that Eve intercepts the single photon transmitted from Alice to Bob, then she prepares another single-photon state and sends it to Bob. However, similar to intercept-resend attack, the state of the detection sequence and Bob's random operations are completely unknown to Eve, and there are many situations in this. Therefore, when conducting a man-in-the-middle attack, the probability that Eve can correctly distinguish the single-photon state that has been measured or reflected is extremely small and can be ignored. In other words, Eve cannot successfully perform a man-in-the-middle attack.

F. The Denial-of-Service Attack

In the denial-of-service attack (DoS) [63], the attacker secretly disrupts communication between the two parties, causing the legitimate communicants to fail to obtain the correct secret message. In this attack, Eve intercepts the particles encoded by Alice, performs his own encoding operations, and sends them to Bob. Therefore, the secret information obtained by Bob through decoding is inconsistent with that sent by Alice. However, this attack cannot be successfully implemented in the DSSQC scheme. Because in Step 5, Eve cannot differentiate the encoding sequence from the detection sequence correctly. When Eve initiates the DoS attack, his behavior will not be able to successfully pass the eavesdropping check. Hence, DoS attacks are invalid for the DSSQC scheme.

G. Information Leakage Problem

Through the analysis of the above several attacks, it can be seen that the eavesdropper cannot obtain any relevant confidential information by attacking the quantum channel. The following mainly analyzes the security of the classical channel disclosed in the DSSQC scheme in combination with the information leakage problem [43]. The information leakage problem means that any other eavesdropper can extract useful information about a secret message without taking active attacks. In Step 5 of the DSSQC scheme, Alice announces the initial degrees of freedom of the single-photon states and the original order of the encoding sequence through the classical channel. However, since Eve does not know the specific encryption mapping relationship, it is impossible to infer any information about the secret message based on the published information. Therefore, there is no information leakage problem in the DSSQC scheme.

5.3.2.4 Qubit Efficiency Analysis

In this subsection, the qubit efficiency of several semi-quantum direct communication (SQDC) schemes will be analyzed and compared, as shown in Table 5.3, including Zou's [64], Zhang's [65], Sun's [66] semi-quantum secure direct

Table 5.3 The comparison of qubit efficiency

	Zou's scheme [64]	Zhang's scheme [65]	Sun's scheme [66]	DSSQC scheme
Quantum resource	Single photon	EPR pair	Bell state	Single photon
Qubit efficiency	*25%*	*33.3%*	*12.5%*	*66.6%*

communication schemes and the proposed DSSQC scheme. It is well known that the qubit efficiency is $\eta = \frac{t}{q}$, where t represents the total number of shared classical bits and q represents the total number of generated qubits in the scheme [67].

In Zou 's scheme [64], the receiver Bob prepares $N = 4n(1 + \delta)$ single photons (δ is small enough and can be neglected), and the sender Alice wants to transmit n-bit classical secret message to Bob. Therefore, the qubit efficiency of Zou's scheme is $\eta = \frac{n}{4n} = \frac{1}{4} = 25\%$.

In Zhang's scheme [65], the sender Alice should prepare $N = 2n(1 + \delta)$ EPR photons and n decoy states. Similarly, the two communicants also want to transmit n-bit classical secret message, so the qubit efficiency of Zhang's scheme is $\eta = \frac{n}{2n+n} = \frac{1}{3} \approx 33.3\%$.

In Sun's scheme [66], the sender Alice has to prepare $N = 4n(1 + \delta)$ Bell states (i.e., 8n qubits), and use them to detect eavesdropping and encode n-bit classical secret message. Hence, the qubit efficiency of Sun's scheme is $\eta = \frac{n}{8n} = \frac{1}{8} = 12.5\%$.

In the proposed DSSQC scheme, Alice and Bob want to transmit n-bit classical secret message M. Alice first prepares n single photons as detection particles, and then prepares $n/2$ single photons to encode n-bit secret message, where each single photon carries two bits of secret message. Therefore, without considering quantum channel noise and resource consumption during eavesdropping check, the qubit efficiency of the DSSQC scheme can be expressed as $\eta = \frac{n}{n+\frac{n}{2}} = \frac{2}{3} \approx 66.6\%$. It can be see that this is the highest qubit efficiency among all semi-quantum direct communication schemes.

5.3.2.5 Summary

This section proposes an efficient and novel deterministic secure semi-quantum communication scheme by using single photons in both polarization and spatial-mode degrees of freedom as the initial quantum resource. In the communication process, communicants adopt single-photon measurement and introduce two unitary operations in each degree of freedom of single photon to independently encrypt secret messages without consuming other quantum resources. Compared with other semi-quantum direct communication schemes, the proposed DSSQC scheme increases the amount of secret information to be transmitted because each single photon state can carry two bits of secret message in a communication process, which means the communication capacity and qubit efficiency of the scheme can be enhanced. At the same time, the proposed DSSQC scheme also has high security and can resist multiple types of attacks. Finally, since the preparation and measurement of a single-photon quantum state in both polarization and spatial-mode degrees of freedom can be easily achieved by employing existing quantum technologies, the DSSQC scheme is feasible in reality.

5.3.3 CDSSQC Scheme

A new controlled deterministic secure semi-quantum communication (CDSSQC) scheme based on GHZ-like states is proposed [68]. In this scheme, the sender Bob introduces the Huffman compression coding technology [69] to compress secret message, and the quantum receiver Alice can only recover the secret message transmitted by the classical sender Bob with the permission of the classical controller Charlie. When the transmission of high security is completed, the receiver Alice can perform error-free decompression to recover the secret message. It is mentioned that GHZ-like states utilized to perform eavesdropping check and message transmission can show certain regularity when doing single-particle measurement under Z or X basis, which has the ability to avoid attacks. At the same time, using the three particle sequences contained in the GHZ-like states as quantum carriers does not require extra insertion of decoy states, which can save quantum resources and improve transmission efficiency.

5.3.3.1 Huffman Coding Technology

The Huffman coding technology [70, 71] mainly uses variable-length coding to encode source symbols. The variable-length coding table is obtained through a method of evaluating the occurrence probability of source symbols. Symbols with high occurrence probability use a shorter code, while those with a low probability of occurrence use a longer code. This reduces the average length and expected value of the encoded string, thereby achieving the purpose of lossless data compression, so it is sometimes called the optimum coding. The traditional Huffman coding algorithm is constructed based on a binary tree. For the convenience of calculation, it is mainly limited to the problem of binary encoding. The specific steps of its encoding are as follows:

1. Arrange the source symbols in descending order of probability, so

$$p(s_1) \geq p(s_2) \geq \cdots \geq p(s_q) \tag{5.23}$$

2. Assign a symbol "0" and "1" to the two source symbols s_{n-1} and s_n with the smallest probability, and merge the two source symbols into a new symbol. After that, the sum of these two smallest probabilities is used as the probability of the new symbol, and as a result, a new source containing only $n - 1$ source symbols is obtained. This is called the first reduced source of the source, denoted by S_1.
3. The symbols of the reduced source S_1 are still arranged in descending order of probability, and Step (2) is repeated to obtain the reduced source S_2 containing only $n - 2$ symbols.
4. Repeat the above steps (1–3) until there are only two symbols left in the reduced source. At this time, the sum of the probabilities of the remaining two symbols

Message Probabilities				Code
Original Message Ensemble	Auxiliary Message Ensembles			
	S_1	S_2	S_3	
S_1: 0.4	0.4	0.4	0.6 0	00
S_2: 0.2	0.2	0.4 0	0.4 1	10
S_3: 0.2	0.2 0	0.2 1		11
S_4: 0.1 0	0.2 1			010
S_5: 0.1 1				011

Fig. 5.4 Huffman binary coding example based on binary tree

must be *1*. Then starting from the last-level reduction of the signal source, and returning forward along the encoding path, the code word corresponding to each source symbol is obtained.

Let's analyze through a specific example.

Consider the EIS $(\{S_1, S_2, S_3, S_4, S_5\}, (0.4, 0.2, 0.2, 0.1, 0.1))$. From Fig. 5.4 below, we can get the corresponding code.

In order to increase decoding speed and obtain the secret message quickly, the receiver will use the Huffman coding technology based on the quaternary tree instead of the binary tree, and its construction principle is the same as that of binary tree [71]. The decoding algorithm is faster than the basic Huffman technology in case of speed, while the compression performance remains almost unchanged, and in terms of decoding time, the quadtree-based Huffman technology is generally more than 50% faster than the conventional Huffman techniques based on the binary tree.

5.3.3.2 CDSSQC Scheme

In the proposed CDSSQC scheme, the three legitimate participants are: the receiver Alice, the sender Bob, and the controller Charlie, and the mission of the CDSSQC scheme is to make Bob transmit secret message *M* to Alice with high security. Among participants, Alice has quantum capabilities and can perform quantum operations, such as preparing quantum states. Conversely, Bob and Charlie have classical abilities, so they are restricted to implement the following operations: (1) Measurement: measure qubits in the computational basis $\{|0\rangle, |1\rangle\}$ and resend the results to Alice; (2) Reflection: reflect qubits back to Alice without disturbance; (3) Reordering: reorder the qubits through different delay lines. During the communication process, Bob plays the role of providing Huffman compression coding to compress secret message. Moreover, Alice and Bob discuss the encoding method of the secret message before the CDSSQC scheme begins. What needs to be declared is that the proposed CDSSQC scheme is carried out in an ideal environment. The main procedures of the CDSSQC scheme are described in the following steps.

Step 1: Prepare Quantum Channel

The quantum Alice prepares 2 *N* ordered three-particle GHZ-like states and divides the first, second, and third particles of each GHZ-like state to form three ordered sequences Q_A, Q_B, and Q_C. Alice retains the Q_A sequence and sends Q_B and Q_C sequences to Bob and Charlie, respectively. To resist the two kinds of Trojan horse attacks [53], Bob and Charlie need to place a wavelength filter and a photon number splitter (PNS) before the particles arrive. The GHZ-like state prepared by Alice can be denoted as

$$\begin{aligned} |Q\rangle_{ABC} &= \frac{1}{2}(|000\rangle + |011\rangle + |110\rangle + |101\rangle)_{ABC} \\ &= \frac{\sqrt{2}}{2}(||+++\rangle + |---\rangle)_{ABC} \end{aligned} \tag{5.24}$$

Step 2: Eavesdropping Checking

After Bob and Charlie receive the Q_B and Q_C sequences, they select the first *N* batches of particles in the 2 *N* GHZ-like states sequence as detection particles and process them as follows:

1. First, rearrange these detection particles using quantum delay lines.
2. Second, randomly perform Measurement or Reflection operation on the rearranged particles. Specifically, in Measurement, Bob and Charlie use the Z basis $\{|0\rangle, |1\rangle\}$ to measure detection particles and keep results. Meantime, they prepare the same states according to the measurement results and send them back to Alice; in th Reflection, Bob and Charlie reflect detection particles back to Alice without disturbance.

Once Alice confirms that all detection particles have been received, Bob and Charlie would publish the initial order of detection particles and the corresponding operations. After the above operations have performed, Alice need to perform eavesdropping checking:

2–1: If Bob and Charlie's choice is Measurement, Alice will perform Z basis measurement on the corresponding particles in the Q_A sequence and then compare the measurement results of the three particles.

2–2: If Bob and Charlie's choice is Reflection, Alice will perform X basis to measure the received detection particles and the corresponding particles in Q_A, respectively and then compare the three measurement results.

The corresponding relationships of the obtained results are shown in Table 5.4. If Alice gets an error rate, which is higher than the security threshold, then the scheme will be terminated; otherwise, Alice discards these detection particles and continues to the next step. Correspondingly, Bob and Charlie also discard the particles which are generated in the Measurement operation. At this time, the sequences owned by Alice, Bob and Charlie are q_A, q_B, and q_C, respectively, and the length of each sequence is *N*.

Table 5.4 The measurement basis and results corresponding to the eavesdropping checking

Measurement basis	Q_A	Q_B	Q_C
X	$\vert+\rangle$	$\vert+\rangle$ (R)	$\vert+\rangle$ (R)
	$\vert-\rangle$	$\vert-\rangle$ (R)	$\vert-\rangle$ (R)
Z	$\vert 0\rangle$	$\vert 0\rangle$	$\vert 0\rangle$
	$\vert 0\rangle$	$\vert 1\rangle$	$\vert 1\rangle$
	$\vert 1\rangle$	$\vert 1\rangle$	$\vert 0\rangle$
	$\vert 1\rangle$	$\vert 0\rangle$	$\vert 1\rangle$

Step 3: Secret Message Compression and Encryption

After passing the eavesdropping checking, Bob compresses the secret message M with a length of N by using a quadtree-based Huffman coding to obtain a binary string $M^{'}$. Subsequently, Bob divides q_B sequence into q_B^1 and q_B^2 sequences, where q_B^1 is the encoding sequence used to encode $M^{'}$, which has been performed Z basis measurement, and q_B^2 is the detection sequence.

3–1: Encoding: Corresponding to the particles in q_B^1, if $M^{'} = 0$, the particle remains unchanged; if $M^{'} = 1$, the particle is flipped. After the encoding operation is complete, encoding sequence $q_B^1 m$ is obtained.

3–2: Detection: Bob randomly selects some qubits in q_B^2 and performs Z basis measurement to obtain $q_B^2 d$. The remaining qubits of q_B^2 are denoted as $q_B^1 f$ and keep unchanged. At this time, $q_B^2 d$ and $q_B^2 f$ are used as detection particles.

Summarize the above two steps, Bob finally sends sequences $q_B^1 m$, $q_B^2 d$, and $q_B^2 f$ to Alice out of order.

Step 4: Secret Message Confirmation

After receiving Alice's confirmation, Bob publishes the initial positions of the detection particles and the corresponding operations, that is, all particles except $q_B^1 m$. At the same time, Charlie divides q_C into encoding sequence q_C^1 and detection sequence q_C^2 according to the information informed by Bob. After performing the same operations as q_B^2 on the corresponding particles in q_C^2 to obtain $q_C^2 d$ and $q_C^2 f$, Charlie shuffles the order of these particles and sends them to Alice. Once Alice confirms the receipt, Charlie will announce the original positions of detection particles and the corresponding operations. Based on the information of the detection particles provided by Bob and Charlie, Alice finds the corresponding particles in q_A and performs the eavesdropping checking similar to Step 2. That is:

4–1: For sequences $q_B^2 d$ and $q_C^2 d$, Alice performs Z basis measurement on the corresponding particles of q_A and compares the measurement results of every three corresponding particles in the three sequences in turn.

4–2: For sequences $q_B^2 f$ and $q_C^2 f$ and the corresponding particles of q_A, Alice implements X basis to measure them, respectively. Then she compares the three measurement results.

Table 5.5 Relationship between measurement results

Sequence	q_A^1	q_C^1	q_B^1	$q_B^1 m$	$M^{'}$
Measurement result	0	0	0	1	1
	1	0	1	1	0
	0	1	1	0	1
	1	1	0	0	0

To sum it up, similarly, according to the results in Table 5.4, the error rate is obtained. If the error rate is higher than security threshold, the communication will be terminated; otherwise, continue to the next step.

Step 5: Secret Message Decoding
After Alice confirms that there is no eavesdropping, Bob and Charlie publish the initial order of $q_B^1 m$ and q_C^1. Alice finds the corresponding particles of sequence q_A^1 in q_A, then performs Z basis measurement on q_A^1 and q_C^1 like q_B^1. According to the states of the three particles, Alice can deduce the states of q_B^1. Then Alice compares the results of q_B^1 and $q_B^1 m$ to get the information of $M^{'}$. Assuming that the results of $q_B^1 m$, q_A^1, and q_C^1 are considered as 1100, 0101, and 0011, the corresponding results of q_B^1 and $M^{'}$ are shown in Table 5.5.

Step 6: Decompression
Finally, Alice performs quadtree-based Huffman decompression on sequence $M^{'}$ to recover the secret message M.

Summarizing all above steps, a flow chart can be used to describe this proposed CDSSQC scheme. The specific implementation process is shown in Fig. 5.5.

In proposed CDSSQC scheme, the first eavesdropping checking in the shared quantum channel phase is to ensure that there is no eavesdropping on the communication channel between Alice and Bob and Alice and Charlie. Similarly, the second eavesdropping checking in Step 4 is to determine whether the eavesdropper has destroyed the coherence of quantum state during the information transmission process. The process of these two security checks establishes a secure environment for the subsequent communication between the three-party.

5.3.3.3 Security Analysis

This subsection is the security analysis of the proposed CDSSQC scheme. A secure CDSSQC scheme should be able to satisfy the following conditions: (1) the receiver can recover the secret message only with the permission of the controller; (2) except for the receiver, no other internal or external eavesdroppers can obtain the secret message. What can be proved is that the CDSSQC scheme has the ability to prevent Trojan horse attack by placing a wavelength filter and a photon number splitter (PNS). Therefore, it is necessary to analyze other common external eavesdropping attacks, such as intercept-resend attack, measure-resend attack [72], CNOT attack, and entangle-measure attack [73], to ensure the security of the CDSSQC scheme.

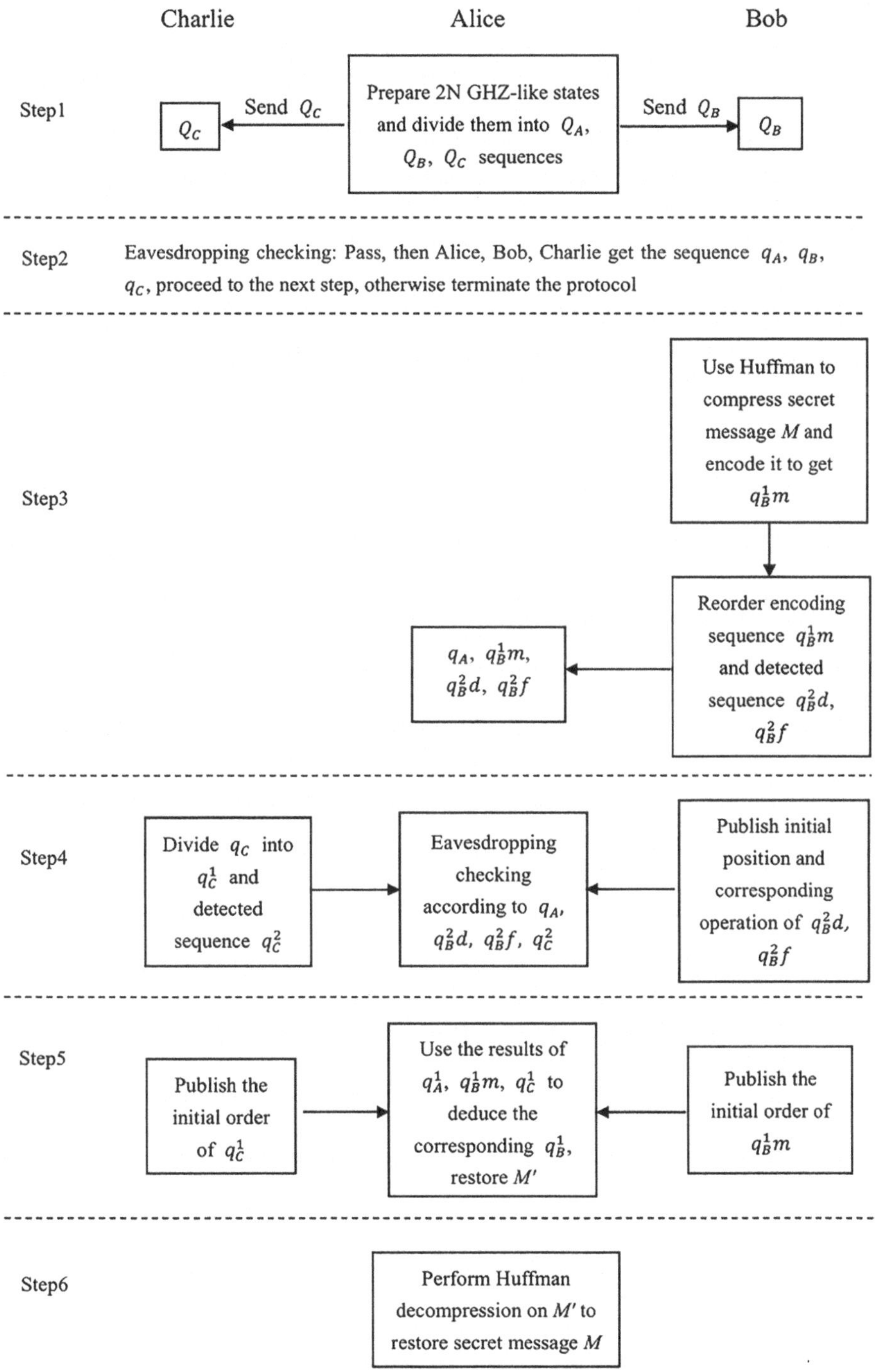

Fig. 5.5 The flow chart of the CDSSQC scheme

A. Intercept-Resend Attack

In the CDSSQC scheme, if the attacker Eve wants to perform an intercept-resend attack, she can intercept some particles from the sequence when Alice sends Q_B or Q_C to Bob or Charlie and then resends a series of particles that she has prepared in advance to Bob or Charlie. However, Alice will find errors once eavesdropping checking is performed, since Eve does not know the initial information of the detection particles when she launches intercept-resend attack. Hence, this attack will be announced to be failed.

B. Measure-Resend Attack

In this attack, Eve captures and measures detection particles in the quantum channel, then prepares the corresponding particles according to the measurement results and sends them to Bob or Charlie. Similar to intercept-resend attack, measure-resend attack will be failed because Eve does not know the initial information of detection particles. Specifically, here is an example to analyze the measure-resend attack of a two-party communication between Alice and Bob (the analysis between Alice and Charlie is similar). If the qubit sent by Alice to Bob is $|0\rangle_B$, Eve intercepts this qubit and measures it in Z basis, then this qubit will collapse into $|0\rangle$. Subsequently, Eve resends the result to Bob without introducing errors. But if Eve uses the X basis to measure, the qubit will collapse into $|+\rangle$ or $|-\rangle$ with each probability of 1/2. Then Eve retransmits the state to Bob. Bob randomly chooses Measurement or Reflection operations. Alice will get the incorrect result with a probability of 1/2 whatever the operation is performed by Bob. Therefore, the probability of Eve not being detected is $\frac{1}{2}\times 1+\frac{1}{2}\times\left(\frac{1}{2}\times\frac{1}{2}+\frac{1}{2}\times\frac{1}{2}\right)=\frac{3}{4}$. Finally, for N detection particles, if Eve performs measure-resend attack, the probability of being detected can be expressed as $1-\left(\frac{3}{4}\right)^N$, and when N is large enough, the error rate of detection will be *100%*. So Eve will be detected, and this attack will be failed ultimately.

C. CNOT Attack

1. Single CNOT Attack.

If Eve performs a single CNOT attack on transmitted sequence Q_C during the phase of establishing quantum channel, the following result will be obtained:

$$\begin{aligned}
|Q\rangle' &= CNOT_{CE}\left(|Q\rangle_{ABC}|0\rangle_E\right)\\
&=\frac{1}{2}(|0000\rangle+|0111\rangle+|1100\rangle+|1011\rangle)_{ABCE}\\
&=\frac{1}{2}(|0000\rangle+|0111\rangle+|1100\rangle+|1011\rangle)_{ABCE}\\
&=\frac{1}{2}\left[|0\rangle_A(|000\rangle+|111\rangle)_{BCE}+|1\rangle_A(|100\rangle+|011\rangle)_{BCE}\right]\\
&=\frac{1}{2}\left[|+\rangle_A|+\rangle_B(|++\rangle+|--\rangle)_{CE}+|-\rangle_A|-\rangle_B(|+-\rangle+|-+\rangle)_{CE}\right],
\end{aligned} \tag{5.25}$$

where Q_C is used as control qubit and the preprepared qubit $|0\rangle_E$ is used as target qubit.

Table 5.6 Measurement basis and corresponding results

MB	Q_A	Q_B	Q_C	Eve with Q_C
X	$\|+\rangle$	$\|+\rangle$ (R)	$\|+\rangle$ (R)	$\|+\rangle_A\|+\rangle_B(\|++\rangle+\|--\rangle)_{CE}$
	$\|-\rangle$	$\|-\rangle$ (R)	$\|-\rangle$ (R)	$\|-\rangle_A\|-\rangle_B(\|+-\rangle+\|-+\rangle)_{CE}$
Z	$\|0\rangle$	$\|0\rangle$	$\|0\rangle$	$(\|000\rangle+\|111\rangle)_{BCE}$
	$\|0\rangle$	$\|1\rangle$	$\|1\rangle$	
	$\|1\rangle$	$\|1\rangle$	$\|0\rangle$	$(\|100\rangle+\|011\rangle)_{BCE}$
	$\|1\rangle$	$\|0\rangle$	$\|1\rangle$	

Comparing the results in Table 5.4, the relationship between the measurement basis and the measurement results of the attacked sequence can be obtained, as shown in Table 5.6. When Alice performs eavesdropping checking, Q_C will give an incorrect result and make the error rate higher than the security threshold. Therefore, Eve cannot successfully perform a single CNOT attack. Similarly, Eve cannot successfully perform this attack on Q_B sequence.

2. Double CNOT Attack.

If Eve performs a double CNOT attack on transmitted sequences Q_B and Q_C during the phase of establishing quantum channel, the following result will be obtained:

$$\begin{aligned}|Q\rangle' &= CNOT_{BE}CNOT_{CE}\left(|Q\rangle_{ABC}|0\rangle_E\right)\\ &= \frac{1}{2}(|0000\rangle+|0110\rangle+|1101\rangle+|1011\rangle)_{ABCE}\\ &= \frac{1}{2}\left[|0\rangle_A(|000\rangle+|110\rangle)_{BCE}+|1\rangle_A(|101\rangle+|011\rangle)_{BCE}\right]\\ &= \frac{1}{2}(|++++\rangle+|+---\rangle+|---+\rangle+|-++-\rangle)_{ABCE},\end{aligned} \tag{5.26}$$

in which Q_B and Q_C are used as control qubits and the preprepared qubit $|0\rangle_E$ is used as target qubit.

Similar to the single CNOT attack, errors will occur when Alice performs eavesdropping checking, which have been exhibited in Table 5.4. So this attack is demonstrated to be failed.

C. Entangle-Measure Attack

In general, Eve's entangle-measure attack is composed of two operations: U_E and U_F. U_E attacks qubits as they go from Alice to Bob (Charlie) and U_F as they go back from Bob (Charlie) to Alice. Note that U_E and U_F share a common probe space [52].

Next, suppose that Eve performs operation U_E on the system composed of the transmitted sequence and the auxiliary photon, which is prepared by Eve as $|\varepsilon\rangle$. Then the state of composite system can be described as

$$U_E|0\rangle|\varepsilon\rangle = a|0\rangle|\varepsilon_{00}\rangle + b|1\rangle|\varepsilon_{01}\rangle \tag{5.27}$$

$$U_E|1\rangle|\varepsilon\rangle = c|0\rangle|\varepsilon_{10}\rangle + d|1\rangle|\varepsilon_{11}\rangle, \tag{5.28}$$

where $|\varepsilon_{00}\rangle$, $|\varepsilon_{01}\rangle$, $|\varepsilon_{10}\rangle$, and $|\varepsilon_{11}\rangle$ belong to the Hilbert space of Eve's probe states and $|a|^2 + |b|^2 = |c|^2 + |d|^2 = 1$, $|a|^2 = |d|^2 = F$, $|c|^2 = |b|^2 = D$.

If Eve executes operation U_E in the quantum channel when the sequence Q_B sent from Alice to Bob (the analysis of sequence Q_C sent from Alice to Charlie is similar), the whole GHZ-like state can be shown as:

$$\begin{aligned}|\varphi\rangle_{Eve} =& \frac{1}{2}[|0\rangle_A(a|0\rangle|\varepsilon_{00}\rangle + b|1\rangle|\varepsilon_{01}\rangle)_{BE}|0\rangle_C + |0\rangle_A(c|0\rangle|\varepsilon_{10}\rangle + d|1\rangle|\varepsilon_{11}\rangle)_{BE}|1\rangle_C \\ &+ |1\rangle_A(c|0\rangle|\varepsilon_{10}\rangle + d|1\rangle|\varepsilon_{11}\rangle)_{BE}|0\rangle_C + |1\rangle_A(a|0\rangle|\varepsilon_{00}\rangle + b|1\rangle|\varepsilon_{01}\rangle)_{BE}|1\rangle_C] \\ =& \frac{1}{2}(a|0\rangle|0\rangle|\varepsilon_{00}\rangle|0\rangle + b|0\rangle|1\rangle|\varepsilon_{01}\rangle|0\rangle + c|0\rangle|0\rangle|\varepsilon_{10}\rangle|1\rangle + d|0\rangle|1\rangle|\varepsilon_{11}\rangle|1\rangle \\ &+ c|1\rangle|0\rangle|\varepsilon_{10}\rangle|0\rangle + d|1\rangle|1\rangle|\varepsilon_{11}\rangle|0\rangle + a|1\rangle|0\rangle|\varepsilon_{00}\rangle|1\rangle + b|1\rangle|1\rangle|\varepsilon_{01}\rangle|1\rangle)_{ABCE} \\ =& \frac{1}{2}[(a|0\rangle|\varepsilon_{00}\rangle + c|1\rangle|\varepsilon_{10}\rangle)_{AE}|0\rangle_B|0\rangle_C + (c|0\rangle|\varepsilon_{10}\rangle + a|1\rangle|\varepsilon_{00}\rangle)_{AE}|0\rangle_B|1\rangle_C \\ &+ (b|0\rangle|\varepsilon_{01}\rangle + d|1\rangle|\varepsilon_{11}\rangle)_{AE}|1\rangle_B|0\rangle_C + (d|0\rangle|\varepsilon_{11}\rangle + b|1\rangle|\varepsilon_{01}\rangle)_{AE}|1\rangle_B|1\rangle_C]\end{aligned} \tag{5.29}$$

During the eavesdropping checking in Step 2, Bob chooses randomly either Measurement or Reflection operation. If Bob selects the Measurement operation, $|\varphi\rangle_{Eve}$ will collapse to $(a|0\rangle|\varepsilon_{00}\rangle + c|1\rangle|\varepsilon_{10}\rangle)_{AE}|0\rangle_B|0\rangle_C$, $(c|0\rangle|\varepsilon_{10}\rangle + a|1\rangle|\varepsilon_{00}\rangle)_{AE}|0\rangle_B|1\rangle_C$, $(b|0\rangle|\varepsilon_{01}\rangle + d|1\rangle|\varepsilon_{11}\rangle)_{AE}|1\rangle_B|0\rangle_C$, or $(d|0\rangle|\varepsilon_{11}\rangle + b|1\rangle|\varepsilon_{01}\rangle)_{AE}|1\rangle_B|1\rangle_C$ $|0\rangle|\varepsilon_{11}\rangle + b|1\rangle|\varepsilon_{01}\rangle)_{AE}|1\rangle_B|1\rangle_C$ each with probability of 1/4. If Bob performs Reflection operation, according to the above,

$$\begin{aligned}|\varphi\rangle_{Eve} =& \frac{1}{2}(a|0\rangle|0\rangle|\varepsilon_{00}\rangle|0\rangle + b|0\rangle|1\rangle|\varepsilon_{01}\rangle|0\rangle + c|0\rangle|0\rangle|\varepsilon_{10}\rangle|1\rangle + d|0\rangle|1\rangle|\varepsilon_{11}\rangle|1\rangle \\ &+ c|1\rangle|0\rangle|\varepsilon_{10}\rangle|0\rangle + d|1\rangle|1\rangle|\varepsilon_{11}\rangle|0\rangle + a|1\rangle|0\rangle|\varepsilon_{00}\rangle|1\rangle + b|1\rangle|1\rangle|\varepsilon_{01}\rangle|1\rangle)_{ABCE}\end{aligned} \tag{5.30}$$

It can be seen that when Alice performs X basis measurement operation on the detection particles, the probability for Eve not to be detected is

$$P = \frac{1}{4}\left(|a|^2 + |d|^2 + |a|^2 + |d|^2\right) = \frac{1}{4}(F + F + F + F) = F \tag{5.31}$$

Similarly, when the operation U_F is performed on the transmitted qubits, which are sent from Bob (Charlie) to Alice, it is also analyzed according to the above method. Now, it is assumed that Eve launches an entangle-measure attack, I_E and I_F represent the maximum amount of information that Eve can obtain when performing operations U_E and U_F, respectively, and d represents the probability that she will be detected by the communicants. Since all the transmitted qubits are in the states $|0\rangle$ or $|1\rangle$, the probability of $|0\rangle$ and $|1\rangle$ is considered to be equal for convenience, that is, $P_0 = P_1 = 0.5$.

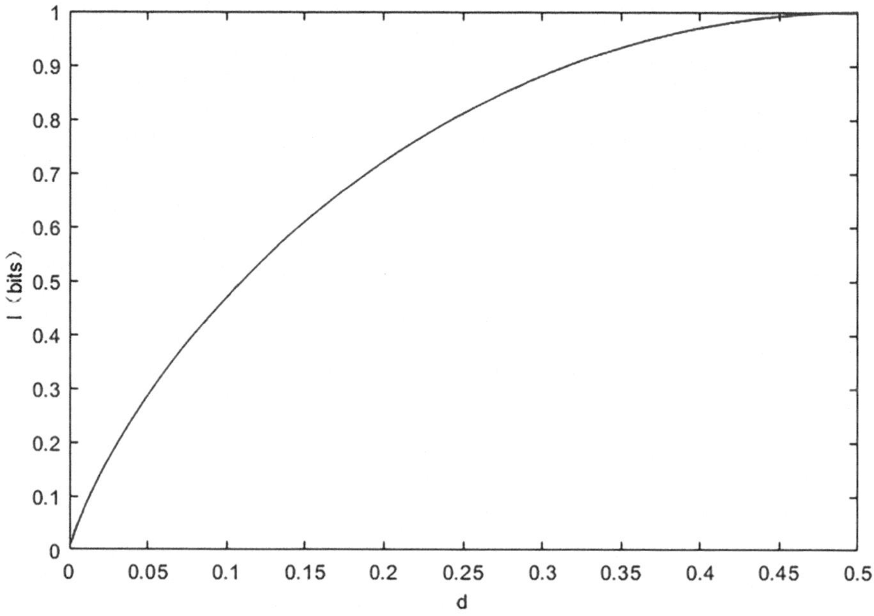

Fig. 5.6 The relationship between the information I and the probability d

According to the above analysis, the probability for Eve to be detected is

$$d = 1 - P = 1 - F \tag{5.32}$$

Combining these two cases, it can be seen that in the case of $P_0 = P_1 = 0.5$, the information Eve obtained between Alice and Bob (Charlie) is equal [65], which can be represented as

$$I_E = I_F = -d \log_2 d - (1 - d) \log_2 (1 - d) \tag{5.33}$$

Figure 5.6 shows the relationship between the information $I_E(I_F)$ that Eve can eavesdrop on and the probability d of the eavesdropping being detected. As can be seen from Fig. 5.6, if Eve wants to obtain one bit of information through the entangle-measure attack under noiseless environment, the probability of being detected is $d = 50\%$. Therefore, if Eve wants to eavesdrop on more information, the corresponding probability of being detected will be greater.

For internal attacks, it is assumed that Charlie is an untrusted third party, and he also wants to obtain Bob's secret message. But during the eavesdropping checking, Charlie does not know all the initial information of detection particles in advance, so he cannot successfully initiate the attack. In addition, the secret message cannot be transferred to Charlie during the transmission, so Charlie will be incapable of inferring the information of the secret message only through the Q_C sequence which he owns.

D. Man-in-the-Middle Attack

Man-in-the-middle attack [74] means that Eve intercepts sequences S_B or S_C transmitted from Alice to Bob or Charlie, and then she prepares another GHZ-like state $|\varphi_e\rangle$, keeps the first qubit in this state, and sends the second and third qubits to Bob and Charlie, respectively. After that, Bob and Charlie perform their operations and send the results back to Alice, which will be intercepted by Eve. However, the order of the returned sequence is rearranged, which is completely unknown to Eve. Therefore, Eve cannot distinguish which photon has been performed the Measurement or Reflection operation. In this situation, even if Eve intercepts these qubits, the attack cannot pass the eavesdropping check and cannot be successfully executed.

E. Denial-of-Service Attack

Denial-of-Service attack (DoS) [63] means that the attacker secretly destroys the communication between the two parties, so that the legitimate communicants cannot obtain the correct secret message. In the DoS attack, Eve intercepts the particles encrypted by Bob and performs his own encoding operation on them and then sends them to Alice. As a result, the secret message obtained by Alice through decoding is inconsistent with the secret message sent by Bob. However, this attack could not be successfully executed in the CDSSQC scheme. Because in Step 3, Eve cannot distinguish the encoding sequence from the detection sequence correctly. If Eve launches a DoS attack, it cannot pass the eavesdropping check. Therefore, for the CDSSQC scheme, the DoS attack is invalid.

5.3.3.4 Efficiency Analysis

The performance of the communication can be expressed in qubit efficiency $\eta = \frac{t}{q}$, where t represents the total number of shared classical bits and q represents the total number of generated qubits in the scheme [67]. (The classical bits used for eavesdropping checking are not considered).

In the CDSSQC scheme, the classical secret message successfully transmitted from Bob to Alice is coded by N bits, that is, $c = N$. Alice prepares $2N$ GHZ-like states as transmission carriers, so a total of $6N$ qubits are consumed, that is, $q = 6N$. In summary, the qubit efficiency of this CDSSQC scheme can be expressed as $\eta = \frac{t}{q} = \frac{N}{6N} = \frac{1}{6}$.

5.3.3.5 Summary

In this section, a controlled deterministic secure semi-quantum communication scheme based on GHZ-like states is proposed. It shows that with assistance of classical controller, classical sender can safely send the secret message to quantum receiver. The proposed CDSSQC scheme mainly contains two phases: establishing quantum communication channel and transmitting secret message. In the first phase, according to the properties of each particle in GHZ-like state, quantum resources consumed in channel can be saved. Because in the case of single-particle measurement, these particles can be directly used for eavesdropping check without inserting additional decoy state. In the second phase, the second particles of the initial GHZ-like states are used to encode the binary sequence that is obtained by compressing the secret message through the Huffman compression coding technology, which improves the confidentiality of communication. What's more, according to the characteristics of Huffman compression coding, the compression ratio will become smaller with the increasing of the length of the secret message, signifying that the higher communication efficiency can be achieved. Eventually, a security analysis of this proposed CDSSQC scheme is carried out on the basis of six different types of attacks, and the results prove that it has strong resistance and high security.

5.4 Quantum Secure Direct Communication

5.4.1 Ping-Pong Scheme

Based on the idea of dense coding, the Ping-Pong scheme [75] encodes information by local operation of EPR pairs. The secret messages can be decoded during the transmission without additional information. The communication parties are in one of two communication modes according to probability P_C, namely, "message mode" and "control mode."

There are two honest parties, Alice and Bob. At the beginning of the scheme, they are set to message mode. Suppose the message sequence is $x^N = (x_1, \cdots, x_N)$, where $x_n \in \{0, 1\}$.

Step 1: Bob prepares an EPR photon pair with state $|\psi^+\rangle = \frac{1}{\sqrt{2}}(|01\rangle + |10\rangle)$. He stores one of the photon pair and then sends the other photon to Alice.

Step 2:

A. Control mode: Alice switches to control mode with probability P_C. In this mode, Alice performs the σ_Z operation on the travel qubit. The measurement outcomes are shown in Table 5.7. Alice sends the outcome to Bob through the public channel.
B. Message mode: Alice holds to message mode with probability $1 - P_C$. Alice performs an encoding operation on the travel qubit. The encoding rules are shown in Table 5.8. Alice sends the travel qubit back to Bob.

Step 3:

A. Control mode: Bob receives the outcomes i from the public channel and switches to control mode. He performs the Z of the particles in the hand,

Table 5.7 The measurement outcomes of Ping-Pong scheme

Travel qubit	i
$\|0\rangle$	0
$\|1\rangle$	1

Table 5.8 The encoding rules of Ping-Pong scheme

Secret message	Travel qubit	Alice's operation	Travel qubit
0	$\vert 0\rangle$	I	$\vert 0\rangle$
	$\vert 1\rangle$	I	$\vert 1\rangle$
1	$\vert 0\rangle$	σ_Z	$\vert 0\rangle$
	$\vert 1\rangle$	σ_Z	$-\vert 1\rangle$

Table 5.9 The decoding rules of Ping-Pong scheme

State	Secret message
$\vert\psi^+\rangle$	0
$\vert\psi^-\rangle$	1

and the value of outcome is j. If $i = j$, there is an eavesdropper in the channel and the communication aborts. If $i \neq j$, the channel is secure and proceed to Step 1.

B. Message mode: Bob receives travel qubits from the quantum channel, holding the message mode. He performs the Bell measurement on two qubits and decodes the message. The decoding rules are shown in Table 5.9. If Bob receives N bits, the communication ends. Otherwise, proceed to Step 1.

The scheme is quasi-secure for direct secret communication if perfect quantum channel is used. However, the capacity of the Ping-Pong scheme is restricted, and the entangled state (EPR pair) carries only one bit of classical information.

5.4.2 Two-Step QSDC Scheme

Based on the two-step QKD scheme, the two-step QSDC scheme uses a set of ordered N EPRs for secure direct communication of blocks [10]. In this scheme, eavesdropping check and information coding are divided into two steps. The party transmits secret messages while ensuring that eavesdroppers cannot obtain valid information.

This QSDC scheme has two honest parties, Alice and Bob, both of quantum parties. A set of N ordered EPR pairs is used as the quantum message carrier, and each EPR pair is randomly in one of the four Bell states. The scheme steps are as follows:

Step 1: Alice prepares an ordered N-th EPR pair sequence with $[(C_1, M_1), \cdots, (C_N, M_N)]$, where C and M represent two particles in an entangled pair and subscripts indicate the position of the particles in the sequence and encode $|\psi^-\rangle, |\psi^+\rangle, |\phi^-\rangle, |\phi^+\rangle$ as 00, 01, 10, and 11, respectively.

Table 5.10 The measurement outcomes of Two-step QSDC scheme

The measurement	Alice	Bob
σ_Z	$\|0\rangle$	$\|1\rangle$
	$\|1\rangle$	$\|0\rangle$
σ_X	$\|+\rangle$	$\|+\rangle$
	$\|-\rangle$	$\|-\rangle$

$$|\psi^-\rangle_{CM} = \frac{1}{\sqrt{2}}\left(|0\rangle_C|1\rangle_M - |1\rangle_C|0\rangle_M\right) \tag{5.34}$$

$$|\psi^+\rangle_{CM} = \frac{1}{\sqrt{2}}\left(|0\rangle_C|1\rangle_M + |1\rangle_C|0\rangle_M\right) \tag{5.35}$$

$$|\phi^-\rangle_{CM} = \frac{1}{\sqrt{2}}\left(|0\rangle_C|0\rangle_M - |1\rangle_C|1\rangle_M\right) \tag{5.36}$$

$$|\phi^+\rangle_{CM} = \frac{1}{\sqrt{2}}\left(|0\rangle_C|0\rangle_M + |1\rangle_C|1\rangle_M\right) \tag{5.37}$$

Step 2: Alice takes one particle from each entangled pair and forms the sequence $S_C = [C_1, \cdots C_N]$, which is called the C-sequence. The remaining particles form the sequence $S_M = [M_1, \cdots M_N]$, which is called the M-sequence.

Step 3: Alice sends the sequence S_C to Bob. Bob randomly selects some particles from S_C, performs local measurement on the selected particles, and the measurement basis randomly selects σ_Z or σ_x . Bob tells Alice which particles he has chosen and what the measurement basis is. Alice chooses the same measurement base for the particles in the same position in S_M, and the two share the measurement outcomes. If there is no eavesdropping, the measurement outcomes are shown in Table 5.10.

Ideally, the two outcomes should be opposite. Based on the measurement results, the QBER can be estimated, and if the bit error rate is less than the threshold, the scheme continues, otherwise the scheme is terminated.

Step 4: Alice encodes a message on the sequence S_M remaining particles and sends it to Bob. To prevent eavesdropping during this transmission, Alice randomly selects some particles in the M-sequence whose locations are known only to her and performs one of the following four operations on them at random:

$$U_0 = I = |0\rangle\langle 0| + |1\rangle\langle 1| \tag{5.38}$$

$$U_1 = \sigma_z = |0\rangle\langle 0| - |1\rangle\langle 1| \tag{5.39}$$

$$U_2 = \sigma_x = |1\rangle\langle 0| + |0\rangle\langle 1| \tag{5.40}$$

$$U_3 = i\sigma_y = |0\rangle\langle 1| - |1\rangle\langle 0| \tag{5.41}$$

The remaining M-sequence particles use intensive encoding during the coding process, transmitting the secret information directly. Alice defines the unitary

Table 5.11 The encoding rules of two-step QSDC scheme

Original sequence	Secret message	Alice's operation	Quantum message
$\lvert\psi^-\rangle$	00	U_0	$\lvert\psi^-\rangle$
$\lvert\psi^-\rangle$	01	U_1	$\lvert\psi^+\rangle$
$\lvert\psi^-\rangle$	10	U_2	$\lvert\phi^-\rangle$
$\lvert\psi^-\rangle$	11	U_3	$\lvert\phi^+\rangle$

operation U according to Eqs. 5.38, 5.39, 5.40, and 5.41, and encodes messages by adding U to the particles of the M sequence. and the encoding rules are shown in Table 5.11. Alice then sends S_M to Bob.

Step 5: After receiving S_M, Bob, combined with S_C, can decode the specific operation by Alice and the corresponding classic bits. Then Alice announces the position and specific value of the random particles, and both sides use this information to estimate the bit error rate of the second transmission and judge the integrity of the message. If the bit error rate of this part is too high, the next step is discarded and the scheme is terminated. Otherwise, the error correction process is performed to preserve the integrity of the message and complete the communication.

The scheme ensures that an eavesdropper can be detected during the transmission of the C sequence. Because the data block is introduced, the security of the message is guaranteed. Even if Eve captured one of the two sequences, she would not be able to read useful information. In addition, the scheme has a high capacity because each EPR pair carries two bits of classical information. The disadvantages of this scheme are the need for quantum memory and the need to transmit the particle sequence twice.

5.4.3 DL04 Scheme

The DL04 scheme [11] is inspired by the classical one-time pad, so it is also called the quantum one-time pad QSDC. The scheme uses a single photon to transmit messages, and if the encoding of the single-photon string is completely random to the eavesdropper, the scheme is unconditionally secure. Below we briefly describe the specific steps of the DL04 scheme.

Step 1: First, Bob prepares a series of single-photon polarization states, and sends all the single photons to Alice by blocks. Each single photon is randomly in one of the following four states: $\lvert H\rangle$, $\lvert V\rangle$, $\lvert +\rangle$, $\lvert -\rangle$. The first two and the last two are the eigenstates of the measured basis σ_Z and σ_x respectively.

Step 2: Alice performs security detection after receiving all single photons. Alice randomly extracts a part of the photons and randomly uses the σ_Z or σ_x for measurement, and informs Bob of the selected measurement base and measurement results through the public channel. The two sides compare the measurement results to detect whether there is eavesdropping, if the bit error rate is below the security threshold, the communication continues, otherwise the communication is terminated immediately.

Step 3: If the security detection result of Step 2 is qualified, Alice performs U operations on the remaining single-photon states in the hand to complete the encoding of secret messages. The encoding mode is as follows: unitary operation $U_0 = I$ represents classical bit 0, and unitary operation $U_3 = \sigma_y$ represents classical bit 1. Namely,

$$U_0 = I = |0\rangle\langle 0| + |1\rangle\langle 1| \tag{5.42}$$

$$U_3 = i\delta_y = |0\rangle\langle 1| - |1\rangle\langle 0| \tag{5.43}$$

$$U_3|0\rangle = -|1\rangle, U_3|1\rangle = |0\rangle, U_3|+\rangle = |-\rangle, U_3|-\rangle = -|+\rangle \tag{5.44}$$

After the encoding is complete, Alice sends a single photon to Bob by blocks. In addition, Alice will randomly extract some of them as security detection photons.

Step 4: Bob uses the true measurement basis to measure the single-photon sequence with secret messages. If the single-photon state remains unchanged, it means that the coding operation is U_0, that is, the classical bit 0; If the single-photon state is flipped between $|H\rangle$ and $|V\rangle$ (or $|+\rangle$ and $|-\rangle$), the encoding operation is U_3, which represents the classical bit 1, completing the reading of the secret messages.

Alice tells Bob which security detection photons she has selected and the unitary operations she has taken on those photons through the public channel. By comparing the results of the two sides, the security of information sequence transmission is judged by the bit error rate, which is the second security detection.

Since the receiver Bob knows the information of the preparation of the quantum state of the encoded messages during the DL04 scheme, the whole decoding process is not classical messages. Compared with the Ping-Pong scheme and the two-step scheme, this scheme uses single-photon state to transmit secret messages and has a wider application prospect.

5.4.4 LM05 Scheme

Lucamarini [76] proposes a QSDC scheme that does not make of entanglement. The scheme uses two sets of conjugate orthogonal bases to attain secure direct communication. The flowchart is shown in Fig. 5.7.

Step 1: Bob randomly produces a qubit at $|0\rangle$, $|1\rangle$, $|+\rangle$ or $|-\rangle$and sends it to Alice.

Step 2: With probability C, Alice switches to control mode. X and Z are randomly selected to measure the received qubits. Or, with probability $1 - C$, the unitary

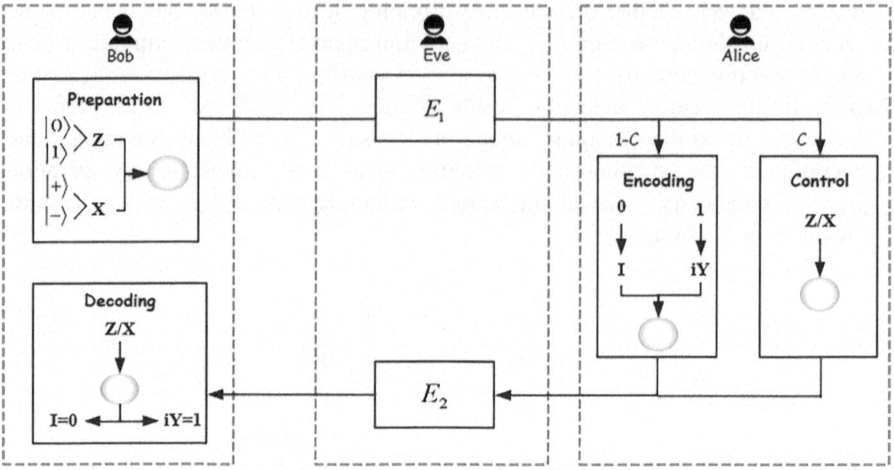

Fig. 5.7 The flowchart of LM05 scheme

operation is performed on the qubit according to the secret message in message mode. The encoding rules: The operation I encodes 0, and the operation iY encodes 1.

Step 3: Bob measures the returned particle. Note that since this qubit is prepared by Bob, Bob can choose the true operation to measure it.

Control mode: Alice tells Bob through the classical channel whether she is measuring and which basis to use. Bob compares the measurement results with his own to detect whether there is eavesdropping in the channel.

Message mode: Bob chooses the true measurement base to measure and decodes the secret message.

The LM05 scheme is easier to prepare, operate, and measure experimentally than entanglement. The scheme can also resist individual attacks in noisy channels. In contrast to the previous schemes, this scheme does not require quantum memory.

5.4.5 MDI-QSDC Scheme

QSDC systems can also be attacked in the apparatus devices, like QKD systems. And the researchers introduced measurement-device-independent (MDI) to resist possible eavesdropping attacks on measuring devices, thus proposing MDI-QSDC. This section introduces the MDI-QSDC scheme based on entanglement [4].

Alice and Bob are the legitimate sender and receiver of the messages, and Charlie is an untrusted third party performing the Bell state measurement. The entanglement pair used in the scheme is in the Bell state and the definition as

$$|\phi^+\rangle = (|00\rangle + |11\rangle)/\sqrt{2} \tag{5.45}$$

$$|\phi^-\rangle = (|00\rangle - |11\rangle)/\sqrt{2} \tag{5.46}$$

$$|\psi^+\rangle = (|01\rangle + |10\rangle)/\sqrt{2} \tag{5.47}$$

$$|\psi^-\rangle = (|01\rangle - |10\rangle)/\sqrt{2} \tag{5.48}$$

The specific steps of the scheme are as follows:

Step 1: Alice prepares n entangled photon pairs of EPR and makes them randomly in $|\psi^+\rangle$ and $|\psi^-\rangle$. Alice takes one photon from each EPR photon pair to form an ordered photon sequence S_A. Alice then prepares m single photons, each of which is randomly in one of four quantum states $\left\{|0\rangle, |1\rangle, |+\rangle = \frac{1}{\sqrt{2}}(|0\rangle + |1\rangle), |-\rangle = \frac{1}{\sqrt{2}}(|0\rangle - |1\rangle)\right\}$. These single photons are randomly inserted into EPR photon pairs to remove another sequence in S_A sequence and form a new sequence C_A. The sequence S_A has n photons and C_A has $n + m$ photons. Similarly, Bob prepares two ordered sequences, S_B with n photons and C_B with $n + m$ photons.

Step 2: Alice and Bob send the sequences C_A and C_B to Charlie, respectively, and keep the S_A and S_B sequences in their hands.

Step 3: After Charlie receives the sequence C_A and C_B, BSM is performed on each pair and the measurement results are announced. The measurement results are in the following three cases:

Cases 1: Both Alice and Bob are photons in the partner of EPR pair. Due to the quantum entanglement swapping,

$$|\psi_{AC}^+\rangle \otimes |\psi_{BD}^+\rangle = \frac{1}{2}\left(|\psi_{AB}^+\rangle|\psi_{CD}^+\rangle - |\psi_{AB}^-\rangle|\psi_{CD}^-\rangle + |\phi_{AB}^+\rangle|\phi_{CD}^+\rangle - |\phi_{AB}^-\rangle|\phi_{CD}^-\rangle\right) \tag{5.49}$$

$$|\psi_{AC}^-\rangle \otimes |\psi_{BD}^+\rangle = \frac{1}{2}\left(|\psi_{AB}^-\rangle|\psi_{CD}^+\rangle - |\psi_{AB}^+\rangle|\psi_{CD}^-\rangle + |\phi_{AB}^-\rangle|\phi_{CD}^+\rangle - |\phi_{AB}^+\rangle|\phi_{CD}^-\rangle\right) \tag{5.50}$$

$$|\psi_{AC}^+\rangle \otimes |\psi_{BD}^-\rangle = \frac{1}{2}\left(|\psi_{AB}^-\rangle|\psi_{CD}^+\rangle - |\psi_{AB}^+\rangle|\psi_{CD}^-\rangle + |\phi_{AB}^+\rangle|\phi_{CD}^-\rangle - |\phi_{AB}^-\rangle|\phi_{CD}^+\rangle\right) \tag{5.51}$$

$$|\psi_{AC}^-\rangle \otimes |\psi_{BD}^-\rangle = \frac{1}{2}\left(|\psi_{AB}^+\rangle|\psi_{CD}^+\rangle - |\psi_{AB}^-\rangle|\psi_{CD}^-\rangle + |\phi_{AB}^+\rangle|\phi_{CD}^+\rangle + |\phi_{AB}^-\rangle|\phi_{CD}^-\rangle\right) \tag{5.52}$$

Table 5.12 Bell measurement on different combinations of photons and their functions

C_A	C_B	Functions
E	E	Entanglement swapping for message transmission
S	S	Security check
E	S	Teleportation that is discarded for simplicity
S	E	

E means a photon from an entangled pair, and *S* stands for a single photon

BSM will entangle photons in the corresponding EPR pairs in S_A and S_B.

Cases 2: Alice and Bob, one of whom is a single photon and the other is a photon in the EPR pair. In principle, these photons could be used to send information. The number of single photons m is usually small; these photons are discarded in order to simplify the implementation of the scheme.

Cases 3: Both Alice and Bob being single photons, and they are used for eavesdropping check. The on different combinations of Bell measurement for photons in Table 5.12.

Step 4: Security detection. After Charlie announces the BSM results, Alice and Bob announce the positions of single photons in C_A and C_B, respectively. For the case that both C_A and C_B are single photons, Alice and Bob exchange the base information of single photons, and only the case that both sides have the same single-photon base vector is used for safety detection:

$$|00\rangle = (|\phi^+\rangle + |\phi^-\rangle)/\sqrt{2} \tag{5.53}$$

$$|01\rangle = (|\psi^+\rangle + |\psi^-\rangle)/\sqrt{2} \tag{5.54}$$

$$|{+}{+}\rangle = (|\phi^+\rangle + |\psi^+\rangle)/\sqrt{2} \tag{5.55}$$

$$|{+}{-}\rangle = (|\phi^-\rangle + |\psi^-\rangle)/\sqrt{2} \tag{5.56}$$

Step 5: Encode the messages. After security detection, Alice and Bob discard the non-entangled photons in the S_A and S_B sequences and the remaining photons in the S_A and S_B sequences form the sequences M_A and M_B, respectively. To send a message, Alice first performs a δ_z operation on a photon in her hand with an initial state of $|\psi^+\rangle$, which is equivalent to all the initial states prepared by Alice being $|\psi^-\rangle$. Then the M_A sequence contains only the photons of the initial state $|\psi^-\rangle$, and for M_B, it contains random photons of the initial state $|\psi^+\rangle$ and $|\psi^-\rangle$. Due to entanglement swapping, photon pairs (M_A, M_B) entangled and only Bob knows these states (M_A, M_B). Alice encodes photons on M_A using four intensive encoding operations, namely, $U_0 = I$, $U_1 = \delta_x$, $U_2 = i\delta_y$, and $U_3 = \delta_z$ for information 00,01,10 and 11, respectively. To prevent Eve from intercepting retransmission attacks, Bob performs random δ_z and I operations on photons of the M_B sequence. In order to ensure the integrity of the information, Alice also performs random check number encoding for photons at random positions in the M_A sequence.

Step 6: Alice and Bob send the encoded M_A and MB sequences to Charlie, who performs the BSM and publishes the measurements. Bob decodes the information based on the measurement results and encodes Alice's random check number. Alice then announces the location and value of the random check number, which Bob compares with Alice to check the integrity of the information. If the error rate of the random number sequence is higher than a certain threshold, it indicates that there is eavesdropping; otherwise, continue to the next step of communication.

The MDI-QSDC scheme requires complete Bell state measurements to distinguish the four Bell states. However, it is impossible to achieve complete Bell measurements with linear optical instruments.

5.5 Quantum Dialogue

5.5.1 QD Scheme

In 2004, the first QD scheme NBA [37] was proposed, which allows the legitimate parties Alice and Bob to exchange secret information directly like a dialogue. The main steps of the scheme are as follows:

Suppose Alice has 2*N* bits secret information $\{(i_1,j_1),(i_2,j_2),\cdots,(i_n,j_n)\}$ and Bob has 2*M* bits secret information $\{(k_1,l_1),(k_2,l_2),\cdots,(k_n,l_n)\}$, where $i_n,j_n,k_n,l_n\in\{0,1\}$, $N=M^2$. First, Bob prepares enough states $|\psi^+\rangle_{AB}=\frac{1}{\sqrt{2}}(|01\rangle+|10\rangle)_{AB}$. Then follow these steps:

Step 1: Initialize, set$n=0$.

Step 2: Let $n=n+1$. Bob selects $C^t_{k_n,l_n}$ to be performed on state $|\psi^+\rangle_{A_nB_n}$ based on the value of (k_n,l_n). Then particle A_n is left and particle B_n is sent to Alice.

Step 3: Alice tells Bob that it's received, and operation $C^t_{i_n,j_n}$ is selected based on the value of (i_n,j_n) to execute on the received particle, and then returns the particle to Bob.

Step 4: Bob tells Alice that the Bell measurement is received and gets the result $\left|\psi^+_{x_n,y_n}\right\rangle_{A_nB_n}$ and then selects the next step based on the operation MM or CM that Alice chooses:

1. If Alice chooses MM, then the decryption phase begins. Bob gets Alice's secret message by calculating in $i_n=|x_n-k_n|,j_n=|y_n-l_n|$. Then Bob tells Alice the value of(x_n,y_n) and Alice gets Bob's secret message by calculating $k_n=|x_n-i_n|,l_n=|y_n-j_n|$. Finally, if$n=N$, the conversation completes and the communication ends; if$n<N$, go to Step 2.

2. If Alice chooses CM, then the eavesdropping check phase begins. Alice announces the value of (i_n, j_n), and Bob computes $|x_n - k_n|$, $|y_n - l_n|$. If $i_n = |x_n - k_n|$, $j_n = |y_n - l_n|$, then the channel is safe, Bob sets $n = n - 1$ and returns to Step 2; otherwise the channel is unsafe, returns to Step 1.

5.5.2 SQD Scheme

The semi-quantum dialogue schemes that have been proposed are based on quantum dialogue between two-party, and there is no semi-quantum dialogue scheme for the three-party currently. With regards to this, we propose a three-party deterministic secure semi-quantum dialogue (DSSQD) scheme based on GHZ states [77], which includes one quantum communicant and two classical communicants. The proposed DSSQD scheme can be divided into two stages: eavesdropping check and encrypted message. The communicants complete two eavesdropping check by introducing decoy state technology and design a classical one-time key encryption scheme to encode secret messages. Compared with the existing quantum dialogue scheme, the proposed DSSQD scheme has the following advantages:

1. Compared with the traditional quantum dialogue scheme, the proposed DSSQD scheme exceeds them in that it does not require all three communicants to possess quantum capabilities. This feature simplifies the measurement and calculation of quantum superposition states involved in the communication process and saves quantum resources.
2. Compared with the existing semi-quantum dialogue scheme, the advantage of the proposed DSSQD scheme is that it uses three-particle GHZ states instead of two-photon entangled states or single-photon states as the quantum carriers, thereby increasing communication party and realizing a three-party secure quantum dialogue;
3. Compared with the existing semi-quantum dialogue scheme [78, 79], the proposed DSSQD scheme has higher security. While avoiding typical information leakage problems, it can resist multiple types of attacks and has good confidentiality and security.

In this section, in order to achieve the purpose of exchanging secret message between Alice, Bob, and Charlie, we employ the GHZ states to construct a three-party DSSQD scheme.

5.5.2.1 GHZ State

Eight GHZ states can be denoted as

$$|\psi_{000}\rangle_{ABC} = \frac{1}{\sqrt{2}}(|000\rangle + |111\rangle)_{ABC} \tag{5.57}$$

$$|\psi_{001}\rangle_{ABC} = \frac{1}{\sqrt{2}}(|000\rangle - |111\rangle)_{ABC} \tag{5.58}$$

$$|\psi_{010}\rangle_{ABC} = \frac{1}{\sqrt{2}}(|100\rangle + |011\rangle)_{ABC} \tag{5.59}$$

$$|\psi_{011}\rangle_{ABC} = \frac{1}{\sqrt{2}}(|100\rangle - |011\rangle)_{ABC} \tag{5.60}$$

$$|\psi_{100}\rangle_{ABC} = \frac{1}{\sqrt{2}}(|010\rangle + |101\rangle)_{ABC} \tag{5.61}$$

$$|\psi_{101}\rangle_{ABC} = \frac{1}{\sqrt{2}}(|010\rangle - |101\rangle)_{ABC} \tag{5.62}$$

$$|\psi_{110}\rangle_{ABC} = \frac{1}{\sqrt{2}}(|110\rangle + |001\rangle)_{ABC} \tag{5.63}$$

$$|\psi_{111}\rangle_{ABC} = \frac{1}{\sqrt{2}}(|110\rangle - |001\rangle)_{ABC} \tag{5.64}$$

5.5.2.2 Three-Party DSSQD Scheme

In the proposed three-party DSSQD scheme, we assume that Alice, Bob, and Charlie each have N-bit secret message respectively, which are expressed as $m_A=\{m_{A_1},m_{A_2},\cdots,m_{A_N}\}$, $m_B=\{m_{B_1},m_{B_2},\cdots,m_{B_N}\}$, and $m_C=\{m_{C_1},m_{C_2},\cdots,m_{C_N}\}$, where the subscript can be represented by i and $i = 1, 2, \cdots, N$. Among them, Alice is a powerful quantum participant who can prepare any quantum state and perform corresponding quantum operations on it, such as measuring and storing qubits. For the other two participants, Bob and Charlie, they are classical and restricted to perform the following operations: (1) M**:** measure qubits on the computational basis $Z = \{|0\rangle, |1\rangle\}$, and prepare qubits in the opposite state of the measurement results. For example, if Bob or Charlie measure the received qubit and get the result $|0\rangle$ ($|1\rangle$), they will prepare the corresponding opposite state $|1\rangle(|0\rangle)$; (2) R**:** reflect the received particles back without any interference; (3) Reordering: reorder the received quantum sequence and temporarily store qubits through different delay lines. It can be seen that at the beginning of the scheme, quantum delay lines [17, 80, 81] are installed in Bob and Charlie respectively to help complete the two eavesdropping check.

The purpose of the proposed three-party DSSQD scheme is to allow any communicants to send his own secret message to the other two parties to complete the mutual dialogue between the three-party without causing information leakage problem. After the communication is completed, each communicant can obtain three

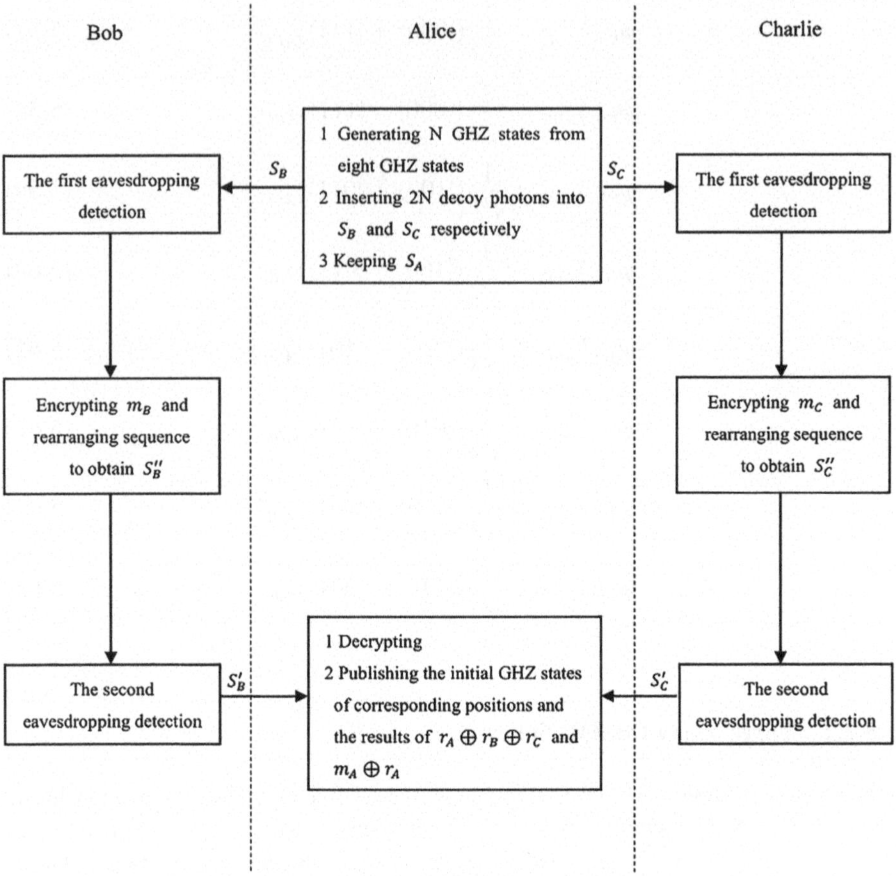

Fig. 5.8 The flow chart of three-party DSSQD scheme

secret messages m_A, m_B and m_C. In addition, the DSSQD scheme assumes that the quantum channel is ideal (i.e., nonlossy and noiseless), and the authenticated classical channel is shared between three legitimate participants.

Alice divides the first, second and third particles in these GHZ states into three ordered sequences S_A, S_B, S_C, and keeps S_A for herself, then sends S_B and S_C to Bob and Charlie, respectively.

The following descriptions are the specific steps of three-party DSSQD scheme, corresponding to Fig. 5.8.

Step 1: Preparation and Transmission of Quantum States

Alice randomly generates N GHZ states from Eqs. 5.57, 5.58, 5.59, 5.60, 5.61, 5.62, 5.63, and 5.64, where N is also the length of secret message. Then she divides the three entanglement particles in the N GHZ states into three sequences that are $S_A = [P_1(A), P_2(A), \ldots, P_N(A)]$, $S_B = [P_1(B), P_2(B), \ldots, P_N(B)]$, and $S_C = [P_1(C),$

$P_2(C), \ldots, P_N(C)]$, respectively, where $P(A)$, $P(B)$, and $P(C)$ represent the first, second, and third particles in GHZ state separately. Next, Alice randomly generates $4N$ decoy states from $\{|0\rangle, |1\rangle, |+\rangle, |-\rangle\}$ and inserts $2N$ decoy states into S_B and S_C, respectively. These decoy states are used to detect whether there is an eavesdropper in the quantum channel, ensuring the security of the communication process. Finally, Alice keeps sequence S_A and transmits the sequences S_B and S_C to Bob and Charlie, respectively.

Step 2: The First Eavesdropping Check

In order to prevent Trojan horse attack [53], Bob and Charlie place a wavelength filter and a photon number splitter (PNS) [54] in front of the receiving device separately. After Bob and Charlie received sequences S_B and S_C, Alice announces the corresponding positions of the $4N$ decoy states. Then Bob and Charlie use quantum delay line to temporarily store sequences and choose the first N from their own $2N$ decoy states separately and randomly perform the following two operations:

1. Reflection: Reflect the photon back to Alice without any disturbance.
2. Measurement: Measure the photon with Z basis, then prepare a fresh one in the same state according to the measurement result and send it back to Alice.

After confirming that Alice has received return decoy particles, Bob and Charlie announce the positions, operations and measurement results of the corresponding N decoy states sequence, respectively. According to these, Alice performs the first eavesdropping check:

1. For the reflected particles, Alice measures them with her preparation basis and calculates the error rate by comparing the measurement results with her prepared initial states.
2. For the measured particles, Alice also measures them with the Z basis and observes whether the results are correct.

Finally, Alice counts the total error rate. If the total error rate is less than the security threshold, the communication will be continued; otherwise, it indicates that there is an eavesdropping in quantum channel and the communication is abandoned.

Step 3: Encryption of Bob and Charlie

First, Bob and Charlie separately perform Z basis measurement on the received sequences S_B and S_C, which are not for the remaining N decoy states and record the measurement results to prepare for the final decryption. Next, they encrypt the initial sequences S_B and S_C in order according to the secret messages $m_B = \{m_{B_1}, m_{B_2}, \cdots, m_{B_N}\}$ and $m_C = \{m_{C_1}, m_{C_2}, \cdots, m_{C_N}\}$, respectively. The specific encryption rules are as follows, where M and R operations correspond to M and R,

1. R: If the bit of secret message is 0, they perform $\boldsymbol{R}$ operation on the particles $P_i(B)$ and $P_i(C)$ which are in the sequences S_B and S_C, respectively.
2. M: If the bit of secret message is 1, they implement M operation on the corresponding particles of sequences S_B and S_C. For example, If the measurement result is $|0\rangle$, they replace the state with $|1\rangle$; otherwise, if it is $|1\rangle$, they substitute the state with $|0\rangle$.

After the encryption of Bob and Charlie is completed, sequences S_B and S_C are turned into sequences S'_B and S'_C. Moreover, after passing the first eavesdropping check, there are still N remaining decoy states located in the sequences S'_B and S'_C, respectively. At this time, Bob and Charlie perform operations on the remaining N decoy states for the second eavesdropping check. The specific operations are as follows:

1. Do nothing.
2. Measure a decoy state with Z basis and prepare a same state according to the measurement result.

At last, they insert these result states after the operations into sequences S'_B and S'_C, then rearrange the order of them using quantum delay line to obtain sequences S''_B and S''_C, respectively. Subsequently, Bob and Charlie send the sequences S''_B and S''_C to Alice.

Step 4: The Second Eavesdropping Check
Similarly, Alice also places a photon number separator (PNS) and a wavelength filter in front of the receiving device to resist Trojan horse attack. After confirming Alice has received sequences S''_B and S''_C, Bob and Charlie announce the positions, operations and measurement results of the corresponding N decoy states in the sequences, respectively. At this time, Alice performs the same operation on these decoy states as the first eavesdropping check to obtain the total error rate. If the total error rate is less than the security threshold, continue to next step; otherwise, terminate this communication.

Step 5: Decryption
After passing the second eavesdropping check, sequences S''_B and S''_C are restored to S'_B and S'_C. Then Bob and Charlie announce to Alice the correct ordering of S'_B and S'_C. According to this, Alice records the state of each group of three parties by measuring the photons $P_i(A)$, $P'_i(B)$, and $P'_i(C)$ in the sequences S_A, S'_B, and S'_C with Z basis, which are denoted as r_{A_i}, r_{B_i}, and r_{C_i}, respectively. Afterward, Alice publishes the results of $r_{A_i} \oplus r_{B_i} \oplus r_{C_i}$ and $m_{A_i} \oplus r_{A_i}$ ($\oplus$ is the XOR operation). In addition, she also announces the initial GHZ states corresponding to each particle in the sequences S_B and S_C. Thus, Bob and Charlie can infer secret message of each other and Alice from the corresponding results of $r_{A_i} \oplus r_{B_i} \oplus r_{C_i}$ and $m_{A_i} \oplus r_{A_i}$, respectively. As for Alice, she also can infer the secret messages m_B and m_C from measuring the sequences S'_B and S'_C. Here is an example that can more clearly illustrate the principle of our decryption. Suppose $m_{A_1}=1$, $m_{B_1}=1$, $m_{C_1}=0$, and the initial GHZ state of $|\psi_{000}\rangle_{ABC} = \frac{1}{\sqrt{2}}(|000\rangle + |111\rangle)_{ABC}$ is used to carry secret message. After Alice sends sequences S_B and S_C to Bob and Charlie separately, Bob and Charlie use the Z basis to measure the photons they hold. Then, when Alice also performs the Z basis measurement on the photons in sequence S_A at the end, she will find that $|\psi_{000}\rangle_{ABC}$ has collapsed. If $|\psi_{000}\rangle_{ABC}$ collapses to $|000\rangle$, the measurement results of Alice, Bob and Charlie are $|0\rangle$, $|0\rangle$, $|0\rangle$, respectively. At this time, the encryption operation is performed, because of $m_{B_1}=1$, $m_{C_1}=0$, Bob replaces the

Table 5.13 The initial GHZ state is $|\psi_{000}\rangle_{ABC}$ and $|\psi_{001}\rangle_{ABC}$

Measurement result	m_A	m_B	m_C	Encoded state	$r_A \oplus r_B \oplus r_C$	$m_A \oplus r_A$
$\|000\rangle$	0	0	0	$\|000\rangle$	0	0
	0	0	1	$\|001\rangle$	1	0
	0	1	0	$\|010\rangle$	1	0
	0	1	1	$\|011\rangle$	0	0
	1	0	0	$\|000\rangle$	0	1
	1	0	1	$\|001\rangle$	1	1
	1	1	0	$\|010\rangle$	1	1
	1	1	1	$\|011\rangle$	0	1
$\|111\rangle$	0	0	0	$\|111\rangle$	1	1
	0	0	1	$\|110\rangle$	0	1
	0	1	0	$\|101\rangle$	0	1
	0	1	1	$\|100\rangle$	1	1
	1	0	0	$\|111\rangle$	1	0
	1	0	1	$\|110\rangle$	0	0
	1	1	0	$\|101\rangle$	0	0
	1	1	1	$\|100\rangle$	1	0

result $|0\rangle$ state with $|1\rangle$ state and sends it back to Alice, while Charlie reflects the $|0\rangle$ state back to Alice without any disturbance. Alice uses the Z basis to measure them after receiving the encrypted photons, according to the results of measurement, she can infer the bit of secret messages about Bob and Charlie that are 1 and 0. When Alice publishes the value of $r_{A_1} \oplus r_{B_1} \oplus r_{C_1}$, which is 1, Bob can calculate that $r_{C_1} = 0$, $m_{C_1} = 0$ by announcing the initial GHZ state through Alice, for the same reason Charlie calculates that $r_{B_1} = 1$, $m_{B_1} = 1$. And when Alice announces the value of $m_{A_1} \oplus r_{A_1}$, which is 1, Bob and Charlie can infer the value of m_{A_1} is 1. Finally, Alice, Bob, and Charlie all infer the secret message of each other, a three-party semi-quantum deterministic secure dialogue scheme is finished. Different initial GHZ states in the proposed three-party DSSQD scheme will get different encryption results, Tables 5.13, 5.14, 5.15, and 5.16 correspond to different decryption results.

Finally, it is worth mentioning that the proposed DSSQD scheme involves two eavesdropping checks which has quantum memory problem. In the proposed DSSQD scheme, due to the limitations of the semi-quantum environment, classical communicants Bob and Charlie do not have quantum registers, and the quantum operations that can be performed by them are also limited, so they can only use quantum delay line technology to temporarily store qubits. However, in recent years, with the in-depth research on quantum-memory-free, in theory, some quantum secure direct communication schemes [6, 7, 82] using quantum-memory-free technology have been designed, and related security analysis has been completed using Wyner's wiretap theory [5, 83, 84]; in practice, some experiments are also achieved through optical fiber [5] and free space [85]. Therefore, with the development of science and technology in the future, quantum-memory-free technology will

Table 5.14 The initial GHZ state is $|\psi_{010}\rangle_{ABC}$ and $|\psi_{011}\rangle_{ABC}$

Measurement result	m_A	m_B	m_C	Encoded state	$r_A \oplus r_B \oplus r_C$	$m_A \oplus r_A$
$\|100\rangle$	0	0	0	$\|100\rangle$	1	1
	0	0	1	$\|101\rangle$	0	1
	0	1	0	$\|110\rangle$	0	1
	0	1	1	$\|111\rangle$	1	1
	1	0	0	$\|100\rangle$	1	0
	1	0	1	$\|101\rangle$	0	0
	1	1	0	$\|110\rangle$	0	0
	1	1	1	$\|111\rangle$	1	0
$\|011\rangle$	0	0	0	$\|011\rangle$	0	0
	0	0	1	$\|010\rangle$	1	0
	0	1	0	$\|001\rangle$	1	0
	0	1	1	$\|000\rangle$	0	0
	1	0	0	$\|011\rangle$	0	1
	1	0	1	$\|010\rangle$	1	1
	1	1	0	$\|001\rangle$	1	1
	1	1	1	$\|000\rangle$	0	1

Table 5.15 The initial GHZ state is $|\psi_{100}\rangle_{ABC}$ and $|\psi_{101}\rangle_{ABC}$

Measurement result	m_A	m_B	m_C	Encoded state	$r_A \oplus r_B \oplus r_C$	$m_A \oplus r_A$
$\|010\rangle$	0	0	0	$\|010\rangle$	1	1
	0	0	1	$\|011\rangle$	0	1
	0	1	0	$\|000\rangle$	0	1
	0	1	1	$\|001\rangle$	1	1
	1	0	0	$\|010\rangle$	1	0
	1	0	1	$\|011\rangle$	0	0
	1	1	0	$\|000\rangle$	0	0
	1	1	1	$\|001\rangle$	1	0
$\|101\rangle$	0	0	0	$\|101\rangle$	0	1
	0	0	1	$\|100\rangle$	1	1
	0	1	0	$\|111\rangle$	1	1
	0	1	1	$\|110\rangle$	0	1
	1	0	0	$\|101\rangle$	0	0
	1	0	1	$\|100\rangle$	1	0
	1	1	0	$\|111\rangle$	1	0
	1	1	1	$\|110\rangle$	0	0

gradually replace the quantum delay line and get greater applications. In the following research, we will also focus on the related technologies of quantum-memory-free, explore the characteristics of the semi-quantum environment at a deeper level, and try to combine the two to achieve the purpose of further improving the semi-quantum communication system.

Table 5.16 The initial GHZ state is $|\psi_{110}\rangle_{ABC}$ and $|\psi_{111}\rangle_{ABC}$

Measurement result	m_A	m_B	m_C	Encoded state	$r_A \oplus r_B \oplus r_C$	$m_A \oplus r_A$
$\|110\rangle$	0	0	0	$\|110\rangle$	0	1
	0	0	1	$\|111\rangle$	1	1
	0	1	0	$\|100\rangle$	1	1
	0	1	1	$\|101\rangle$	0	1
	1	0	0	$\|110\rangle$	0	0
	1	0	1	$\|111\rangle$	1	0
	1	1	0	$\|100\rangle$	1	0
	1	1	1	$\|101\rangle$	0	0
$\|001\rangle$	0	0	0	$\|001\rangle$	1	0
	0	0	1	$\|000\rangle$	0	0
	0	1	0	$\|011\rangle$	0	0
	0	1	1	$\|010\rangle$	1	0
	1	0	0	$\|001\rangle$	1	1
	1	0	1	$\|000\rangle$	0	1
	1	1	0	$\|011\rangle$	0	1
	1	1	1	$\|010\rangle$	1	1

5.5.2.3 Correctness Analysis

When analyzing the correctness of the DSSQD scheme, the description of eavesdropping check is ignored. In Step 1, Alice has the initial GHZ state $|\psi\rangle_{ABC}$, secret message $m_A = \{m_{A_1}, m_{A_2}, \cdots, m_{A_N}\}$, and the sequence S_A, and Bob and Charlie have secret messages $m_B = \{m_{B_1}, m_{B_2}, \cdots, m_{B_N}\}$ and $m_C = \{m_{C_1}, m_{C_2}, \cdots, m_{C_N}\}$ and the sequences S_B and S_C, respectively.

In Step 3, Bob and Charlie encode the secret messages m_B and m_C on the initial sequences S_B and S_C; therefore, they form two new qubit sequences S'_B and S'_C:

$$S'_B = E_{m_B}(S_B), S'_C = E_{m_C}(S_C), \tag{5.65}$$

where E_{m_B} and E_{m_C}, respectively, represent the encryption operation on secret messages m_B and m_C.

In Step 5, after Alice receives sequences S'_B and S'_C, she can get the original sequences S_B and S_C by decrypting and deduce the secret messages m_B and m_C to be transmitted by Bob and Charlie according to the initial GHZ state $|\psi\rangle_{ABC}$:

$$S_B = D_{m_B}\left(S'_B\right), S_C = D_{m_C}\left(S'_C\right), \tag{5.66}$$

where D_{m_B} and D_{m_C}, respectively, represent the decryption operation on secret messages m_B and m_C.

By measuring S_A, S'_B and S'_C with correct bases, Alice obtains r_A, r_B and r_C,

$$\mathrm{r}_A = M_Z(S_A), \mathrm{r}_B = M_Z(S'_B), \mathrm{r}_C = M_Z(S'_C) \tag{5.67}$$

Where M_Z denotes measuring sequences with Z basis.

Finally, Alice publishes the results of $r_A \oplus r_B \oplus r_C$ and $m_A \oplus r_A$ ($\oplus$ is the XOR operation) and the corresponding initial GHZ states. Bob and Charlie can infer secret message of Alice from the corresponding results of $m_A \oplus r_A$ and deduce secret message of each other from the corresponding results of $r_A \oplus r_B \oplus r_C$.

Obviously, the three-party DSSQD scheme is correct in the ideal scenario.

5.5.2.4 Security Analysis

Security is the most important aspect of quantum direct communication schemes. In this section, we analyze the security of the proposed three-party DSSQD scheme.

A. Information Leakage Problem

Information leakage problem [43] means that any other eavesdropper can extract some useful information about a secret message without taking an active attack. To demonstrate this point, we should examine what eavesdropper Eve can obtain from the public declaration. Next, it will be proved that the proposed three-party DSSQD scheme can avoid information leakage problem. Suppose the initial GHZ state prepared by Alice is $|\psi_{010}\rangle_{ABC}$ (other situations are similar). In Step 5, Alice first obtains the state of each encoded single photon carrying a secret message bit in sequences S'_B and S'_C by using Z basis measurement. Next, Alice performs the calculation of $r_{A_i} \oplus r_{B_i} \oplus r_{C_i}$ and $\mathrm{m}_{A_i} \oplus r_{A_i}$. Finally, Alice announces the value of $r_{A_i} \oplus r_{B_i} \oplus r_{C_i}$ and $\mathrm{m}_{A_i} \oplus r_{A_i}$ instead of the measurement results. Therefore, eavesdropper Eve has no chance to know the measurement result of each corresponding encoded photon from the public statement of Alice or Bob or Charlie and can only make inferences based on the initial GHZ state and the calculation results disclosed by Alice. In this way, Eve knows that the secret message bits transmitted by Alice, Bob, and Charlie must be one of $\{(0,0,0), (0,0,1), (0,1,0), (0,1,1), (1,0,0), (1,0,1), (1,1,0), (1,1,1)\}_{ABC}$ randomly, where the subscripts A, B, and C stand for Alice, Bob, and Charlie, respectively. These results occur with equal probabilities. In summary, according to Shannon's information theory, the information can be concealed completely only if the key uncertainty is at least log N [86]. For each photon that encodes one bit of secret message, there are totally eight kinds of uncertainty, so the information entropy of secret message is $-\sum_{i=1}^{8} P_i \, log_2 P_i = -8 \times \frac{1}{8} \, log_2 \frac{1}{8} = 3$ bits of information, which exactly equal to the total number of secret messages to be transmitted between Alice, Bob, and Charlie. As a result, it can be considered that none of the three secret message bits have been leaked to Eve. Hence, the proposed three-party DSSQD scheme is believed to avoid information leakage problem.

B. Intercept-Measure-Resend Attack

In the proposed three-party DSSQD scheme, Alice inserts $2N$ decoy states into the sequences S_B and S_C each for two eavesdropping check, respectively. When the eavesdropper Eve intercepts the transmitted photon sequences S_B and S_C, she will randomly select the X or Z basis to measure the intercepted photons. Therefore, she has a 50% probability to choose the correct measurement basis to obtain the correct measurement results, of course, there is also a 50% possibility to choose the wrong measurement basis. Here is the analysis of the transmission between Alice and Bob as an example (the analysis between Alice and Charlie is similar).

It is assumed that Eve intercepts the photon sent by Alice to Bob is $|+\rangle$. On the one hand, if Eve performs a measurement on it along the X basis, this photon will be collapsed into $|+\rangle$. Eve then resends this result state $|+\rangle$ to Bob. Consequently, Eve induces no error. On the other hand, if Eve performs single-photon measurement on it with the Z basis, $|+\rangle$ will collapse into $|0\rangle$ or $|1\rangle$, each with a probability of 1/2. Then, Eve retransmits this result state to Bob. According to the Step 2 of the proposed DSSQD scheme, Bob randomly chooses Measurement or Reflection operation. If it is a Measurement, the particle will be kept and induces no error; if it is a Reflection, Alice performs X basis measurement and obtains $|+\rangle$ or $|-\rangle$ with the same probability of 1/2. Thus, for a decoy state prepared with X basis, the probability for Eve to pass the security checking is $\frac{1}{2}\times 1+\frac{1}{2}\times\left(\frac{1}{2}\times 1+\frac{1}{2}\times\frac{1}{2}\right)=\frac{7}{8}$.

Now suppose Eve intercepts the photon sent by Alice to Bob is $|0\rangle$. The analysis process of the first case is similar to the above. However, in the second case, if Eve performs single photon measurement on it with the X basis, it will be collapsed into $|+\rangle$ or $|-\rangle$, each with a probability of 1/2. Eve then resends the result state to Bob. Accordingly, if Bob selects Reflection, Alice will get the incorrect result with a probability of 1/2. If Bob chooses Measurement, Alice will obtain the incorrect result with a probability of 1/2 as well. Therefore, for a decoy state prepared with Z basis, the probability of not being detected is $\frac{1}{2}\times 1+\frac{1}{2}\times\left(\frac{1}{2}\times\frac{1}{2}+\frac{1}{2}\times\frac{1}{2}\right)=\frac{3}{4}$.

Finally, for $2N$ decoy states, if Eve initiates intercept-measure-resend attack, the probability of being detected is $d=1-\left(\frac{3}{4}\times\frac{7}{8}\right)^{2N}=1-\left(\frac{21}{32}\right)^{N}$. When N is large enough, the detection rate will be approximate to 1. In addition, since Eve does not know the specific location information of the decoy state in the proposed three-party DSSQD scheme, her intercept-measure-resend attack will inevitably be detected.

C. Flip Attack

In the flip attack, Eve modifies the contents of the transmitted particles to make the receiver obtain wrong secret messages. When the sequences S''_B and S''_C are transmitted from Bob and Charlie to Alice, Eve can cause Alice to decode the wrong bit value of the secret message by flipping some particles in sequences, so that the entire communication will be destroyed. We suppose that Alice generates GHZ state $|\psi_{010}\rangle_{ABC}$and sends the second particle to Bob (the analysis between Alice and Charlie is similar). Then Eve intercepts and flips the particle when it is sent back to Alice. According to the proposed three-party DSSQD scheme, Bob randomly selects to perform the M or R operation on the photon that encrypts secret message. Specifically, if the bit of secret message is 1, Bob decides to perform M operation. At this time, suppose the particle collapses into $|1\rangle$ ($|0\rangle$) (the corresponding complete state is $|011\rangle$ ($|100\rangle$). Then Bob sends the corresponding opposite state $|0\rangle$ ($|1\rangle$) to Alice. At the same time, Eve intercepts this particle and performs a flip attack on it so that the final state Alice received is $|1\rangle$ ($|0\rangle$), which cause Alice to deduce the original GHZ state as $|\psi_{010}\rangle_{ABC}$ or $|\psi_{011}\rangle_{ABC}$ each with probability of 1/2. In other words, Alice finds the flip attack with a probability of 1/2. Similarly, Bob chooses R operation when the bit of secret message is 0. However, in this case, Eve launches the flip attack without introducing any errors.

In addition, for N decoy states, similar to the analysis in the intercept-measure-resend attack, Alice can find the flip attack with the help of the announcement of Bob during the eavesdropping check phase. Thus, the error rate introduced by Eve is 100%.

In summary, if Eve performs the flip attack on a particle encoding secret message of sequences S''_B (S''_C), the probability for she to pass the security checking is $\frac{1}{2}\times 1+\frac{1}{2}\times\frac{1}{2}=\frac{3}{4}$. Therefore, for N-qubit particle encoding secret messages, the probability of being detected is $d=1-\left(\frac{3}{4}\right)^N$. When N is large enough, the probability of Eve being discovered will increase accordingly.

D. Man-in-the-Middle Attack

Man-in-the-middle attack means that Eve intercepts the photon sequences S_B or S_C transmitted from Alice to Bob or Charlie, then she prepares another GHZ state $|\varphi_e\rangle$, reserves the first qubit, and sends the second and third qubits to Bob and Charlie respectively. After that, Bob and Charlie perform encryption operation and send the result back to Alice, which will be intercepted by Eve. However, the order of the resulting sequence is rearranged, which is completely unknown to Eve. Therefore, Eve cannot distinguish which photon has been performed the M or R operation, that is, cannot obtain confidential information. In addition, decoy state technology is used in the proposed three-party DSSQD scheme, even if Eve intercepts these qubits, her attack can eventually be detected by decoy state technology.

E. Trojan Horse Attack

Trojan horse attack includes a multi-photon-signal attack [60], an invisible-photon attack [61] or a delay-photon attack [54]. For this, receiver not only has to estimate the number of photons in each legitimate single-photon signal in the received sequence, but also filter out illegal photon signals. In the proposed three-party DSSQD scheme, Alice, Bob, and Charlie all place a wavelength filter and a photon number splitter (PNS) in front of the receiving device. The wavelength filter is mainly used to filter out single photons with illegitimate wavelengths and throw it away before reaching the receiving device. And the photon number splitter (PNS) is used in each path to divide each signal into two pieces, each of which is performed single-photon measurement. Then, Alice, Bob, and Charlie can distinguish whether there is a multi-photon signal in the received sequence through the measurement results. Hence, the proposed three-party DSSQD scheme is safe against Trojan horse attack in principle.

F. Entangle-Measure Attack

In general, Eve's entangle-measure attack is composed of two operations U_E and U_F. U_E is applied on the composed state of Eve's probe and the travel qubit as they go from Alice to Bob or Charlie (forward channel); U_F is applied on the composed state of Eve's probe and travel qubit as they go back from Bob or Charlie to Alice (backward channel) [65], corresponding to Fig. 5.9.

Here, we think that Eve utilizes the same attack strategy in the forward and backward channels, that is, U_E=U_F. Therefore, we can only analyze the impact of the U_E operation. Suppose Eve performs operation U_E on the system composed of decoy states $\{|0\rangle, |1\rangle, |+\rangle, |-\rangle\}$ and the auxiliary photon which is prepared by Eve as $|\varepsilon\rangle$. We can get

$$U_E|0\rangle|\varepsilon\rangle = a|0\rangle|\varepsilon_{00}\rangle + b|1\rangle|\varepsilon_{01}\rangle \tag{5.68}$$

$$U_E|1\rangle|\varepsilon\rangle = c|0\rangle|\varepsilon_{10}\rangle + d|1\rangle|\varepsilon_{11}\rangle \tag{5.69}$$

$$\begin{aligned} U_E|+\rangle|\varepsilon\rangle &= \frac{1}{\sqrt{2}}[|0\rangle(a|\varepsilon_{00}\rangle + c|\varepsilon_{10}\rangle) + |1\rangle(b|\varepsilon_{01}\rangle + d|\varepsilon_{11}\rangle)] \\ &= \frac{1}{2}[|+\rangle(a|\varepsilon_{00}\rangle + b|\varepsilon_{01}\rangle + \mathrm{c}|\varepsilon_{10}\rangle + d|\varepsilon_{11}\rangle) \\ &\quad +|-\rangle(a|\varepsilon_{00}\rangle - b|\varepsilon_{01}\rangle + \mathrm{c}|\varepsilon_{10}\rangle - d|\varepsilon_{11}\rangle)] \end{aligned} \tag{5.70}$$

$$\begin{aligned} U_E|-\rangle|\varepsilon\rangle &= \frac{1}{\sqrt{2}}[|0\rangle(a|\varepsilon_{00}\rangle - c|\varepsilon_{10}\rangle) + |1\rangle(b|\varepsilon_{01}\rangle - d|\varepsilon_{11}\rangle)] \\ &= \frac{1}{2}[|+\rangle(a|\varepsilon_{00}\rangle + b|\varepsilon_{01}\rangle - \mathrm{c}|\varepsilon_{10}\rangle - d|\varepsilon_{11}\rangle) \\ &\quad +|-\rangle(a|\varepsilon_{00}\rangle - b|\varepsilon_{01}\rangle - \mathrm{c}|\varepsilon_{10}\rangle + d|\varepsilon_{11}\rangle)], \end{aligned} \tag{5.71}$$

where $|\varepsilon_{00}\rangle$, $|\varepsilon_{01}\rangle$, $|\varepsilon_{10}\rangle$ and $|\varepsilon_{11}\rangle$ belong to the Hilbert space of Eve's probes and $|a|^2 + |b|^2 = |c|^2 + |d|^2 = 1$.

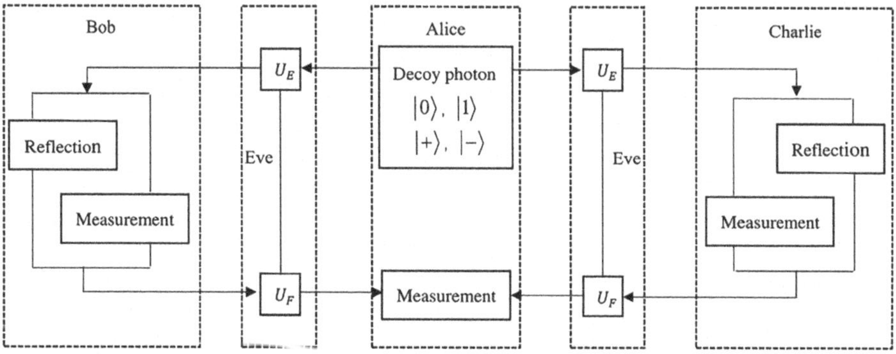

Fig. 5.9 The flow chart of the entangle-measure attack

At this time, if Eve wants to pass the eavesdropping check in Step 2, she should perform the following actions. When the decoy state is $|0\rangle$ or $|1\rangle$, Eve sets $b = c = 0$; in addition, when the decoy state is $|+\rangle$ or $|-\rangle$, she should set $a|\varepsilon_{00}\rangle - b|\varepsilon_{01}\rangle + c|\varepsilon_{10}\rangle - d|\varepsilon_{11}\rangle = a|\varepsilon_{00}\rangle + b|\varepsilon_{01}\rangle - c|\varepsilon_{10}\rangle - d|\varepsilon_{11}\rangle = \vec{0}$, where $\vec{0}$ represents a zero vector. However, when $b = c = 0$, it means $a|\varepsilon_{00}\rangle = d|\varepsilon_{11}\rangle$. It can be seen from this that the price of Eve evading eavesdropping check is not being able to distinguish between $a|\varepsilon_{00}\rangle$ and $d|\varepsilon_{11}\rangle$. Hence, the proposed three-party DSSQD scheme can resist the entangle-measure attack.

5.5.2.5 Efficiency Analysis

The performance of the communication can be expressed in qubit efficiency $\xi = \frac{c}{q+b}$, where c represents the total number of shared classical secret message bits, q represents the total number of generated qubits in the scheme, and b represents the number of exchanged classical bits used for decrypting [87] (the qubits used for eavesdropping check are not considered).

In three-party DSSQD scheme, Alice, Bob, and Charlie can all get mutual secret message, so the classical secret message that is finally successfully transmitted has a total of $6N$ bits, that is, $c = 6N$. The scheme involves N GHZ states as transmission carriers, which includes a total of $3N$ qubits. In addition, Alice also prepares $4N$ decoy states for eavesdropping check. Considering these two situations, we can get the total number of qubits that need to be transmitted, that is, $q = 3N + 4N = 7N$. Finally, Alice needs $3N$, N, and N classical bits, respectively, to announce the corresponding initial GHZ state, the results of $r_A \oplus r_B \oplus r_C$ and $m_A \oplus r_A$. Thus, the total number of classical bits consumed for decrypting is $b = 3N + N + N = 5N$. In summary, the qubit efficiency of the proposed three-party DSSQD scheme can be expressed as $\xi = \frac{c}{q+b} = \frac{6N}{7N+5N} = 50\%$.

5.5.2.6 Summary

Due to the high cost of quantum equipment and the shortage of quantum resources, more and more scholars pay more attention to the field of semi-quantum communication. Based on this, this section proposes a three-party deterministic secure semi-quantum dialogue scheme based on GHZ states. After security analysis, the scheme can successfully overcome the problem of information leakage and resist intercept-measure-resend attack, flip attacks, man-in-the-middle attack, Trojan horse attack, etc. Compared with the traditional quantum dialogue schemes, the proposed three-party DSSQD scheme only requires partial communicants to have quantum capabilities, which saves quantum resources and has good implementation feasibility. So it is a very meaningful and efficient scheme. Since various noises inevitably exist in actual quantum channels, the current research on the influence of noise in the semi-quantum environment is still in its infancy. Therefore, in future research, it is possible to increase noise-related analysis to make the proposed three-party DSSQD scheme more complete and to promote its practical application.

5.6 Brief Summary

In this chapter, based on the basic principle of quantum direct communication, some classical communication schemes are introduced and reviewed. In addition, considering the semi-quantum environment, and corresponding to the proposed three schemes, an efficient and novel semi-quantum deterministic secure quantum communication scheme, controlled deterministic secure semi-quantum communication scheme, and three-party deterministic secure semi-quantum dialogue scheme based on the GHZ states. The anti-attack and transmission efficiency of the three schemes have been analyzed in detail in each section. In a word, all three schemes save quantum resource and communication cost.

References

1. J.-Y. Hu, B. Yu, M.-Y. Jing, L.-T. Xiao, S.-T. Jia, G.-Q. Qin, G.-L. Long, Experimental quantum secure direct communication with single photons. Light Sci. Appl. **5**(9), e16144 (2016)
2. F. Zhu, W. Zhang, Y. Sheng, Y. Huang, Experimental long-distance quantum secure direct communication. Sci.Bull. **62**(22), 1519–1524 (2017)
3. Z.R. Zhou, Y.B. Sheng, P.H. Niu, L.G. Yin, G.L. Long, L. Hanzo, Measurement-device-independent quantum secure direct communication. Sci. China Phys. Mech. Astron. **63**(3), 230362 (2019)
4. P.-H. Niu, Z.-R. Zhou, Z.-S. Lin, Y.-B. Sheng, L.-G. Yin, G.-L. Long, Measurement-device-independent quantum communication without encryption. Sci. Bull. **63**(20), 1345–1350 (2018)
5. R.Y. Qi, Z. Sun, Z. Lin, P. Niu, W. Hao, L.Y. Song, Q. Huang, J. Gao, L.G. Yin, G.L. Long, Implementation and security analysis of practical quantum secure direct communication. Light Sci. Appl. **8**(1), 22 (2019)
6. Z. Sun, R.Y. Qi, Z. Lin, Design and implementation of a practical quantum secure direct communication system, in *2018 IEEE Globecom Workshops (GC Wkshps)*, (IEEE, 2018), pp. 1–6
7. Z. Sun, L.Y. Song, Q. Huang, L.G. Yin, G.-L. Long, J.H. Lu, L. Hanzo, Toward practical quantum secure direct communication: a quantum memory-free scheme and code design. IEEE Trans. Commun. **68**(9), 5778–5792 (2020)
8. H. Zhang, Z. Sun, R. Qi, L. Yin, G.-L. Long, J. Lu, Realization of quantum secure direct communication over 100 km fiber with time-bin and phase quantum states. Light Sci. Appl. **11**(1), 83 (2022)
9. G.L. Long, X.S. Liu, Theoretically efficient high-capacity quantum-key-distribution scheme. Phys. Rev. A **65**(3), 032302 (2002)
10. F.G. Deng, G.L. Long, X.S. Liu, Two-step quantum direct communication scheme using the Einstein-Podolsky-Rosen pair block. Phys. Rev. A **68**(4), 042317 (2003)

11. F.G. Deng, G.L. Long, Secure direct communication with a quantum one-time pad. Physical Review A—Atomic, Molecular, and Optical Physics **69**(5), 052319 (2004)
12. A.D. Zhu et al., Secure direct communication based on secret transmitting order of particles. Phys. Rev. A **73**(2), 457–460 (2006)
13. J. Wang, Q. Zhang, C.J. Tang, Quantum secure direct communication based on order rearrangement of single photons. Phys. Lett. A **358**(4), 256–258 (2006)
14. Z. Rong, D. Qiu, X. Zou, Two single-state semi-quantum secure direct communication schemes based on single photons. Int. J. Mod. Phys. B **34**(11), 2050106 (2020)
15. D. Fu-Guo, L. Xi-Han, L. Chun-Yan, Z. Ping, Z. Hong-Yu, Economical quantum secure direct communication network with single photons. Chin. Phys. **16**(12), 3553 (2007)
16. Z. Rong, D. Qiu, X. Zou, Semi-quantum secure direct communication using entanglement. Int. J. Theor. Phys. **59**, 1807–1819 (2020)
17. C.W. Yang, C.W. Tsai, Advanced semi-quantum secure direct communication protocol based on bell states against flip attack. Quantum Inf. Process **19**, 1–13 (2020)
18. D. Wu, H.J. Lv, G.J. Xie, Robust anti-collective noise quantum secure direct dialogue using logical bell states. Int. J. Theor. Phys. **55**, 457–469 (2016)
19. J.M. Qi, G. Xu, X.B. Chen, T.Y. Wang, X.Q. Cai, Y.X. Yang, Two authenticated quantum dialogue protocols based on three-particle entangled states. Quantum Inf. Process **17**, 1–19 (2018)
20. N.R. Zhou et al., Secure quantum dialogue scheme based on W states without information leakage. Int. J. Theor. Phys. **52**(9), 3204–3211 (2013)
21. C.H. Chang et al., Quantum dialogue schemes over collective noise using entanglement of GHZ state. Quantum Inf. Process **15**(7), 2971–2991 (2016)
22. X.R. Jin et al., Three-party quantum secure direct communication based on GHZ states. Phys. Lett. A **1**(2), 354 (2006)
23. M. Zhong-Xiao, X. Yun-Jie, Improvement of security of three-party quantum secure direct communication based on GHZ states. Chin. Phys. Lett. **24**(1), 15 (2007)
24. W. Mei-Yu, Y. Feng-Li, Three-party simultaneous quantum secure direct communication scheme with EPR pairs. Chin. Phys. Lett. **24**(9), 2486 (2007)
25. L.Y. Wang et al., Information leakage in three-party simultaneous quantum secure direct communication with EPR pairs. Opt. Commun. **284**(7), 1719–1720 (2011)
26. S.K. Chong, T. Hwang, The enhancement of three-party simultaneous quantum secure direct communication scheme with EPR pairs. Opt. Commun. **284**(1), 515–518 (2011)
27. X. Yin et al., Efficient three-party quantum secure direct communication with EPR Pairs. J. Quantum Inf. Sci. **03**(1), 1–5 (2013)
28. L.L. Wang et al., Three-party quantum secure direct communication with single photons in both polarization and spatial-mode degrees of freedom. Int. J. Theor. Phys. **55**(5), 2490–2499 (2016)
29. Y.F. He, W.P. Ma, Three-party quantum secure direct communication against collective noise. Quantum Inf. Process **16**(10), 252 (2017)
30. K. Shimizu, N. Imoto, Communication channels secured from eavesdropping via transmission of photonic Bell states. Phys. Rev. A **60**(1), 157–166 (1999)
31. A. Beige, B.G. Englert, C. Kurtsiefer, H. Weinfurter, Secure communication with single-photon two-qubit states. J. Phys. A **35**(28), L407–L413 (2002)
32. Z.X. Man, Z.J. Zhang, L. Yong, Deterministic secure direct communication by using swapping quantum entanglement and local unitary operations. Chin. Phys. Lett. **22**, 18 (2005)
33. X.H. Li, F.G. Deng, C.Y. Li, Y.J. Liang, P. Zhou, H.Y. Zhou, Deterministic secure quantum communication without maximally entangled states. J. Korean Phys. Soc. **49**, 1354 (2006)
34. G.L. Long, F.G. Deng, C. Wang, X.H. Li, K. Wen, W.Y. Wang, Quantum secure direct communication and deterministic secure quantum communication. Front. Phys. China **2**(3), 251–272 (2007)
35. C. Shukla, A. Pathak, Orthogonal-state-based deterministic secure quantum communication without actual transmission of the message qubits. Quantum Inf. Process **13**, 2099–2113 (2014)

36. D. Joy, S.P. Surendran, M. Sabir, Efficient deterministic secure quantum communication schemes using multipartite entangled states. Quantum Inf. Process **16**, 157 (2017)
37. B.A. Nguyen, Quantum dialogue. Phys. Lett. A **328**(1), 6–10 (2004)
38. X. Ji, S. Zhang, Secure quantum dialogue based on single-photon. Chin. Phys. **15**, 1418–1420 (2006)
39. Y.G. Yang, Q.Y. Wen, Quasi-secure quantum dialogue using single photons. Sci. China Ser. G-Phys. Mech. Astron. **50**(5), 558–562 (2007)
40. Y. Xia, C.B. Fu, S. Zhang, et al., Quantum dialogue by using the GHZ state. J. Korean Phys. Soc. **48**(1), 24–27 (2006)
41. L. Dong, X.M. Xiu, Y.J. Gao, et al., Quantum dialogue scheme using a class of three-photon W states. Commun. Theor. Phys. **52**(5), 853–856 (2009)
42. W. Li, X. Zha, Y. Yu, Secure quantum dialogue scheme based on four-qubit cluster state. Int. J. Theor. Phys. **57**, 371–380 (2018)
43. F. Gao, F. Guo, Q. Wen, et al., Revisiting the security of quantum dialogue and bidirectional quantum secure direct communication. Sci. China Ser. G-Phys. Mech. As **51**, 559–566 (2008)
44. G.F. Shi, X.Q. Xi, M.L. Hu, et al., Quantum secure dialogue by using single photons. Opt. Commun. **283**(9), 1984–1986 (2010)
45. Y. Luo, C. Lin, T. Hwang, Efficient quantum dialogue using single photons. Quantum Inf. Process **13**, 2451–2461 (2014)
46. N. Zhou, T. Hua, G. Wu, et al., Single-photon secure quantum dialogue scheme without information leakage. Int. J. Theor. Phys. **53**, 3829–3837 (2014)
47. C. Tsai, C. Hsieh, T. Hwang, Dense coding using cluster states and its application on deterministic secure quantum communication. Eur. Phys. J. D **61**, 779–783 (2011)
48. D. Liu, J.L. Chen, W. Jiang, High-capacity quantum secure direct communication with single photons in both polarization and spatial-mode degrees of freedom. Int. J. Theor. Phys. **51**(9), 2923–2929 (2012)
49. L.L. Wang, W.P. Ma, D.S. Shen, et al., Efficient bidirectional quantum secure direct communication with single photons in both polarization and spatial-mode degrees of freedom. Int. J. Theor. Phys. **54**, 3443–3453 (2015)
50. X. Zhang, R.G. Zhou, An efficient and novel semi-quantum deterministic secure quantum communication scheme. Int. J. Theor. Phys. **61**, 94 (2022)
51. F.G. Deng, G.L. Long, Secure direct communication with a quantum one-time pad. Phys. Rev. A **69**, 521–524 (2004)
52. M. Boyer, D. Kenigsberg, T. Mor, Quantum key distribution with classical Bob. Phys. Rev. Lett. **99**, 140501 (2007)
53. M. Boyer et al., Semi-quantum key distribution, in *Quantum Information*, (Springer, 2007), pp. 235–274
54. X.H. Li, F.G. Deng, H.Y. Zhou, Improving the security of secure direct communication based on the secret transmitting order of particles. Phys. Rev. A **74**, 054302 (2006)
55. Z. Liu, H. Chen, D. Wang, et al., Cryptanalysis and improvement of three-particle deterministic secure and high bit-rate direct quantum communication scheme. Quantum Inf. Process **13**(6), 1345–1351 (2014)
56. Y.S. Zhang, C.F. Li, G.C. Guo, Comment on "Quantum key distribution without alternative measurements". Phys. Rev. A **61**, 052312 (2000) Phys. Rev. A 6303(3), 036301 (2001)
57. F. Gao, Q.Y. Wen, F.C. Zhu, Comment on: "Quantum exam". Phys. Lett. A **350**, 174 (2006) Phys. Lett. A 360(6), 748–750 (2007)
58. S.J. Qin, F. Gao, F.Z. Guo, Q.Y. Wen, Comment on "Two-way schemes for quantum cryptography with a nonmaximally entangled qubit pair". Phys. Rev. A **82**(3), 036301 (2010)
59. Z. Liu, H. Chen, W. Liu, et al., Analyzing and revising a two-way scheme for quantum cryptography with a nonmaximally entangled qubit pair. Int. J. Quantum. Inf. **9**(5), 1329–1339 (2011)

60. F.G. Deng, X.H. Li, H.Y. Zhou, Z.J. Zhang, Erratum: Improving the security of multiparty quantum secret sharing against Trojan horse attack. Phys. Rev. A **72**, 044302 (2005) Phys. Rev. A 73(4), 049901 (2006)
61. Q.Y. Cai, Eavesdropping on the two-way quantum communication schemes with invisible photons. Phys. Lett. A **351**(1–2), 23–25 (2006)
62. T. Ye, H. Li, J. Hu, Semi-quantum key distribution with single photons in both polarization and spatial-mode degrees of freedom. Int. J. Theor. Phys. **59**, 2807–2815 (2020)
63. Q.Y. Cai, The ping-pong scheme can be attacked without eavesdropping. Phys. Rev. Lett. **91**, 109801 (2003)
64. X.F. Zou, D.W. Qiu, Three-step semi-quantum secure direct communication scheme. Sci. China Phys. Mech. Astron. **57**(9), 1696–1702 (2014)
65. M.-H. Zhang, H.-F. Li, Z.-Q. Xia, X.-Y. Feng, J.-Y. Peng, Semi-quantum secure direct communication using EPR pairs. Quantum Inf. Process **16**(5), 117 (2017)
66. Y. Sun, L. Yan, Y. Chang, S. Zhang, T. Shao, Y. Zhang, Two semi-quantum secure direct communication schemes based on Bell states. Mod. Phys. Lett. A **34**(1), 10 (2019)
67. C. Yang, C. Tsai, T. Hwang, Fault tolerant deterministic quantum communications using GHZ states over collective-noise channels. Quantum Inf. Process **12**(9), 3043–3055 (2013)
68. R.G. Zhou, X. Zhang, Controlled deterministic secure semi-quantum communication. Int. J. Theor. Phys. **60**, 1767–1782 (2021)
69. D.A. Huffman, A method for the construction of minimum-redundancy codes. Reson **11**, 91–99 (2006)
70. D. Sridhara, Efficient coding of information: Huffman coding. Reson **11**, 51–73 (2006)
71. A. Habib, M.S. Rahman, Balancing decoding speed and memory usage for Huffman codes using quaternary tree. Appl. Inform. **4**, 5 (2017)
72. E. Biham, M. Boyer, G. Brassard, J. Van de Graaf, T. Mor, Security of quantum key distribution against all collective attacks. Algorithmica **34**(4), 372–388 (2002)
73. J. Wang, Q. Zhang, C.J. Tang, Multiparty controlled quantum secure direct communication using Greenberger-Horne-Zeilinger state. Opt. Commun. **266**(2), 732–737 (2006)
74. L.L. Yan, Y.H. Sun, Y. Chang, S.B. Zhang, G.G. Wan, Z.W. Sheng, Semi-quantum scheme for deterministic secure quantum communication using Bell states. Quantum Inf. Process **17**(11), 315 (2018)
75. K.J. Boström, T. Felbinger, Deterministic secure direct communication using entanglement. Phys. Rev. Lett. **89**(18), 187902 (2002)
76. M. Lucamarini, S. Mancini, Secure deterministic communication without entanglement. Phys. Rev. Lett. **94**(14), 140501 (2005)
77. R.G. Zhou, X. Zhang, F. Li, Three-party semi-quantum scheme for deterministic secure quantum dialogue based on GHZ states. Quantum Inf. Process **20**, 153 (2021)
78. T.Y. Ye, C.Q. Ye, Semi-quantum dialogue based on single photons. Int. J. Theor. Phys. **57**(5), 1440–1454 (2018)
79. H.-M. Pan, Semi-quantum dialogue with bell entangled states. Int. J. Theor. Phys. **59**, 1–8 (2020)
80. J. Gu, P.-H. Lin, T. Hwang, Double C-NOT attack and counterattack on 'Three-step semi-quantum secure direct communication scheme'. Quantum Inf. Process **17**(7), 1–8 (2018)
81. K.-F. Yu, J. Gu, T. Hwang, P. Gope, Multi-party semi-quantum key distribution-convertible multiparty semi-quantum secret sharing. Quantum Inf. Process **16**(8), 194 (2017)
82. D. Pan et al., Single-photon-memory two-step quantum secure direct communication relying on Einstein-Podolsky-Rosen pairs. IEEE Access **8**, 121146–121161 (2020)
83. J. Wu et al., Security of quantum secure direct communication based on Wyner's wiretap channel theory. Quantum Eng. **1**(4), e26 (2019)
84. P.-H. Niu et al., Security analysis of measurement-device-independent quantum secure direct communication. Quantum Inf. Process **19**(10), 1–14 (2020)

85. D. Pan, Z. Lin, J. Wu, H. Zhang, Z. Sun, D. Ruan, et al., Experimental free-space quantum secure direct communication and its security analysis. Photonics Res. **8**(9), 1522–1531 (2020)
86. C.E. Shannon, Communication theory of secrecy system. Bell Syst. Tech. J. **28**, 656–715 (1949)
87. A. Cabello, Quantum key distribution in the Holevo limit. Phys. Rev. Lett. **85**, 5635 (2000)

Chapter 6
Quantum Key Agreement

Abstract The security of classical key agreement schemes must be based on the assumption of computational complexity, which is challenged by the enhancement of computing power or algorithm. Some mathematical problems, in particular, are vulnerable to quantum computers. Quantum key agreement schemes can obtain unconditional security through quantum mechanics, in theory. Through the quantum key agreement scheme, two or more parties can establish a classical shared key equally and fairly based on quantum mechanics for secure communication.

6.1 Introduction

As part of quantum cryptography, the quantum key agreement (QKA) scheme not only ensures the quantum cryptography security of the key shared by the participants but also ensures fairness in the key agreement process, that is, all participants contribute the same to the final shared key, and the key cannot be independently controlled by a single individual or part of the participants. Taking advantage of the characteristics of this scheme key, participants can establish a key in a public but insecure channel without prior sharing of the key. Therefore, key agreement is widely used in open networks, and it is one of the important techniques to achieve secure communication. It can meet the needs of website user authentication and temporary user secure session. QKA is an important part of quantum cryptography, which plays a similar role in application to traditional key agreement and plays an important role in their respective fields. The QKA scheme is a secure scheme based on quantum channel, which requires participants to establish a secure shared key through negotiation, and does not allow any party to control the key alone. In addition, such schemes must be secure and efficient to ensure that participants can establish a shared key between security and efficiency. The security of the scheme requires it to have enough defense ability, resist the threat from the participants and external attackers, and ensure the security and reliability of the scheme. High efficiency means that the scheme design should have good performance under

R.-G. Zhou et al., *Design and Analysis of Secure Quantum Communication Schemes*,
https://doi.org/10.1007/978-3-031-78428-6_6

other conditions. Unlike quantum cryptography, the session key for QKA is generated by all participants through negotiation. Therefore, the design of such schemes is very important in both theoretical and practical sense.

6.2 Principle and Research Progress of Quantum Key Agreement

In an open, dynamic network, to achieve secure end-to-end communication, a shared key must be established between two nodes to meet basic security requirements, such as encryption and authentication. In a classical environment, there are two main ways for two or more parties to get the shared key: key distribution and key agreement. In key distribution, a key distributor or key distribution center generates a shared key in advance and sends it to all other participants, who trust the key distributor or key distribution center. In key agreement, all participants participate in the generation and distribution of the shared key, and the contribution to the negotiated shared key is equal. No participant in the scheme can control the shared key, and no trivial subset of all participants can independently determine the shared key. QKA and QKD solve the problem of how to generate a quantum cryptography key between two parties to a communication, the difference is between "distribution" and "agreement." The QKA emphasizes that the key is generated by one party and distributed to the other. For example, in BB84, the resulting key must be contained in Alice's bit sequence and does not have the right to determine the contents of the key. QKA means that the final key is negotiated by two or more parties, and it emphasizes that the participants of the key have equal contribution to the final key generation.

QKA scheme satisfies the following four characteristics: correctness, fairness, security, and privacy. Correctness means that the result of the operation of the scheme is that the participants of the key agreement obtain the same key, that is, the logic of the scheme is correct. Fairness is the characteristic of the QKA scheme, which shows that every negotiator has the same contribution to the generation of the final key, and neither party can decide any part of the final key. Security and confidentiality define the security of the scheme from external and internal aspects. External security means that any eavesdropper cannot get the final key, while internal security means that any participant only knows his or her contribution to the final key, and the contribution of other participants is unknown to him or her.

In recent years, the QKA scheme has attracted a lot of researchers' interest. In 2004, Zhou et al. [1] first proposed a QKA scheme based on quantum teleportation. In the same year, Hsueh and Chen et al. [2] developed a QKA scheme with maximum entanglement to meet the security requirements at that time. However, in 2009, Tsai and Hwang's research results showed that the QKA scheme designed by Zhou et al. in 2004 cannot meet the security requirements, because it cannot effectively resist participant attacks. This allows certain participants to control the generation of shared keys without detection through malicious operations. The next year, Tsai et al. found that the QKA scheme established by Hsueh and Chen could not resist participant and external attacks effectively, and did not achieve the expected effect. In 2010, Chong et al. [3] successfully developed a two-party QKA scheme using unitary transformation and delay measurement technology, which is the first successful QKA scheme in history, but its quantum bit efficiency is not satisfactory. In 2011, Chong et al. [4] established a two-party QKA scheme using Bell State, which greatly improved the utilization efficiency of quantum bits and the performance of QKA scheme. However, researchers have found that although the QKA scheme meets the requirements of fairness, security, and efficiency, it can only achieve security in theory. Because the influence of noise cannot be ignored in quantum state transmission, in order to resist the influence of noise effectively, the QKA scheme should be implemented by immune noise. Therefore, the immune noise QKA scheme [5–14] emerged. In 2014, Huang et al. [5] first realized a two-party QKA scheme on noisy channel by using EPR pair and single-photon measurement technique. Subsequently, Huang et al. [6] proposed a QKA scheme that can resist collective noise by using logical quantum states and single photon measurement technology. However, the traditional method of designing a QKA scheme through logical quantum states can greatly reduce the quantum bit efficiency of the scheme. With the development of technology, until 2017, He et al. [8] designed two QKA schemes with excellent performance by utilizing five-particle logic quantum states.

However, these schemes apply only to cases where both parties are involved. With the rapid development of science and technology, more and more people need a multiparty QKA scheme. Therefore, in 2013, Shi et al. [15] proposed a multiparty QKA scheme based on Bell measurement and Bell state entanglement swapping to achieve a more effective key agreement. However, in the same year, Liu et al. [16] proved that the scheme designed by Shi et al. cannot resist the attack of participants effectively, that is, the legitimate participants can control the generation of shared keys through malicious operations. Liu et al. further designed a secure multiparty QKA scheme using single-particle technology, which can effectively resist the threat of participants and external attackers. Then Sun et al. [17] developed a ring QKA scheme, which can improve the performance of the scheme and protect data security better.

In the abovementioned QKA scheme, the use of quantum states is mostly single-particle or Bell state. Subsequently, researchers began to explore the QKA scheme based on multiparticle entanglement and achieved fruitful results [14, 18–22]. In 2018, Min et al. [19] proposed a multiparty QKA scheme based on G-like and Bell states. In the same year, Cai et al. [20] constructed another multiparty QKA scheme using the five-particle Brown state. Subsequently, Yang et al. [21] proposed a novel multiparty QKA scheme in 2019, which is implemented with a four-particle cluster state and a ring structure.

At present, a large number of excellent QKA schemes have emerged. However, these schemes require the participants to have full quantum capabilities, which means that the participants have to bear excessive communication costs. To solve this problem, Boyer et al. first proposed the semi-quantum concept in 2007. Then researchers apply the semi-quantum idea to other fields of quantum cryptography and propose a variety of quantum communication schemes under the semi-quantum system.

In the field of QKA, researchers have also introduced the concept of semi-quantum to simplify the quantum capability of participants and achieved some outstanding results. In 2017, Shukla et al. [23] used Bell states to develop a two-party semi-quantum key agreement (SQKA) scheme with a quantum bit efficiency of 9.09%. In the same year, Liu et al. [24] also proposed a multiparty SQKA scheme by using the entanglement of Bell states to meet the needs of different applications. In 2019, Yan et al. [25] first developed an SQKA scheme using Bell states and then used Bell states to optimize the scheme for multiple parties. The following year, Li et al. [26] designed a new SQKA scheme using high-dimensional single-particle states, which is the first time that such a scheme is implemented by high-dimensional particle states. In the same year, Zhou et al. [27] designed a three-party SQKA scheme that was realized by using four-particle cluster states and extended the original scheme to multiparty [28] to meet different security requirements. In 2021, Guo et al. [29] proposed a novel three-party SQKA scheme based on the three-particle maximally entangled state GHZ; therefore, an SQKA scheme is realized that is more suitable for reality. In 2022, Liu et al. [30] designed a two-party SQKA scheme by using the chi-square entangled state to provide the security required by the two parties in the key agreement. In the same year, Zhou et al. [31] developed a two-party SQKA scheme that can resist untrusted third parties by using G-like states and entrustment quantum computing technology.

Since its birth, SQKA has developed rapidly and become an important part of quantum cryptography. At present, the SQKA scheme designed by scholars cannot only resist the attack of participants and external attackers but also ensure that each participant contributes the same shared key. However, some existing SQKA schemes cannot meet the actual requirements in terms of quantum bit efficiency, and rely too much on trusted third parties for key agreement. Therefore, it is necessary to further study the SQKA scheme.

6.3 Typical Quantum Key Agreement Scheme

6.3.1 MQKA Scheme

The Liu2013 MQKA scheme [16] is a fully graphical multiparty QKA scheme based on a single photon. The core of the scheme is that each participant encodes its key as a quantum bit to transmit to all other participants and monitors the eavesdropper by randomly inserting decoy states. Suppose there are N participants who need to negotiate a key together. Details of the Liu2013 scheme are as follows:

Step 1: Each node P_i generates its key string K_i randomly and encodes K_i into $N - 1$ identical qubit sequences S_i (0 encodes as $|0\rangle$, 1 encodes as $|1\rangle$).

Step 2: Each node randomly generates a decoy state for each string, and each decoy state is randomly selected between $\{|+\rangle, |-\rangle, |+y\rangle, |-y\rangle\}$. The decoy states are randomly inserted into each string of S_i to obtain $S_{i\to 0}, S_{i\to 1}, \cdots, S_{i\to N-1}$.

Step 3: Each node sends the generated sequence of qubits to all other nodes.

Step 4: After each node receives the sequence sent by all other nodes and stores it, it announces the position and base of the decoy state and measures the decoy state of the corresponding position with the corresponding base according to the information published by the other nodes. Half of the measurements are randomly published, the other half are checked for accuracy, and the other half are published for initial state of decoy. Each node compares its measurement results with the initial state published by the remaining nodes. Each node simultaneously judges the security of the transmission process by the accuracy of both sides. If the bit error rate is too high, then the transmission process has the possibility of eavesdropping, terminating the scheme.

Step 5: If the transmission process is secure, each node measures the quantum states at the remaining positions on the $\{|0\rangle, |1\rangle\}$ basis and obtains the key of the remaining nodes.

Step 6: Each node does XOR operations on its own key and the keys of all other nodes it obtains. The result is the final key, $K_f = K_0 \oplus K_1 \oplus \cdots \oplus K_{N-1}$.

The security of the Liu2013 scheme is mainly guaranteed by the decoy state. The so-called decoy state is that the eavesdropper changes the initial state of quantum information by intercepting and measuring the quantum information and then finds the eavesdropper by comparing the measurement results. Since the position of the decoy state is random, eavesdropper Eve does not know the position of the decoy state and the signal state, so it can only randomly select the signal eavesdropping in some positions. When the decoy state is measured, the uncertainty principle determines that it cannot be measured accurately on both the $\{|+\rangle, |-\rangle\}$ and $\{|+y\rangle, |-y\rangle\}$ bases and that only one base can be chosen for the measurement. When the selected base is different from the prepared base, the measurement result is random, which interferes with the original state of the decoy state, which is monitored in Step 4. The Liu2013 scheme is therefore externally secure. Liu also points out that the scheme can withstand attacks from internal players.

6.3.2 MQKA Scheme Based on GHZ-Like States and χ^+ States

In order to enrich the types of multiparty QKA schemes, this section proposes a novel multiparty quantum key agreement (MQKA) scheme based on entanglement swapping of GHZ-like states and χ^+ states [32]. The MQKA scheme designs encoding rules based on Bell state and $\{|\chi^\pm\rangle, |\omega^\pm\rangle\}$ state, in which each state corresponds to three classical bits. Compared with the QKA scheme proposed by Min [19], the proposed two-party quantum key agreement (TQKA) scheme improves the quantum efficiency by about *30%*. In addition, in the multiparty quantum key agreement scheme, as the number of participants grows, the efficiency will gradually increase to nearly *50%*.

6.3.2.1 Entanglement Swapping

The Bell state is the maximum entangled state of two qubits. The four Bell states are expressed as

$$|\phi^{\pm}\rangle = \frac{1}{\sqrt{2}}(|00\rangle \pm |11\rangle) \tag{6.1}$$

$$|\psi^{\pm}\rangle = \frac{1}{\sqrt{2}}(|01\rangle \pm |10\rangle) \tag{6.2}$$

For convenience, we define the $Y = \{|\chi^{\pm}\rangle, |\omega^{\pm}\rangle\}$ as Y basis, and the ground state $|\chi^{\pm}\rangle$ and $|\omega^{\pm}\rangle$ can be obtained by performing the Hadamard gate operation on the second qubit of the Bell state [33] as follows:

$$\begin{aligned}
|\chi^{+}\rangle &= \frac{|\phi^{-}\rangle + |\psi^{+}\rangle}{\sqrt{2}} = \frac{|00\rangle - |11\rangle + |01\rangle + |10\rangle}{2} \\
|\chi^{-}\rangle &= \frac{|\phi^{+}\rangle + |\psi^{-}\rangle}{\sqrt{2}} = \frac{|00\rangle + |11\rangle + |01\rangle - |10\rangle}{2} \\
|\omega^{+}\rangle &= \frac{|\phi^{+}\rangle - |\psi^{-}\rangle}{\sqrt{2}} = \frac{|00\rangle + |11\rangle - |01\rangle + |10\rangle}{2} \\
|\omega^{-}\rangle &= \frac{|\phi^{-}\rangle - |\psi^{+}\rangle}{\sqrt{2}} = \frac{|00\rangle - |11\rangle - |01\rangle - |10\rangle}{2}
\end{aligned} \tag{6.3}$$

6.3.2.2 GHZ-Like State and χ + State

The GHZ-like state is composed of a single photon state and an EPR state, and it can also be expressed as the tensor product of $|\varphi\rangle_1$ and $|\psi^{+}\rangle_{23}$ states. The detailed process is as follows:

Table 6.1 Encoding rules

State	ϕ^+	ϕ^-	χ^+	χ^-	ψ^+	ψ^-	ω^+	ω^-
Binary code	000	001	010	011	100	101	110	111

$$
\begin{aligned}
|G_0\rangle_{123} = |\varphi\rangle_1 \otimes |\psi^+\rangle_{23} &= \frac{1}{\sqrt{2}}(|0\rangle_1 + |1\rangle_1) \otimes \frac{1}{\sqrt{2}}(|01\rangle + |10\rangle)_{23} \\
&= \frac{1}{2}(|001\rangle + |010\rangle + |101\rangle + |110\rangle)_{123}
\end{aligned}
\tag{6.4}
$$

Next, use the first particle as the control qubit and the third particle as the target qubit to perform the CNOT gate operation, and can get

$$
\begin{aligned}
|G_1\rangle_{123} &= \frac{1}{\sqrt{2}}(|001\rangle + |010\rangle + |100\rangle + |111\rangle)_{123} \\
&= \frac{1}{\sqrt{2}}(|0\rangle_1 |\psi^+\rangle_{23} + |1\rangle_1 |\phi^+\rangle_{23})
\end{aligned}
\tag{6.5}
$$

At this time, the entanglement swapping principle of GHZ-like state and χ^+ state can be expressed as follows [34]:

$$
\begin{aligned}
&|G_1\rangle_{123} \otimes |\chi^+\rangle_{45} \\
&= \frac{1}{\sqrt{2}}(|0\rangle_1 |\psi^+\rangle_{23} + |1\rangle_1 |\phi^+\rangle_{23}) \otimes |\chi^+\rangle_{45} \\
&= \frac{1}{4}|0\rangle_1 (|01\rangle_{23} + |10\rangle_{23})(|00\rangle_{45} - |11\rangle_{45} + |01\rangle_{45} + |10\rangle_{45}) \\
&\quad + \frac{1}{4}|1\rangle_1 (|00\rangle_{23} + |11\rangle_{23})(|00\rangle_{45} - |11\rangle_{45} + |01\rangle_{45} + |10\rangle_{45}) \\
&= \frac{1}{4}|0\rangle_1 (|0010\rangle - |0111\rangle + |0011\rangle + |0110\rangle + |1000\rangle - |1101\rangle + |1001\rangle + |1100\rangle)_{2435} \\
&\quad + \frac{1}{4}|1\rangle_1 (|0000\rangle - |0101\rangle + |0001\rangle + |0100\rangle + |1010\rangle - |1111\rangle + |1011\rangle + |1110\rangle)_{2435} \\
&= \frac{1}{2\sqrt{2}}(|0\rangle_1 \otimes (|\phi^+\rangle_{24}|\omega^+\rangle_{35} - |\phi^-\rangle_{24}|\omega^-\rangle_{35} + |\psi^+\rangle_{24}|\chi^+\rangle_{35} - |\psi^-\rangle_{24}|\chi^-\rangle_{35}) \\
&\quad + |1\rangle_1 \otimes (|\phi^+\rangle_{24}|\chi^+\rangle_{35} + |\phi^-\rangle_{24}|\chi^-\rangle_{35} + |\psi^+\rangle_{24}|\omega^+\rangle_{35} - |\psi^-\rangle_{24}|\omega^-\rangle_{35})
\end{aligned}
\tag{6.6}
$$

Obviously, according to Table 6.1, the relationship between the possible measurement results and the initial states will be as follows:

$$
R_1 \oplus R_{23} \oplus R_{45} = M_1 \oplus M_{24} \oplus M_{35},
\tag{6.7}
$$

where R_{ij} is the binary code corresponding to the initial state of two particles i and j and M_{st} is the binary code corresponding to the measurement results of two particles s and t. Otherwise, $R_1(M_1)$ represents the first particle; if $R_1(M_1)$ is $|0\rangle$ state, it represents binary information *0*; if $R_1(M_1)$ is $|1\rangle$ state, it represents binary information *1*. For example, if the measurement results of particles 1 and 2,4 and 3,5 arel 0⟩, $|\phi^-\rangle$ and $|\omega^-\rangle$, then there is

$$R_1 \oplus R_{23} \oplus R_{45} = 0 \oplus 100 \oplus 010 = 110 \tag{6.8}$$

$$M_1 \oplus M_{24} \oplus M_{35} = 0 \oplus 001 \oplus 111 = 110 \tag{6.9}$$

6.3.2.3 Two-Party QKA-Scheme

Suppose that U_{00}, U_{01}, U_{10}, U_{11}indicate the four local unitary operations *I, X, Z, iY* respectively, as follows:

$$\begin{aligned} U_{00} &\equiv I = \mid 0\rangle\langle 0| + |1\rangle\langle 1 \mid \\ U_{01} &\equiv X = \mid 0\rangle\langle 1| + |1\rangle\langle 0 \mid \\ U_{10} &\equiv Z = \mid 0\rangle\langle 0| - |1\rangle\langle 1 \mid \\ U_{11} &\equiv iY = \mid 0\rangle\langle 1| - |1\rangle\langle 0 \mid \end{aligned} \tag{6.10}$$

In addition, $|0\rangle$ and $|1\rangle$ form Z basis, $|+\rangle$ and $|-\rangle$ form X basis, while $|\chi^{\pm}\rangle$ and $|\omega^{\pm}\rangle$ form Y basis, where $|\pm\rangle = \frac{1}{\sqrt{2}}(|0\rangle \pm |1\rangle)$. Table 6.2 shows the conversion relationship between Bell state, Y state, and the four unitary operations.

Suppose that the communicants Alice and Bob randomly generate *3n* bit strings K_A and K_B as their respective security keys, and the final shared secret key is expressed as $K = K_A \oplus K_B$, where the details of K_A and K_B are shown as follows ($k_A^t, k_B^t \in \{0, 1\}$):

$$\begin{aligned} K_A &= \left(k_A^1, k_A^2, \ldots, k_A^t, \ldots, k_A^{3n}\right) \\ K_B &= \left(k_B^1, k_B^2, \ldots, k_B^t, \ldots, k_B^{3n}\right) \end{aligned} \tag{6.10}$$

Step 1: Preparation of Quantum States

Alice prepares *n* GHZ-like states and divides them into three ordered sequences a_1, a_2, and a_3. Bob prepares $n\ \chi^+$ states and divides them into two ordered sequences b_1 and b_2 (see Fig. 6.1a).

Table 6.2 The relationship between Bell state, Y state, and four unitary operations

	$I \otimes U_{00}$	$I \otimes U_{01}$	$I \otimes U_{10}$	$I \otimes U_{11}$
$\lvert\phi^{\pm}\rangle$	$\lvert\phi^{\pm}\rangle$	$\lvert\psi^{\pm}\rangle$	$\lvert\phi^{\mp}\rangle$	$\lvert\psi^{\mp}\rangle$
$\lvert\psi^{\pm}\rangle$	$\lvert\psi^{\pm}\rangle$	$\lvert\phi^{\pm}\rangle$	$\lvert\psi^{\mp}\rangle$	$\lvert\phi^{\mp}\rangle$
$\lvert\chi^{\pm}\rangle$	$\lvert\chi^{\pm}\rangle$	$\lvert\chi^{\mp}\rangle$	$\lvert\omega^{\pm}\rangle$	$\lvert\omega^{\mp}\rangle$
$\lvert\omega^{\pm}\rangle$	$\lvert\omega^{\pm}\rangle$	$\lvert\omega^{\mp}\rangle$	$\lvert\chi^{\pm}\rangle$	$\lvert\chi^{\mp}\rangle$

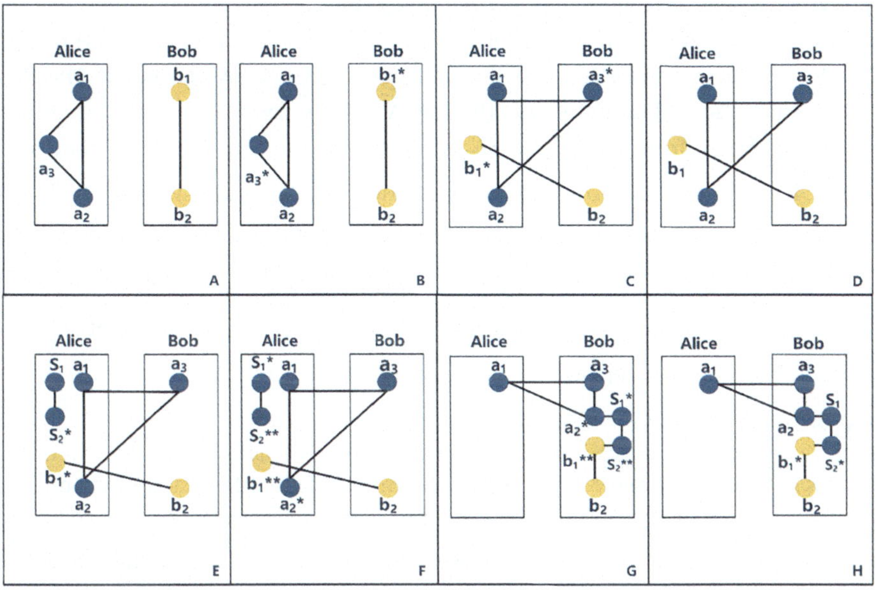

Fig. 6.1 The transmission process of photon in TQKA scheme

Step 2: Preparation of Decoy States
Alice (Bob) randomly selects a group of photons from $|0\rangle$, $|1\rangle$, $|+\rangle$, $|-\rangle$ as decoy states and inserts them into $a_3(b_1)$ randomly to obtain a new sequence $a_3^*(b_1^*)$ (see Fig. 6.1b). Subsequently, Alice (Bob) sends the mixed sequences $a_3^*(b_1^*)$ to Bob (Alice) (see Fig. 6.1c).

Step 3: Eavesdropping Check
After confirming the receipt of all the sequences, Alice (Bob) announces the position of the decoy states and the corresponding measurement base. Then Bob (Alice) measures the decoy states with the correct measurement base and sends the measurement results to Alice (Bob). Finally, Alice and Bob check the security of the quantum channel by comparing the measurement results. If the error rate exceeds the preset value, discard this round of communication; otherwise, continue to the next step.

Step 4: Bell Basis Measurement and Y Basis Measurement
Alice (Bob) removes the decoy states from the received sequence $b_1^*(a_3^*)$ to obtain the sequence $b_1(a_3)$ (see Fig. 6.1d). After that, Alice performs Z basis measurement on sequence a_1 and Bell basis measurement on sequences a_2 and b_1. Meanwhile, Bob performs Y basis measurement on sequences a_3 and b_2 and records the results as M_B, according to Table 6.1. Subsequently, Alice publishes the measurement results of a_1. Thus, Alice and Bob can obtain each other's measurement results based on the correlation of the entanglement exchange (According to Eq. 6.6).

Step 5: Transmission of Encrypted Quantum States
Alice again randomly prepares $\frac{n}{2}$ Bell states from $\{|\phi^{\pm}\rangle, |\psi^{\pm}\rangle\}$, the first and second particles of which form the sequences S_1 and S_2, respectively. At this point, Alice has $\frac{3n}{2}$ Bell states. According to the security key K_A, Alice performs a series of unitary operations chosen from $\{U_{00}, U_{01}, U_{10}, U_{11}\}$ on the j-th ($j = 1,2,..., \frac{3n}{2}$) particle in $b_1\|S_2$ ("||" represents the concatenation operation) to obtain a new sequence $b_1^*\|S_2^*$ (see Fig. 6.1e). Next, Alice randomly selects a set of decoy states and inserts them into $a_2\|S_1$ and $b_1^*\|S_2^*$ to obtain $a_2^*\|S_1^*$ and $b_1^{**}\|S_2^{**}$(see Fig. 6.1f). Finally, Alice sends the new mixed sequences $a_2^*\|S_1^*$ and $b_1^{**}\|S_2^{**}$ to Bob (see Fig. 6.1g). At this point, Alice and Bob perform an eavesdropping check as Step 3.

Step 6: Decrypt and Generate Shared Key
After removing the decoy state (see Fig. 6.1h), Bob tells Alice the value of $K'_B = K_B \oplus M_B$. At the same time, Alice tells Bob the initial Bell states of sequence $\{a_2\|S_1, b_1\|S_2\}$. Then Bob performs the Bell measurement on sequence $\{a_2\|S_1, b_1^*\|S_2^*\}$. According to the measurement results and the initial Bell state information told by Alice, Bob can calculate a series of unitary operations performed by Alice and obtain the key K_A according to these unitary operations. Alice and Bob can obtain mutual keys. From this, the shared key can be calculated: $K = K_A \oplus K_B$.

6.3.2.4 Multiparty QKA Scheme

The circular MQKA scheme is the most popular scheme at present. In order to improve the efficiency, this section improves the TQKA scheme proposed in Sect. 6.3.2.3 and constructs a new loop MQKA with each of the two participants as a group. The specific description is as follows:

Step 1: Preparation Stage
Suppose N communicants are randomly divided into M ($M = \frac{N}{2}$) groups, denoted as A_1, B_1; A_2, B_2;...; A_M, B_M, where each group of A_i and B_i ($i = 1, 2,..., M$) implements a two-party QKA scheme. A_i prepares n GHZ-like states and $n/2$ Bell states, while B_i prepares n χ^+ states. It is worth mentioning that B_i can obtain the secret subkey K_i from its own group that has executed the TQKA scheme. After that, B_i performs the Hadamard-gate operation on the second particle of its $n/2$ Bell states. Therefore, each B_i has $3n/2$ Y states. Finally, B_i performs unitary operation to restore all Y states to χ^+ states, according to Table 6.2.

Step 2: Construction of Circle-Type MQKA Scheme
All participants B_i randomly combine to become P_i ($P_i \in \{B_1, B_2, \ldots, B_M\}$), where each participant does not know who the adjacent participant is, as shown in Fig. 6.2.

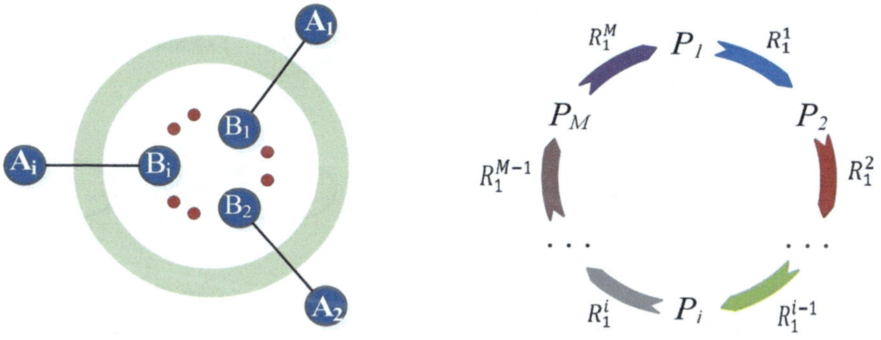

Fig. 6.2 Circle-type MQKA scheme

Step 3: Preparation of Decoy States
P_i forms the first and second particles in the $3n/2\ \chi^+$ states into two sequences U_i and R_i, respectively. In addition, P_i prepares a set of decoy states and randomly inserts them into R_i to obtain R_i^i, and then sends R_i^i to P_{i+1}.

Step 4: Eavesdropping Check
After P_{i+1} confirms that all sequences have been received, P_i announces the position of the decoy states and the corresponding measurement basis. Then P_{i+1} measures the decoy states with the correct measurement basis and sends the measurement results to P_i. P_i judges whether there is eavesdropping in the channel by comparing the measurement results. If the error rate exceeds the preset value, discard this round of communication; otherwise, continue to the next step.

Step 5: Encryption Operation
After removing the decoy particles, According to the security key K_{i+1}, P_{i+1} performs a series of unitary operations chosen from $\{U_{00}, U_{01}, U_{10}, U_{11}\}$ on the j-th ($j = 1, 2, \ldots, \frac{3n}{2}$) particle in R_i^i. For example, if K_{i+1} =00 (01) or 10 (11), then P_{i+1} will perform the unitary operation $U_{00}(U_{01})$ or $U_{10}(U_{11})$ on the particle sequence R_i^i. Then P_{i+1} randomly selects a group of decoy states and randomly inserts them into the sequence R_i^i to obtain R_i^{i+1}. Finally, P_{i+1} sends the mixed sequence R_i^{i+1} to P_{i+2}. At this time, P_{i+1} and P_{i+2} perform eavesdropping check as Step 4.

Step 6: Decrypt and Generate Shared Key
The participants P_{i+2}, P_{i+3},…,P_{i-1} perform Step 5 in the same way as the participant P_{i+1}. As shown in Fig. 6.3. Finally, When P_i receives sequence R_i^{i-1} and removes the decoy states, then P_i measures the sequence $\{U_i, R_i^{i-1}\}$ with Y basis, through Tables 6.3 and 6.4, P_i can get the key $K_i' = K_{i+1} \oplus K_{i+2} \oplus \ldots \oplus K_{i-1}$, so as to calculate the final shared key $K = K_i \oplus K_i'$. Finally, P_i encrypts K with the original group secret key and sends it to the corresponding group member. It is worth noting that each participant P_i ($i = 1, 2, \ldots, M$) must send its own particle sequence to the next recipient P_{i+1} at the same time.

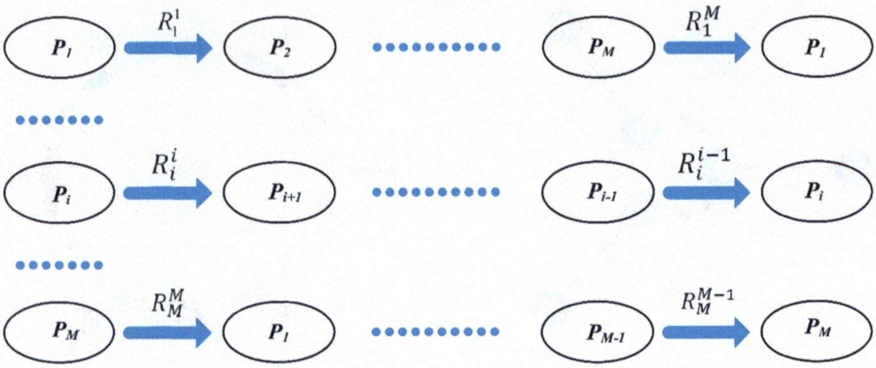

Fig. 6.3 Transmission process of MQKA scheme

Table 6.3 New encoding rules

	$\lvert\chi^+\rangle$	$\lvert\chi^-\rangle$	$\lvert\omega^+\rangle$	$\lvert\omega^-\rangle$
Binary code	00	01	10	11

Table 6.4 The relationship between the state $\{\lvert\chi^\pm\rangle, \lvert\omega^\pm\rangle\}$ and four unitary operations

Initial state	K_{i+1}	Operate on the second particle	The first state	K_{i+2}	Operate on the second particle	Final state	$K_{i+1} \oplus K_{i+2}$
$\lvert\chi^+\rangle_{12}$	00	U_{00}	$\lvert\chi^+\rangle_{12}$	00	U_{00}	$\lvert\chi^+\rangle_{12}$	00
				01	U_{01}	$\lvert\chi^-\rangle_{12}$	01
				10	U_{10}	$\lvert\omega^+\rangle_{12}$	10
				11	U_{11}	$\lvert\omega^-\rangle_{12}$	11
	01	U_{01}	$\lvert\chi^-\rangle_{12}$	00	U_{00}	$\lvert\chi^-\rangle_{12}$	01
				01	U_{01}	$\lvert\chi^+\rangle_{12}$	00
				10	U_{10}	$\lvert\omega^-\rangle_{12}$	11
				11	U_{11}	$\lvert\omega^+\rangle_{12}$	10
	10	U_{10}	$\lvert\omega^+\rangle_{12}$	00	U_{00}	$\lvert\omega^+\rangle_{12}$	10
				01	U_{01}	$\lvert\omega^-\rangle_{12}$	11
				10	U_{10}	$\lvert\chi^+\rangle_{12}$	00
				11	U_{11}	$\lvert\chi^-\rangle_{12}$	01
	11	U_{11}	$\lvert\omega^-\rangle_{12}$	00	U_{00}	$\lvert\omega^-\rangle_{12}$	11
				01	U_{01}	$\lvert\omega^+\rangle_{12}$	10
				10	U_{10}	$\lvert\chi^-\rangle_{12}$	01
				11	U_{11}	$\lvert\chi^+\rangle_{12}$	00

6.3.2.5 Security Analysis

Security analysis is an important part of the MQKA scheme. This section will analyze the security of the proposed TQKA and MQKA schemes from both participant and outsider attacks.

A. Participant Attack

In the TQKA scheme, delay measurement technology [35] is used to ensure security.

In the MQKA scheme, the biggest fear is internal participant attacks. If a dishonest participant P_i gets the final key K first, then he will be able to control everyone's final key without being detected. For example, if P_i will perform a unitary operation according to $K^{*}= K \oplus K'$, where K' is the key determined according to P_i's own will, so other honest participants will only get $K \oplus (K \oplus K') = K'$. In the proposed MQKA scheme, all participants send their own particle sequence R_i^i with decoy states at the same time. Therefore, no participant can obtain the final shared key in advance.

In addition, a collusion attack is the biggest security risk of the MQKA scheme [36], however, In this proposed MQKA scheme, the participants do not know who the neighboring participants are, so it is impossible to find the corresponding dishonest participants to predetermine the secret key.

B. Outsider Attack

1. **Trojan Horse Attack**
   According to Refs. [15, 16], because of the same sequence being transmitted many times in our TQKA and MQKA schemes, it may be attacked by Trojan horse. In order to avoid attack, a wavelength filter and a photon number splitters (PNS) will be installed for each participant in the proposed TQKA and MQKA schemes. PNS can divide each signal into two parts and discover delayed photon attacks. If a multi-photon signal has an unreasonably high rate, the attack will be detected.
2. **Intercept-Resend Attack**
   In the intercept-resend attack, the eavesdropper Eve intercepts the sequence sent by Alice (Bob) to Bob (Alice), and sends the pseudo sequence prepared in advance to Bob (Alice). In the proposed TQKA scheme, Alice and Bob will conduct eavesdropping check, and the possibility of being detected is $1 - \left(\frac{1}{4}\right)^n$ [37], where n denotes the number of decoy states. When n is large enough, the intercept-resend attack must be detected. In the proposed MQKA scheme, each participant P_i ($P_i \in \{B_1, B_2, \ldots, B_M\}$) will perform eavesdropping check to ensure the security of the scheme. The analysis of the MQKA scheme is similar to the above.
3. **Measure-Resend Attack**
   Eve may carry out measure-resend attack on particles transmitted through quantum channels and randomly uses Z basis or X basis to measure the obtained particles. The decoy particle state will collapse when Eve performs the measurement. Therefore, in the proposed TQKA and MQKA schemes, because Eve does not know the position of the decoy state and the corresponding measurement basis, so the attack cannot be successfully executed.

4. **Entangle-Measure Attack**
If an eavesdropper wants to launch an entangle-measure attack, she can use the auxiliary quantum state to entangle sequence transmitted in the quantum channel, and then obtain information by measuring the auxiliary quantum state. Without loss of generality, suppose the eavesdropper wants to perform an entangle-measure attack with the auxiliary particle $|E\rangle$, which can be expressed as

$$\begin{aligned} U_E \mid 0\rangle \mid E\rangle &= a \mid 0\rangle \mid E_{00}\rangle + b \mid 1\rangle \mid E_{01}\rangle \\ U_E \mid 1\rangle \mid E\rangle &= c \mid 0\rangle \mid E_{10}\rangle + d \mid 1\rangle \mid E_{11}\rangle \end{aligned} \tag{6.11}$$

$$\begin{aligned} U_E \mid +\rangle \mid E\rangle &= \frac{1}{\sqrt{2}}(a|0\rangle|E_{00}\rangle + b|1\rangle|E_{01}\rangle + c|0\rangle|E_{10}\rangle + d|1\rangle|E_{11}\rangle) \\ &= \frac{1}{2}(|+\rangle(a|E_{00}\rangle + b|E_{01}\rangle + c|E_{10}\rangle + d|E_{11}\rangle)) \end{aligned} \tag{6.12}$$

$$\begin{aligned} U_E \mid -\rangle \mid E\rangle &= \frac{1}{\sqrt{2}}(a|0\rangle|E_{00}\rangle + b|1\rangle|E_{01}\rangle - c|0\rangle|E_{10}\rangle - d|1\rangle|E_{11}\rangle) \\ &= \frac{1}{2}(|-\rangle(a|E_{00}\rangle - b|E_{01}\rangle - c|E_{10}\rangle + d|E_{11}\rangle)), \end{aligned} \tag{6.13}$$

where $|a|^2 + |b|^2=1$ and $|c|^2 + |d|^2=1$. If the eavesdropper does not want to introduce errors, the above equation must be satisfied. Therefore, it's easy to find out that $a = d = 1$, $b = c = 0$ and $|E_{00}\rangle=|E_{11}\rangle$. So Eq. 6.14 will become

$$\begin{aligned} U_E \mid 0\rangle \mid E\rangle &= \mid 0\rangle \mid E_{00}\rangle \\ U_E \mid 1\rangle \mid E\rangle &= \mid 1\rangle \mid E_{11}\rangle \end{aligned} \tag{6.14}$$

From Eq. 6.14, it means that only when the transmitted qubit is $|0\rangle$ or $|1\rangle$, the eavesdropper will not be detected. In our proposed TQKA and MQKA schemes, because we use four decoy particles $\{|0\rangle,|1\rangle,|+\rangle,|-\rangle\}$ for eavesdropping check. Obviously, it is impossible for eavesdroppers to an implement entangle-measure attack without being detected.

6.3.2.6 Efficiency Analysis

In Cabellon's scheme [38], qubit efficiency in quantum scheme is defined as

$$\eta = \frac{c}{q+b}, \tag{6.15}$$

where c is the number of bits of the generated shared key, q represents the total number of qubits used, and b represents the classical information used to decode. In proposed MQKA scheme, in order to generate 3n-bit shared key, each group needs to prepare n GHZ-like states, $n/2$ Bell states, n χ^+ states and $Nn/2$ decoy states.

Table 6.5 Comparison with similar multiparty QKA schemes

Scheme	Category	Quantum resource	Qubit efficiency
Ref. [15]	Complete-graph-type	Bell states	$\frac{1}{3N(N-1)}$
Ref. [16]	Complete-graph-type	Single photons	$\frac{1}{2N(N-1)}$
Ref. [17]	Circle-type	Single photons	$\frac{1}{N^2}$
Ref. [19]	Circle-type	G-like states and bell states	$\frac{8}{(10+N)N}$
Ours	Circle-type	G-like states and χ^+ states	$\frac{12}{(12+N)N}$

Therefore, the qubit efficiency of proposed MQKA scheme can be calculated as $\eta = \frac{3n}{\left(6n+\frac{N}{2}n\right)\frac{N}{2}} = \frac{12}{(12+N)N}$, where N represents the number of participants and $\frac{N}{2}$ represents the number of groups. Next, the proposed MQKA scheme will be compared with similar MQKA schemes in terms of category, quantum resources, and qubit efficiency, as shown in Table 6.5.

Compared with Min's scheme [19], the efficiency improvement of the proposed MQKA scheme can be expressed as follows:

$$\eta' = \frac{\frac{12}{(12+N)N} - \frac{8}{(10+N)N}}{\frac{8}{(8+N)N}} = \frac{N+6}{2N+24} \tag{6.16}$$

Figure 6.4 shows the relationship between the improved efficiency η' and the number of participants N. It is obvious that when the number of participants N is relatively small, the efficiency η' is about 30%, and when the number of participants is relatively large, the efficiency η' can be nearly 50%.

6.3.2.7 Summary

Based on the entanglement exchange between the GHZ-like state and χ^+state, this section designs a new multiparty quantum key agreement scheme. Different from other similar MQKA schemes, this scheme uses χ^+ state and Bell state to encode three bits of classical information corresponding to two particles, thereby greatly improving the efficiency. The results show that the proposed MQKA scheme can effectively resist internal and external attacks. In addition, this MQKA scheme is simple and has a small amount of calculation, which can be implemented with current technology.

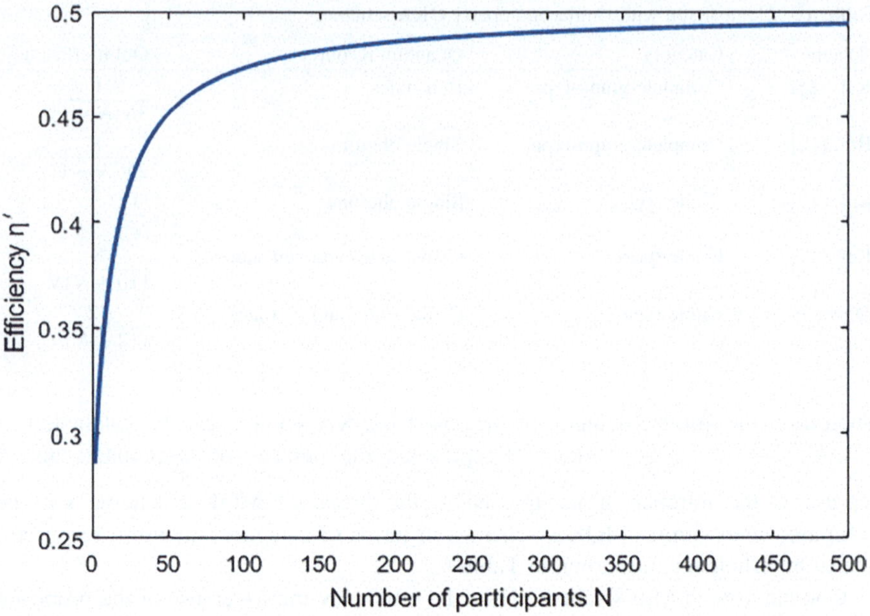

Fig. 6.4 The relationship between number of participants N and efficiency η'

6.3.3 SQKA Scheme

In order to improve the efficiency and practicality of semi quantum key agreement schemes, this section proposes a three-party semi-quantum key agreement (SQKA) scheme based on GHZ states, which includes four participants: Alice, Bob, Charlie, and David. Alice, Bob, and Charlie are classic users who can only perform two semi quantum operation: measurement and reflection. David is a semi-honest fourth party who can perform full quantum operation. With the help of quantum user David, three classic users Alice, Bob, and Charlie were able to participate in the final key negotiation process together. The semi-quantum key agreement scheme proposed not only satisfies the security of the key agreement scheme but also ensures fairness among the three classical participants.

The participants of the proposed three-party SQKA scheme are shown in Fig. 6.5. Alice, Bob, and Charlie are classic users who can only perform semi-quantum operations and only have two basic quantum functions: measurement and reflection. Measurement refers to measuring particles sent from the quantum party using Z basis $\{|0\rangle, |1\rangle\}$ and saving the measurement results. Reflection means that the particles sent by the quantum party are reflected back without any modifications [41]. David is a semi-honest fourth party capable of performing full quantum operations, with advanced quantum functions including preparing quantum states, storing quantum states, and using various basic measurements on quantum states. The semi-honest aspect of David's behavior means that he may leak the secret key during the

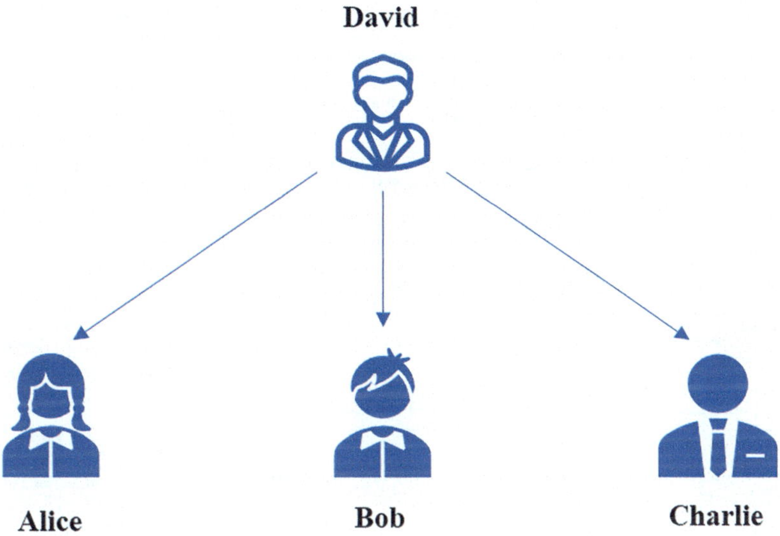

Fig. 6.5 The participants of the proposed three-party SQKA scheme

execution of the scheme, but he must execute the scheme truthfully and cannot publish false results or collude with Alice, Bob, and Charlie.

Before proposing the SQKA scheme, we first describe the GHZ states that will be used in the SQKA scheme. The eight GHZ states can be denoted as:

$$|\varphi_1\rangle = \frac{1}{\sqrt{2}}(|000\rangle + |111\rangle) \tag{6.17}$$

$$|\varphi_2\rangle = \frac{1}{\sqrt{2}}(|000\rangle - |111\rangle) \tag{6.18}$$

$$|\varphi_3\rangle = \frac{1}{\sqrt{2}}(|100\rangle + |011\rangle) \tag{6.19}$$

$$|\varphi_4\rangle = \frac{1}{\sqrt{2}}(|100\rangle - |011\rangle) \tag{6.20}$$

$$|\varphi_5\rangle = \frac{1}{\sqrt{2}}(|010\rangle + |101\rangle) \tag{6.21}$$

$$|\varphi_6\rangle = \frac{1}{\sqrt{2}}(|010\rangle - |101\rangle) \tag{6.22}$$

$$|\varphi_7\rangle = \frac{1}{\sqrt{2}}(|110\rangle + |001\rangle) \tag{6.23}$$

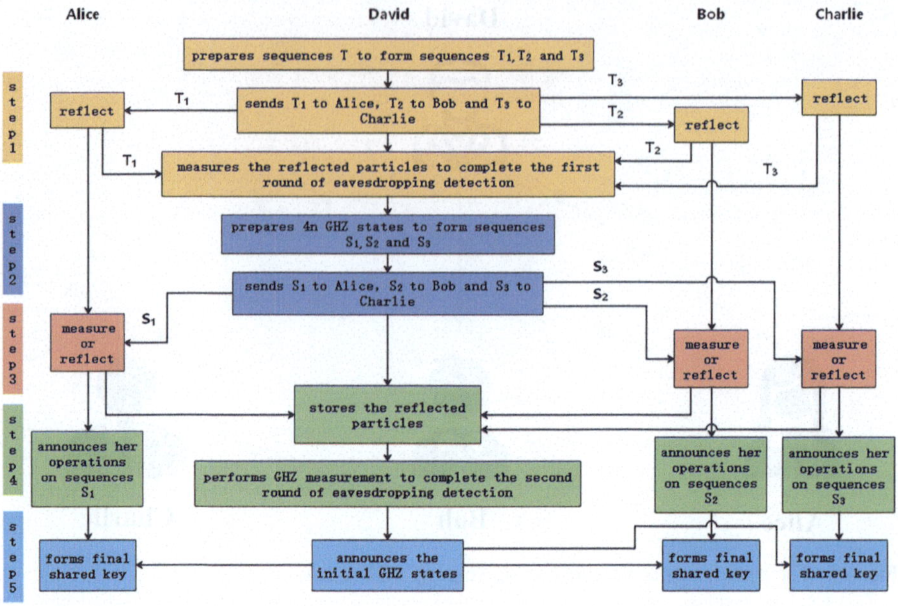

Fig. 6.6 The flow chart of the proposed three-party SQKA scheme

$$|\varphi_8\rangle = \frac{1}{\sqrt{2}}(|110\rangle - |001\rangle) \tag{6.24}$$

It is assumed in the proposed three-party SQKA scheme that the quantum channels are ideal (i.e., nonlossy and noiseless) and an authenticated classical channel is shared among all the legitimate participants. An adversary can eavesdrop on a message communicated through this channel. Based on the above scenario, the implementation process of the scheme will be described in steps. The flow chart of the proposed three-party SQKA scheme based on GHZ states is shown in Fig. 6.6.

6.3.3.1 Three-Party SQKA Scheme

Step 1: David randomly generates a quantum bit string $T = \{t_1, t_2, t_3, \ldots, t_{3n}\}$ containing $3n$ single particles for eavesdropping check, while recording the state of each particle. The single particle $t_i (i = 1, 2, 3, \ldots, 3n)$ is randomly in one of the four states: $|0\rangle$, $|1\rangle$, $|+\rangle$ and $|-\rangle$. David extracts the first n single particles to form a sequence T_1, that is, $T_1 = \{t_1, t_2, t_3, \ldots, t_n\}$, extracts the middle n single particles to form a sequence T_2, that is, $T_2 = \{t_{n+1}, t_{n+2}, t_{n+3}, \ldots, t_{2n}\}$, and extracts the last n single particles to form a sequence T_3, that is, $T_3 = \{t_{2n+1}, t_{2n+2}, t_{2n+3}, \ldots, t_{3n}\}$.

David sends the sequence T_1 to Alice, the sequence T_2 to Bob, and the sequence T_3 to Charlie. Alice, Bob, and Charlie all perform reflection operations. After receiving the reflected particle sequences, David performed Z basis measurements on the previously recorded particles with states $|0\rangle$ or $|1\rangle$ and X basis measurements on the previously recorded particles with states $|+\rangle$ or $|-\rangle$. David compares the measurement results of each particle with the previously recorded states and calculates the error rate. If the error rate is below the predetermined threshold, David concludes that there are no eavesdroppers in the quantum channel and proceeds to Step 2. Otherwise, he infers that the channel has been eavesdropped and terminates the SQKA scheme.

Step 2: After the first round of eavesdropping check, David prepares $4n$ three-particle GHZ states as mentioned above, while recording the state of each three-particle GHZ state. Each is randomly assigned one of eight states $\{|\varphi_1\rangle, |\varphi_2\rangle, |\varphi_3\rangle, |\varphi_4\rangle, |\varphi_5\rangle, |\varphi_6\rangle, |\varphi_7\rangle, |\varphi_8\rangle\}$. The first particle of each GHZ state is extracted to form a sequence S_1 in order; the second particle of each GHZ state is extracted to form a sequence S_2 in order; and the third particle of each GHZ state is extracted to form a sequence S_3 in order. Sequences S_1, S_2, and S_3 are used to generate the final shared key. David uses block transmission technology [26, 42] to send S_1 to Alice, S_2 to Bob, and S_3 to Charlie, where S_1, S_2, and S_3 all contain $4n$ qubits.

Step 3: After receiving the sequences S_1, S_2, and S_3 sent by David, Alice, Bob, and Charlie randomly choose between measurement and reflection operations for each particle in the sequences. Measurement means performing a Z basis $\{|0\rangle, |1\rangle\}$ measurement on the particle in the sequences sent by David and saving the measurement result. Reflection means not modifying the particle in the sequences sent by David and reflecting it directly back.

Step 4: David stores the particles reflected back by Alice, Bob, and Charlie and publicly confirms their receipt. Subsequently, Alice, Bob, and Charlie announced the operations performed on each particle in the sequences S_1, S_2, and S_3. Then David saves the particles that Alice, Bob, and Charlie performed reflection operations on the sequences S_1, S_2, and S_3 at the same position in order of position. David performs joint GHZ basis measurements on the particles in the saved sequence S_1, the corresponding particles in sequence S_2, and the corresponding particles in sequence S_3. David compares all measurement results with the previously recorded three-particle states and calculates the error rate. For instance, David prepares the initial GHZ states $|\varphi_2\rangle$, the measurement result should be $|\varphi_2\rangle$. If the error rate exceeds the predefined threshold, the scheme is terminated. Otherwise, Alice, Bob, and Charlie consider that there are no eavesdroppers in the channel and proceed to Step 5.

Step 5: After passing the second round of eavesdropping check, Alice, Bob, and Charlie can form the final shared key with David's assistance based on their own operations and measurement results for each particle. If Alice, Bob, and Charlie's measurement results are $|0\rangle$, they are decoded as classical bit 0. If Alice, Bob, and Charlie's measurement results are $|1\rangle$, they are decoded as classical bit 1. The decoding rules for the initial GHZ states are specified in Table 6.6. In Step

Table 6.6 The decoding rules for the initial GHZ states

Initial GHZ states	Corresponding classic bits	Initial GHZ states	Corresponding classic bits
$\|\varphi_1\rangle$	*000*	$\|\varphi_5\rangle$	*100*
$\|\varphi_2\rangle$	*001*	$\|\varphi_6\rangle$	*101*
$\|\varphi_3\rangle$	*010*	$\|\varphi_7\rangle$	*110*
$\|\varphi_4\rangle$	*011*	$\|\varphi_8\rangle$	*111*

4, David already knows the operations on each particle of sequences S_1, S_2, and S_3 performed by Alice, Bob, and Charlie. Next, there are *8* different cases to provide the methods for forming the final shared key under each situation.

1. If Alice, Bob, and Charlie perform measurement operations on the particles at the same position in the received sequence, David announces the three-particle states of that position in the three-particle sequences prepared in Step 2. Due to the entanglement of quantum entangled states, Alice can obtain the measurement results of Bob and Charlie's corresponding particles based on her own measurement results and David's published initial GHZ state.

   Alice takes the classical bit corresponding to her measurement result as the first bit of the final shared key, the classical bit corresponding to Bob's measurement result as the second bit of the final shared key, and the classical bit corresponding to Charlie's measurement result as the third bit of the final shared key.

   Similarly, due to the entanglement of quantum entangled states, Bob can obtain the measurement results of Alice and Charlie's corresponding particles based on his own measurement results and David's published initial GHZ state. Bob takes the classical bit corresponding to his measurement result as the second bit of the final shared key, the classical bit corresponding to Alice's measurement result as the first bit of the final shared key, and the classical bit corresponding to Charlie's measurement result as the third bit of the final shared key.

   Similarly, due to the entanglement of quantum entangled states, Charlie can obtain the measurement results of Alice and Bob's corresponding particles based on his own measurement results and David's published initial GHZ state. Charlie takes the classical bit corresponding to his measurement result as the third bit of the final shared key, the classical bit corresponding to Alice's measurement result as the first bit of the final shared key, and the classical bit corresponding to Bob's measurement result as the second bit of the final shared key.

   In the end, Alice, Bob, and Charlie all formed the same final shared key.
2. If Alice performs a measurement operation on a particle in the received sequence S_1, Bob performs a reflection operation on particles at the same position in the received sequence S_2, and Charlie performs a measurement operation on particles at the same position in the received sequence S_3.

   Alice and Charlie share the classic bits corresponding to their measurement results with Bob. David announces the three-particle state at this position in the three-particle quantum sequence prepared in Step 2, which is the initial GHZ states. Alice uses her measured classical bit corresponding to the measurement result as the first bit of the final shared key. She decodes the corresponding classical bits of the initial GHZ state announced by David according to Table 6.6, using them as the second, third, and fourth bits of the final shared key. Based on the entanglement property of quantum entangled states, Charlie's measured classical bit corresponding to the measurement result is used as the fifth bit of the final shared key.

Similarly, Bob decodes the corresponding classical bits from David's initial GHZ state according to Table 6.6 as the second, third, and fourth bits of the final shared key, uses Alice's shared classical bits as the first bit of the final shared key, and uses Charlie's shared classical bits as the fifth bit of the final shared key.

Similarly, Charlie uses his measured classical bit corresponding to the measurement result as the fifth bit of the final shared key. Charlie decodes the corresponding classical bits of the initial GHZ state announced by David according to Table 6.6, using them as the second, third, and fourth bits of the final shared key. Based on the entanglement property of quantum entangled states, Alice's measured classical bit corresponding to the measurement result is used as the first bit of the final shared key.

In the end, Alice, Bob, and Charlie all formed the same final shared key.

3. If Alice performs a measurement operation on a particle in the received sequence S_1, Bob also performs a measurement operation on particles at the same position in the received sequence S_2, and Charlie performs a reflection operation on particles at the same position in the received sequence S_3. Alice and Bob share the classic bits corresponding to their measurement results with Charlie. David announces the initial GHZ state of the particles prepared in Step 2.

   Alice uses her measured classical bit corresponding to the measurement result as the first bit of the final shared key. Based on the entanglement property of quantum entangled states, Bob's measured classical bit corresponding to the measurement result is used as the second bit of the final shared key. She decodes the corresponding classical bits of the initial GHZ state announced by David according to Table 6.6, using them as the third, fourth, and fifth bits of the final shared key.

   Similarly, Bob uses the measured classical bit corresponding to his measurement result as the second bit of the final shared key and uses Alice's measured classical bit corresponding to her measurement result, which is obtained based on the entanglement property of quantum entangled states, as the first bit of the final shared key. Bob also decodes the corresponding classical bits of the initial GHZ state announced by David according to Table 6.6, using them as the third, fourth, and fifth bits of the final shared key.

   Similarly, Charlie decodes the corresponding classical bits of the initial GHZ state announced by David according to Table 6.6, using them as the third, fourth, and fifth bits of the final shared key. He uses the classical bits received from Alice as the first bit of the final shared key, and the classical bits received from Bob as the second bit of the final shared key.

   In the end, Alice, Bob, and Charlie all formed the same final shared key.

4. If Alice performs a measurement operation on a particle in the received sequence S_1, Bob performs a reflection operation on particles at the same position in the received sequence S_2, and Charlie also performs a reflection operation on particles at the same position in the received sequence S_3. Alice shares the classic bits corresponding to their measurement results with Bob and Charlie. David announces the initial GHZ state of the particles prepared in Step 2.

   Alice uses her measured classical bit corresponding to the measurement result as the first bit of the final shared key. She decodes the corresponding classical bits of the initial GHZ state announced by David according to Table 6.6, using them as the second, third, and fourth bits of the final shared key.

   Similarly, Bob and Charlie decode the corresponding classical bits of the initial GHZ state announced by David according to Table 6.6, using them as the second, third, and fourth bits of the final shared key. They use the classical bits received from Alice as the first bit of the final shared key.

   In the end, Alice, Bob, and Charlie all formed the same final shared key.
5. If Alice performs a reflection operation on a particle in the received sequence S_1, Bob performs a measurement operation on particles at the same position in the received sequence S_2, and Charlie also performs a measurement operation on particles at the same position in the received sequence S_3. Bob and Charlie share the classic bits corresponding to their measurement results with Alice. David announces the initial GHZ state of the particles prepared in Step 2.

   Alice decodes the corresponding classical bits of the initial GHZ state announced by David according to Table 6.6, using them as the first, second, and third bits of the final shared key. She uses the classical bits received from Bob as the fourth bit of the final shared key and the classical bits received from Charlie as the fifth bit of the final shared key.

   Similarly, Bob uses the classical bit corresponding to his measurement result as the fourth bit of the final shared key. He derives the classical bit corresponding to Charlie's measurement result based on the entanglement property of the quantum entangled state, using it as the fifth bit of the final shared key. Bob then decodes the corresponding classical bits of the initial GHZ state announced by David according to Table 6.6, using them as the first, second, and third bits of the final shared key.

   Similarly, Charlie uses the classical bit corresponding to his measurement result as the fifth bit of the final shared key. He derives the classical bit corresponding to Bob's measurement result based on the entanglement property of the quantum entangled state, using it as the fourth bit of the final shared key. Charlie then decodes the corresponding classical bits of the initial GHZ state announced by David according to Table 6.6, using them as the first, second, and third bits of the final shared key.

   In the end, Alice, Bob, and Charlie all formed the same final shared key.
6. If Alice performs a reflection operation on a particle in the received sequence S_1, Bob also performs a reflection operation on particles at the same position in the received sequence S_2, and Charlie performs a measurement operation on particles at the same position in the received sequence S_3.

Charlie shares the classic bits corresponding to their measurement results with Alice and Bob. David announces the initial GHZ state of the particles prepared in Step 2.

Alice decodes the corresponding classical bits of the initial GHZ state announced by David according to Table 6.6, using them as the first, second, and third bits of the final shared key. She also uses the classical bit shared by Charlie as the fourth bit of the final shared key.

Similarly, Bob decodes the corresponding classical bits of the initial GHZ state announced by David according to Table 6.6, using them as the first, second, and third bits of the final shared key. He also uses the classical bit shared by Charlie as the fourth bit of the final shared key.

Similarly, Charlie uses the classical bit corresponding to his measurement result as the fourth bit of the final shared key. He decodes the corresponding classical bits of the initial GHZ state announced by David according to Table 6.6, using them as the first, second, and third bits of the final shared key.

In the end, Alice, Bob, and Charlie all formed the same final shared key.

7. If Alice performs a reflection operation on a particle in the received sequence S_1, Bob performs a measurement operation on particles at the same position in the received sequence S_2, and Charlie performs a reflection operation on particles at the same position in the received sequence S_3. Bob shares the classic bits corresponding to their measurement results with Alice and Charlie. David announces the initial GHZ state of the particles prepared in Step 2.

   Alice decodes the corresponding classical bits of the initial GHZ state announced by David according to Table 6.6, using them as the first, second, and fourth bits of the final shared key. She also uses the classical bit shared by Bob as the third bit of the final shared key.

   Similarly, Bob uses the classical bit corresponding to his measurement result as the third bit of the final shared key. He decodes the corresponding classical bits of the initial GHZ state announced by David according to Table 6.6, using them as the first, second, and fourth bits of the final shared key.

   Similarly, Charlie decodes the corresponding classical bits of the initial GHZ state announced by David according to Table 6.6, using them as the first, second, and fourth bits of the final shared key. He also uses the classical bit shared by Bob as the third bit of the final shared key.

   In the end, Alice, Bob, and Charlie all formed the same final shared key.

8. If Alice, Bob, and Charlie all perform reflection operations on particles at the same position in the received sequence. David announces the initial GHZ state of the particles prepared in Step 2.

   Alice, Bob, and Charlie all decode the corresponding classical bits of the initial GHZ state announced by David according to Table 6.6, using them as the final shared key.

   In the end, Alice, Bob, and Charlie all formed the same final shared key.

In summary, the methods for forming the final shared key are provided in 8 different scenarios. The above 5 steps are the entire process of our proposed

Table 6.7 The initial GHZ state is $|\varphi_1\rangle$

Initial GHZ states	Alice's operation	Bob's operation	Charlie's operation	Final shared key
$\lvert\varphi_1\rangle$	M	M	M	000 or 111
	M	R	M	00000 or 10,001
	M	M	R	00000 or 11,000
	M	R	R	0000 or 1000
	R	M	M	00000 or 00011
	R	R	M	0000 or 0001
	R	M	R	0000 or 0010
	R	R	R	000

Table 6.8 The initial GHZ state is $|\varphi_2\rangle$

Initial GHZ states	Alice's operation	Bob's operation	Charlie's operation	Final shared key
$\lvert\varphi_2\rangle$	M	M	M	000 or 111
	M	R	M	00010 or 10,011
	M	M	R	00001 or 11,001
	M	R	R	0001 or 1001
	R	M	M	00100 or 00111
	R	R	M	0010 or 0011
	R	M	R	0001 or 0011
	R	R	R	001

three-party semi-quantum key agreement scheme based on GHZ states. The $4n$ three-particle initial GHZ states prepared by David in Step 2 are one of the eight states $\{|\varphi_1\rangle, |\varphi_2\rangle, |\varphi_3\rangle, |\varphi_4\rangle, |\varphi_5\rangle, |\varphi_6\rangle, |\varphi_7\rangle, |\varphi_8\rangle\}$. In each of the eight possible states, Alice, Bob, and Charlie can perform a measurement or reflection operation, resulting in a total of eight possible scenarios. So there are a total of 64 cases in which the scheme forms the final shared key, all of which are shown in Tables 6.7, 6.8, 6.9, 6.10, 6.11, 6.12, 6.13 and 6.14, where the parameters "M" and "R" denote measurement and reflection.

6.3.3.2 Security Analysis

This subsection assumes that the external attacker is Eve and provides a detailed analysis of the security of the three-party SQKA scheme we proposed under six external attacks: Trojan horse attack, intercept-resend attack, intercept-measure-resend attack, entangle-measure attack, flipping attack, and man-in-the-middle attack, as well as semi-honest fourth party internal attack. This proves that the SQKA scheme is secure under these seven common attacks.

Table 6.9 The initial GHZ state is $|\varphi_3\rangle$

Initial GHZ states	Alice's operation	Bob's operation	Charlie's operation	Final shared key
$|\varphi_3\rangle$	M	M	M	100 or 011
	M	R	M	00101 or 10,100
	M	M	R	10,010 or 01010
	M	R	R	0010 or 1010
	R	M	M	01000 or 01011
	R	R	M	0100 or 0101
	R	M	R	0100 or 0110
	R	R	R	010

Table 6.10 The initial GHZ state is $|\varphi_4\rangle$

Initial GHZ states	Alice's operation	Bob's operation	Charlie's operation	Final shared key
$|\varphi_4\rangle$	M	M	M	100 or 011
	M	R	M	10,110 or 00111
	M	M	R	10,011 or 01011
	M	R	R	0011 or 1011
	R	M	M	01100 or 01111
	R	R	M	0110 or 0111
	R	M	R	0101 or 0111
	R	R	R	011

Table 6.11 The initial GHZ state is $|\varphi_5\rangle$

Initial GHZ states	Alice's operation	Bob's operation	Charlie's operation	Final shared key
$|\varphi_5\rangle$	M	M	M	010 or 101
	M	R	M	01000 or 11,001
	M	M	R	01100 or 10,100
	M	R	R	0100 or 1100
	R	M	M	10,010 or 10,001
	R	R	M	1000 or 1001
	R	M	R	1000 or 1010
	R	R	R	100

A. Trojan Horse Attack

Cai pointed out that bidirectional quantum communication schemes are susceptible to Trojan horse attack [39]. Since the designed three-party SQKA scheme is a bidirectional communication scheme, it is vulnerable to Trojan horse attack from external attackers. Quantum Trojan attacks generally involve two types: one is the use of invisible photons for eavesdropping [39], and the other is the delayed photon Trojan horse attack [40], also known as multi-photon signal attacks.

Table 6.12 The initial GHZ state is $|\varphi_6\rangle$

Initial GHZ states	Alice's operation	Bob's operation	Charlie's operation	Final shared key
$\|\varphi_6\rangle$	M	M	M	010 or 101
	M	R	M	01010 or 11,011
	M	M	R	01101 or 10,101
	M	R	R	0101 or 1101
	R	M	M	10,110 or 10,101
	R	R	M	1010 or 1011
	R	M	R	1001 or 1011
	R	R	R	101

Table 6.13 The initial GHZ state is $|\varphi_7\rangle$

Initial GHZ states	Alice's operation	Bob's operation	Charlie's operation	Final shared key
$\|\varphi_7\rangle$	M	M	M	110 or 001
	M	R	M	11,100 or 01101
	M	M	R	11,110 or 00110
	M	R	R	0110 or 1110
	R	M	M	11,010 or 11,001
	R	R	M	1100 or 1101
	R	M	R	1100 or 1110
	R	R	R	110

Table 6.14 The initial GHZ state is $|\varphi_8\rangle$

Initial GHZ states	Alice's operation	Bob's operation	Charlie's operation	Final shared key
$\|\varphi_8\rangle$	M	M	M	110 or 001
	M	R	M	11,110 or 01111
	M	M	R	11,111 or 00111
	M	R	R	0111 or 1111
	R	M	M	11,110 or 11,101
	R	R	M	1110 or 1111
	R	M	R	1101 or 1111
	R	R	R	111

General detection tools are difficult to resist Trojan horse attack. To ensure communication security, it is necessary to add a photon number separator (PNS) and a wavelength filter (WQF) before Alice, Bob, and Charlie's devices to preprocess the received quantum bits. Among them, the function of wavelength filters is to filter out illegal photons under invisible photon attacks, while photon number separators resist delayed photon Trojan horse attack by detecting the presence of multi-photon signals.

B. Intercept-Resend Attack

Intercept-resend attack [43] refers to the external attacker Eve intercepting the sequences sent by David to Alice, Bob, and Charlie, replacing the intercepted sequences with preforged ones and sending them to Alice, Bob, and Charlie. After Alice, Bob, and Charlie perform a series of operations, Eve intercepts the sequences sent to David again and measures it. Based on the measurement results, she infers the operations performed by Alice, Bob, and Charlie, and applies the inferred operations performed by Alice, Bob, and Charlie to the initial intercepted sequences and sends them to David, thus achieving the purpose of stealing information. In the SQKA scheme we proposed, there are two places where external attackers Eve may implement intercept-resend attack:

1. If the external attacker Eve implements intercept-resend attack during the process of sending sequences T_1, T_2, and T_3 to Alice, Bob, and Charlie, respectively, in Step 1. Eve replaces the intercepted sequences T_1, T_2, and T_3 with prefabricated sequences F_1, F_2, and F_3 and then sends them to Alice, Bob, and Charlie. After Alice, Bob, and Charlie perform a series of operations, Eve intercepts the sequence they send to David again and measures it. The measurement results show that there is no change in the sequence. Thus, it is presumed that Alice, Bob, and Charlie do not make any modifications to the received sequence and return it directly to the sender. Therefore, Eve does not make any changes to the intercepted sequence of T_1, T_2, and T_3 and returns it directly to David. In this case, David cannot detect the presence of the external attacker Eve during the first round of eavesdropping check, but Eve also does not obtain any useful information.
2. If the external attacker Eve implements an intercept-resend attack during the process of sending sequences S_1, S_2, and S_3 to Alice, Bob, and Charlie, respectively, in Step 2. Eve replaces the intercepted sequences S_1, S_2, and S_3 with prefabricated sequences H_1, H_2, and H_3 and then sends them to Alice, Bob, and Charlie. After Alice, Bob, and Charlie perform a series of operations, Eve intercepts the sequence they send to David again and measures it. Because Alice, Bob, and Charlie randomly choose measurement or reflection operations and only send the particles that undergo reflection operations to David, Eve cannot deduce the operations performed by Alice, Bob, and Charlie based on the measurement results. Eve can only make random guesses and operate on the intercepted sequence before sending it to David. In this case, David will inevitably detect the presence of external attacker Eve during the second round of eavesdropping check.

In summary, the three-party SQKA scheme can resist intercept-resend attack.

C. Intercept-Measure-Resend Attack

Intercept-measure-resend attack refers to an external attacker Eve intercepting particles sent by David to Alice, Bob, and Charlie, performing measurement operations, and then preparing new particles based on the measurement results to send them in order to achieve the malicious purpose of the attacker tampering with the transmitted information, causing the information receiver to mistakenly believe that they have received the original information. In the three-party SQKA scheme proposed, there are two places where external attackers Eve may implement intercept-measure-resend attack:

1. If the external attacker Eve implements intercept-measure-resend attack during the process of sending sequences T_1, T_2, and T_3 to Alice, Bob, and Charlie respectively in Step 1. Due to Eve's lack of knowledge of the type of decoy particles, she can only make a blind guess. Even if Eve guesses that the decoy particle is a single particle, she doesn't know what basis to use for measurement. Eve can only choose X basis or Z basis measurements with equal probability. For example, the decoy particle prepared by David is $|0\rangle$ ($|1\rangle$) state. If Eve intercepts this particle and selects the Z basis for measurement, the obtained measurement result will still be $|0\rangle$ ($|1\rangle$) state.

   But if Eve chooses the X basis for measurement, then the particle will collapse into $|+\rangle$ state or $|-\rangle$ state each with the probability of $\frac{1}{2}$. Eve prepared new particles based on this measurement result and sent them to Alice, Bob, and Charlie. In this way, David can detect the presence of Eve during the first round of eavesdropping check. That is to say, if Eve chooses the right measurement basis, he can successfully pass David's first round of eavesdropping check. So, the probability for Eve to pass the eavesdropping check of a single decoy particle is $\frac{1}{2}$. In Step 1, David prepared $3n$ decoy particles for eavesdropping check. So, for Eve's intercept-measure-resend attack, the probability of being detected in the first round of eavesdropping check is $P_1 = 1 - \left(\frac{1}{2}\right)^{3n} = 1 - \left(\frac{1}{8}\right)^{n}$. The relationship between P_1 and n is shown in Fig. 6.7. This probability is approximate to 1 if n is large enough.
2. If the external attacker Eve implements intercept-measure-resend attack during the process of sending sequences S_1, S_2, and S_3 to Alice, Bob, and Charlie respectively in Step 2. In proposed three-party SQKA scheme, Alice, Bob, and Charlie randomly select measurement or reflection operations. However, the external attacker Eve does not know which particles Alice, Bob, and Charlie performed measurement operations or reflection operations on. If Eve implements intercept-measure-resend attack, they will be discovered. For example, if the initial GHZ state of particles prepared by David is $|\varphi_1\rangle$state, and Eve intercepts the particles sent by David to Alice, Bob, and Charlie for Z basis measurement. The $|\varphi_1\rangle$ state will collapse to $|000\rangle$ state or $|111\rangle$ state, and Eve prepares particles based on the measurement results and sends them to Alice, Bob, and Charlie. If Alice, Bob, and Charlie have chosen the measurement operation, Eve will not be discovered. In view of$|000\rangle = \frac{1}{\sqrt{2}}(|\varphi_1\rangle + |\varphi_2\rangle)$

   ($|111\rangle = \frac{1}{\sqrt{2}}(|\varphi_1\rangle - |\varphi_2\rangle)$), if Alice, Bob, and Charlie all choose the reflection

   operation, David performs GHZ basis measurement in Step 4 and then obtains $|\varphi_1\rangle$ or $|\varphi_2\rangle$ each with the probability of $\frac{1}{2}$. So, the probability of not discovering Eve is $\frac{1}{2}$. Therefore, the probability for Eve to pass the eavesdropping check is

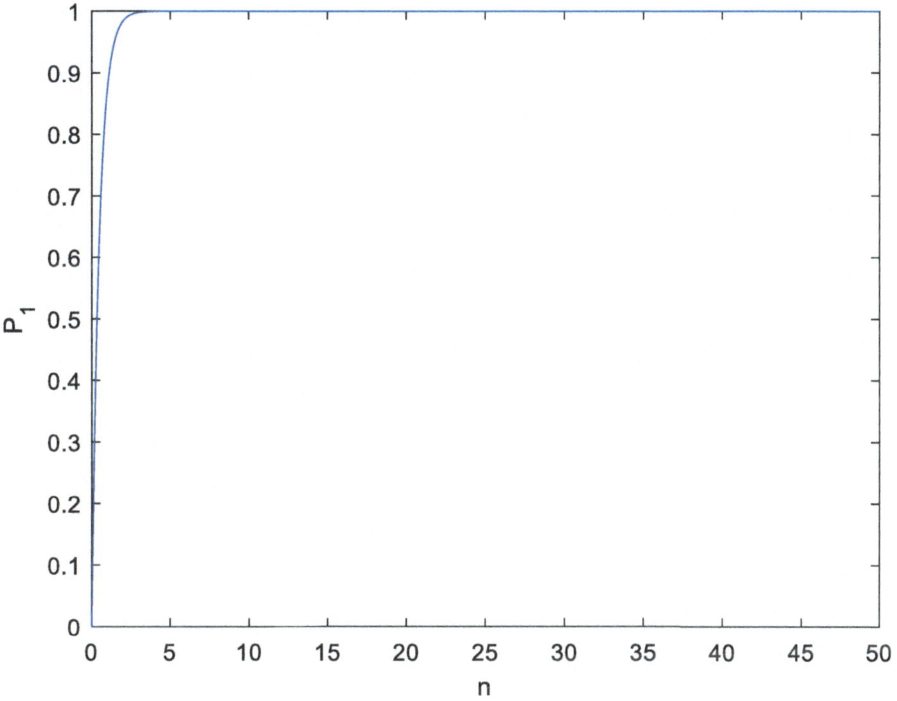

Fig. 6.7 The relationship between P_1 and $\boldsymbol{n}$

$P = \frac{7}{8} \times 1 + \frac{1}{8} \times \frac{1}{2} = \frac{15}{16}$. So, for Eve's intercept-measure-resend attack, the probability of being detected in the second round of eavesdropping check is $P_2 = 1 - \left(\frac{15}{16}\right)^{4n} = 1 - \left(\frac{50625}{65536}\right)^n$.The relationship between P_2 and n is shown in Fig. 6.8. This probability is approximate to 1 if n is large enough.

In summary, three-party SQKA scheme can resist intercept-measure-resend attack.

D. Entangle-Measure Attack

Entangle-measure attack [44], also known as auxiliary particle attack, attacks by intercepting quantum states in transmission. Then the attacker attacks the intercepted particles and the prepared particles, causing them to entangle with each other. Then the attacker sends the intercepted particles back to the receiver and measures their own auxiliary particles to obtain valuable information.

In the proposed three-party SQKA scheme, if an external attacker Eve attempts to use entangle-measure attack. So, he will perform a unitary operation U_E to entangle the intercepted particles sent by David to Alice, Bob, and Charlie with the auxiliary particle $|e\rangle$ he has prepared and finally steal information by measuring the auxiliary particles. Eve's attack operations on particles are as follows:

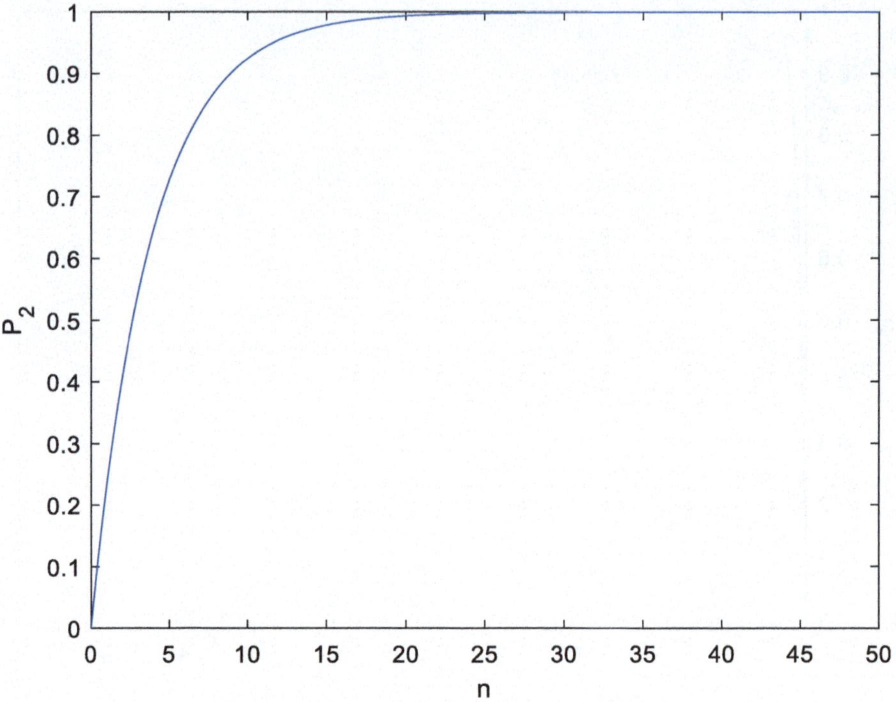

Fig. 6.8 The relationship between P_2 and ***n***

$$U_E|0,e\rangle = \alpha|0,e_{00}\rangle + \beta|1,e_{01}\rangle \tag{6.25}$$

$$U_E|1,e\rangle = \lambda|0,e_{10}\rangle + \mu|1,e_{11}\rangle \tag{6.26}$$

$$\begin{aligned} U_E|+,e\rangle = &\frac{1}{2}|+\rangle(\alpha|e_{00}\rangle + \beta|e_{01}\rangle + \lambda|e_{10}\rangle + \mu|e_{11}\rangle) \\ &+\frac{1}{2}|-\rangle(\alpha|e_{00}\rangle - \beta|e_{01}\rangle + \lambda|e_{10}\rangle - \mu|e_{11}\rangle) \end{aligned} \tag{6.27}$$

$$\begin{aligned} U_E|-,e\rangle = &\frac{1}{2}|+\rangle(\alpha|e_{00}\rangle + \beta|e_{01}\rangle - \lambda|e_{10}\rangle - \mu|e_{11}\rangle) \\ &+\frac{1}{2}|-\rangle(\alpha|e_{00}\rangle - \beta|e_{01}\rangle - \lambda|e_{10}\rangle + \mu|e_{11}\rangle), \end{aligned} \tag{6.28}$$

where $|e_{00}\rangle$, $|e_{01}\rangle$, $|e_{10}\rangle$, and $|e_{11}\rangle$ are pure states determined only by the unitary operation U_E and

$$|\alpha|^2 + |\beta|^2 = 1 \tag{6.29}$$

$$|\lambda|^2 + |\mu|^2 = 1 \tag{6.30}$$

$$\langle e_i|e_i\rangle = 1(i = 00, 01, 10, 11) \tag{6.31}$$

$$\langle e_{00}|e_{01}\rangle = \langle e_{10}|e_{11}\rangle = 0 \tag{6.32}$$

In channel detection, if Eve is not detected to introduce any errors, it can be obtained that

$$U_E|0,e\rangle = \alpha|0,e_{00}\rangle \tag{6.33}$$

$$U_E|1,e\rangle = \mu|1,e_{11}\rangle \tag{6.34}$$

$$\alpha|e_{00}\rangle + \lambda|e_{10}\rangle = \beta|e_{01}\rangle + \mu|e_{11}\rangle \tag{6.35}$$

$$\alpha|e_{00}\rangle + \beta|e_{01}\rangle = \lambda|e_{10}\rangle + \mu|e_{11}\rangle \tag{6.36}$$

From the above four formulas, it can be inferred that

$$\alpha = \mu = 1, \beta = \lambda = 0 \tag{6.37}$$

$$|e_{00}\rangle = |e_{11}\rangle \tag{6.38}$$

Substituting the initial four equations yields

$$U_E|0,e\rangle = |0,e_{00}\rangle \tag{6.39}$$

$$U_E|1,e\rangle = |1,e_{11}\rangle = |1,e_{00}\rangle \tag{6.40}$$

$$U_E|+,e\rangle = |+\rangle(\alpha|e_{00}\rangle + \lambda|e_{10}\rangle) = |+,e_{00}\rangle \tag{6.41}$$

$$U_E|-,e\rangle = |-\rangle(\alpha|e_{00}\rangle - \lambda|e_{10}\rangle) = |-,e_{00}\rangle \tag{6.42}$$

The above four formulas indicate that external attacker Eve cannot obtain any valuable information by initiating entangle-measure attack. In addition, the external attacker Eve initiating an entangle-measure attack will cause the original quantum state to change. As the number of particles used for eavesdropping check increases, Eve's presence will inevitably be discovered in two rounds of eavesdropping check.

In summary, three-party SQKA scheme can resist entangle-measure attack.

E. Flipping Attack

The attack method of flipping attack is to intercept the quantum state in transmission, and then the attacker uses the classical measurement basis to measure it. If the measurement result is $|0\rangle(|1\rangle)$, the attacker flips and prepares $|1\rangle(|0\rangle)$ to send to the receiver, causing the receiver to obtain incorrect information. In the three-party SQKA scheme, there are two places where external attackers Eve may implement flipping attack

1. If the external attacker Eve implements flipping attack during the process of sending sequences T_1, T_2, and T_3 to Alice, Bob, and Charlie, respectively, in Step 1. So, David's calculated error rate during the first round of eavesdropping check will be 100%, which will reveal the presence of external attacker Eve.
2. If the external attacker Eve implements flipping attack during the process of sending sequences S_1, S_2, and S_3 to Alice, Bob, and Charlie, respectively, in Step

2. Assuming the initial GHZ state prepared by David in Step 2 is $|\varphi_1\rangle$ and the measurement result of the sequence S_1 is $|0\rangle(|1\rangle)$ state, Eve performs a flipping attack and sends the $|1\rangle(|0\rangle)$ state to Alice. According to the entanglement of quantum entangled states, the measurement result of sequence S_2 can only be the $|0\rangle(|1\rangle)$ state. Eve performs a flip attack and sends the $|1\rangle(|0\rangle)$ state to Bob. According to the entanglement of quantum entangled states, the measurement result of sequence S_3 can only be the $|0\rangle(|1\rangle)$ state. Eve performs a flip attack and sends the $|1\rangle(|0\rangle)$ state to Charlie.

If Alice, Bob, and Charlie do not all perform reflection operations, external attacker Eve will not be detected. When Alice, Bob, and Charlie all perform reflection operations, David performs GHZ basis measurement on the reflected $|111\rangle(|000\rangle)$ state and then obtains $|\varphi_1\rangle$ or $|\varphi_2\rangle$ each with probability of $\frac{1}{2}$. So, the probability of not discovering Eve is $\frac{1}{2}$. Therefore, the probability for Eve to pass the eavesdropping check is $P=\frac{7}{8}\times 1+\frac{1}{8}\times\frac{1}{2}=\frac{15}{16}$. So, for Eve's flipping attack, the probability of being detected in the second round of eavesdropping check is $P_3=1-\left(\frac{15}{16}\right)^{4n}=1-\left(\frac{50625}{65536}\right)^{n}$. The relationship between P_3 and n is shown in Fig. 6.9. This probability is approximate to 1 if n is large enough.

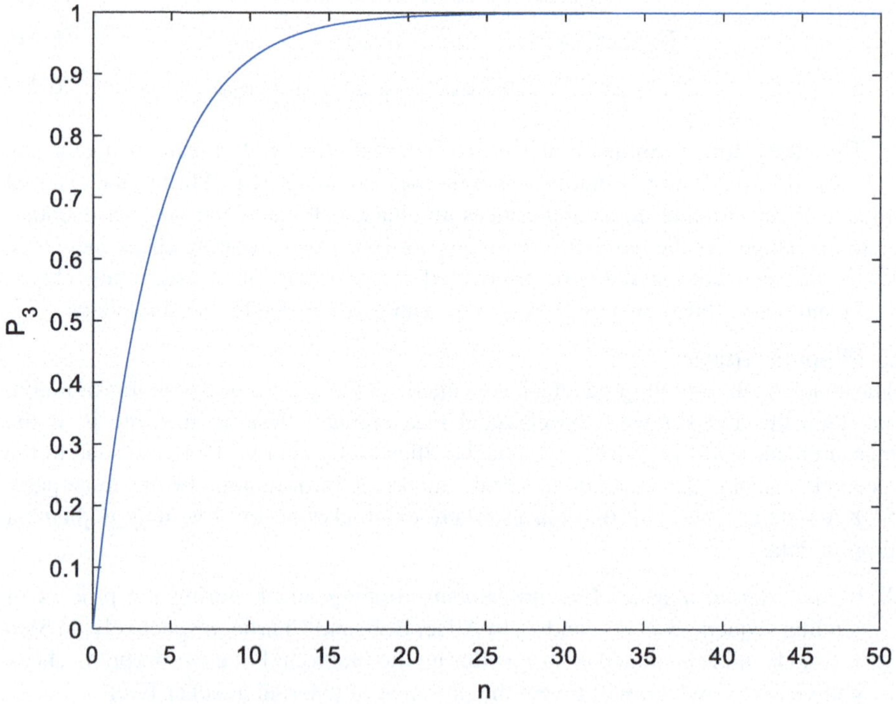

Fig. 6.9 The relationship between P_3 and $\boldsymbol{n}$

In summary, three-party SQKA scheme can resist flipping attack. And external attacker Eve cannot obtain any useful secret information when performing a flipping attack.

F. Man-in-the-Middle Attack

Man-in-the-middle attack [45] refers to the attacker Eve impersonating a legitimate user and initiating a conversation with three legitimate communication parties Alice, Bob, and Charlie to complete the communication process, attempting to obtain confidential information without being discovered by the communication participants.

Assuming that the attacker Eve impersonates the legitimate user David and intercepts the initial GHZ states $|\varphi_1\rangle$ sent by David to Alice, Bob, and Charlie in Step 2. After measurement, prepare the same GHZ states and send the first particle to Alice, the second particle to Bob, and the third particle to Charlie. Then Eve captures the particles reflected back by Alice, Bob, and Charlie and measures them, preparing the same particles based on the measurement results and sending them to David. So, the particle David received is $|000\rangle$ state or $|111\rangle$ state. David performs GHZ basis measurement on it and then obtains $|\varphi_1\rangle$ or $|\varphi_2\rangle$ each with the probability of $\frac{1}{2}$. So, the probability of not discovering Eve is $\frac{1}{2}$. When there are enough particles prepared by David, the attacker Eve will definitely be detected in eavesdropping check. So, the three-party SQKA scheme can resist man-in-the-middle attack.

G. Semi-Honest Fourth Party Internal Attack

In the three-party SQKA scheme, David is a semi-honest fourth party. The semi-honest aspect of David's behavior means that he may leak the secret key during the execution of the scheme, but he must execute the scheme truthfully and cannot publish false results or collude with Alice, Bob, and Charlie.

Assuming that David attempts to infer the final shared key during the execution of the scheme and then leaks it out. In the proposed three-party SQKA scheme, the initial GHZ states were prepared by David. Alice, Bob, and Charlie announced the operations to be performed on each particle in sequences S_1, S_2, and S_3. Based on this information, David can infer the final shared key. When Alice, Bob, and Charlie do not all perform reflection operations, David can correctly guess the measurement results of Alice, Bob, and Charlie with the probability of $\frac{1}{2}$. So, David will have a probability of $\frac{1}{2}$ to get the final shared key. When Alice, Bob, and Charlie all perform reflection operations, as the initial GHZ states particles are prepared by David and David also knows the corresponding decoding rules, it is certain that the final shared key can be inferred. Therefore, the probability that David gets the final shared key from one of these rounds is $P = \frac{7}{8} \times \frac{1}{2} + \frac{1}{8} \times 1 = \frac{9}{16}$. So, the probability that David can get all the final shared keys is $P_4 = \left(\frac{9}{16}\right)^{4n} = \left(\frac{6561}{65536}\right)^n$. The relationship between P_4 and n is shown in Fig. 6.10. This probability is approximate to 0 if n is large enough. That is, when the number of initial GHZ states particles prepared is large enough, the probability that David can get all the final shared keys is negligible.

In summary, the three-party SQKA scheme can resist semi-honest fourth party internal attack.

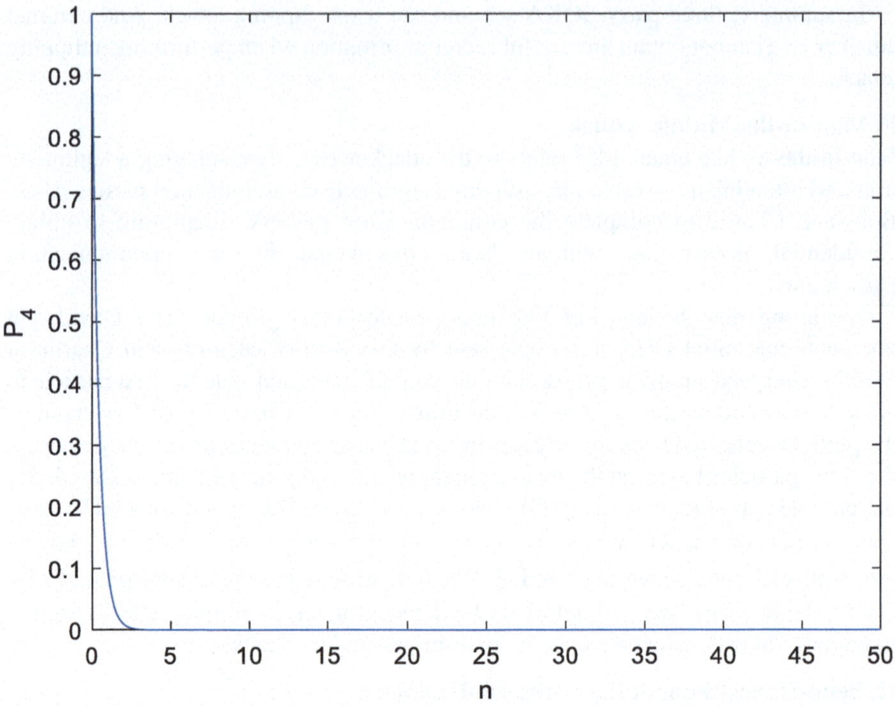

Fig. 6.10 The relationship between P_4 and n

6.3.3.3 Fairness Analysis

This subsection conducts a fairness analysis of the proposed three-party SQKA scheme. In the three-party SQKA scheme, in Step 3, Alice, Bob, and Charlie randomly select one of the measurement and reflection operations for each particle in the sequence after receiving sequences S_1, S_2, and S_3 sent by David. In Step 4, Alice, Bob, and Charlie announce the operations they will perform on each particle in sequences S_1, S_2, and S_3. In Step 5, David only announces the initial GHZ states he prepared in Step 2. Therefore, Alice, Bob, and Charlie cannot modify their operations based on the initial GHZ states. The formation of the final shared key is jointly determined by the random operations carried out by Alice, Bob, and Charlie and the initial GHZ states prepared by David. No participant can generate the final shared key according to their own ideas. By executing this scheme, the three classical users Alice, Bob, and Charlie can jointly participate in the key negotiation process and generate the final shared key.

6.3.3.4 Efficiency Analysis

The efficiency of the proposed three-party SQKA scheme is measured by Cabello's definition of qubit efficiency [38] $\eta = \frac{c}{q+b}$, where q is the number of all qubits used in the semi-quantum key agreement scheme, b is the number of classical bits used for decoding secret information (excluding the number of classical bits used for eavesdropping check), and c is the length of the final shared key generated by the semi-quantum key agreement scheme.

In the three-party SQKA scheme, in Step 1, David randomly generates a sequence containing $3n$ single particles and sends it to Alice, Bob, and Charlie. Alice, Bob, and Charlie then reflect back n single particles for the first round of eavesdropping check. In Step 2, David prepared $4n$ three-particle GHZ states ($12n$ qubits) and sent them to Alice, Bob, and Charlie. Alice, Bob, and Charlie then all reflected back $2n$ single particles for the second round of eavesdropping check. Therefore, total $q = 3n + n + n + n + 12n + 2n + 2n + 2n = 24n$. In the three-party SQKA scheme, in order to decode the secret information and generate the final shared key, in Step 4, Alice, Bob, and Charlie publicly announce the operations performed on each particle in the sequences S_1, S_2, and S_3. For this, Alice, Bob, and Charlie all use 4n classical bits. Since each operation can decode secret information to generate the final shared key, these classical bits need to be calculated in b. In Step 5, if Alice, Bob, and Charlie perform measurement, reflection, and measurement operations, respectively, Alice and Charlie need to share n classical bits corresponding to their measurement results with Bob in total. If Alice, Bob, and Charlie perform measurement, reflection, and reflection operations separately, Alice needs to share the classical bits corresponding to her measurement results with Bob and Charlie, and the number of classical bits required is also n. The other four situations are the same as these two situations. Except for Alice, Bob, and Charlie all performing measurement operations and reflection operations, the remaining 6 cases require the consumption of classical bits, and a total of $6n$ classical bits are required for the 6 cases. Therefore, total $b = 4n + 4n + 4n + 6n = 18n$. The length of the final shared key corresponding to each GHZ states is 3 or 4 or 5. Therefore, in the three-party SQKA scheme, the total length of the final shared key is at least $12n$ and at most $20n$, that is, the minimum is $12n$ and the maximum is $20n$. When c is $12n$, the efficiency of our proposed scheme is $\eta_1 = \frac{c}{q+b} = \frac{12n}{24n+18n} \approx 28.57\%$; when c is $20n$, the efficiency of the three-party SQKA scheme is $\eta_2 = \frac{c}{q+b} = \frac{20n}{24n+18n} \approx 47.62\%$. In summary, the qubit efficiency of the three-party SQKA scheme is 28.57 % ~ 47.62%.

Table 6.15 shows the efficiency comparison between the proposed three-party SQKA scheme and other semi-quantum key agreement schemes. Among them, Shukla's scheme [23], Yan's scheme [46], and Yan's scheme [31] are two-party semi-quantum key agreement schemes, while Zhou's scheme [25] is a three-party semi-quantum key agreement scheme. Compared with some existing semi-quantum key agreement schemes, the three-party SQKA scheme has higher efficiency, as shown in Table 6.15.

Table 6.15 Comparison with related semi-quantum key agreement scheme

Scheme	Quantum resource	Qubit efficiency (%)
Ref. [23]	Bell states	9.09
Ref. [46]	Bell states	6.7
Ref. [31]	Bell states	16.7
Ref. [25]	Cluster states	2.08
Ours	GHZ states	28.57 ~ 47.62

6.3.3.5 Summary

This section proposes a three-party semi-quantum key agreement scheme based on GHZ states. With the help of David, a semi-honest fourth party who can perform full quantum operation, three classical users Alice, Bob, and Charlie can jointly participate in the final key negotiation process, and finally form the final shared key. This three-party SQKA scheme first explained some things that need to be prepared before scheme execution. Then we provided the specific execution steps of the three-party SQKA scheme. Subsequently, a detailed security and fairness analysis was conducted. After detailed analysis, it has been proven that the three-party SQKA scheme can resist common attack methods such as Trojan horse attack, intercept-resend attack, intercept-measure-resend attack, entangle-measure attack, flipping attack, man-in-the-middle attack, and semi-honest fourth party internal attack. In addition, compared with some existing semi-quantum key agreement schemes [23, 25, 31, 46], the three-party SQKA scheme has higher efficiency. In addition, since the three-party SQKA scheme is based on semi-quantum technology, the three classic users Alice, Bob, and Charlie who participate in key agreement only need to have two basic quantum functions: measurement and reflection. They can negotiate the final shared key without having to equip expensive quantum equipment to perform complex quantum operation, effectively reducing the cost of quantum communication. Classical users with limited quantum capabilities also have the opportunity to participate in quantum communication. Compared to the three-party quantum key agreement scheme, under existing technical conditions, the proposed three-party SQKA scheme can be implemented at a lower cost, which is more practical and feasible. This will promote the application of quantum key agreement schemes. To some extent, this will promote the popularization process of quantum communication.

6.4 Brief Summary

Based on the quantum cryptography theory, this chapter studies the theory of quantum key agreement. According to the full quantum and semi-quantum environment, three multiparty quantum key agreement schemes are proposed, respectively. The research of these two improved schemes enriches the theory of quantum key agreement and also provides a new idea for the follow-up research.

References

1. N. Zhou, G. Zeng, J. Xiong, Quantum key agreement scheme. Electron. Lett. **40**(18), 1149 (2004)
2. C.C. Hsueh, C.Y. Chen, Quantum key agreement scheme with maximally entangled states, in *14th Information Security Conference (ISC 2004)*, (National Taiwan University of Science and Technology, Taipei, 2004), pp. 236–242
3. S.K. Chong, T. Hwang, Quantum key agreement scheme based on BB84. Opt. Commun. **283**(6), 1192–1195 (2010)
4. S.K. Chong, C.W. Tsai, T. Hwang, Improvement on "quantum key agreement scheme with maximally entangled states". Int. J. Quantum Inf. **50**(6), 1793–1800 (2011)
5. W. Huang, Q.Y. Wen, B. Liu, et al., Quantum key agreement with EPR pairs and single-particle measurements. Quantum Inf. Process **13**(3), 649–663 (2014)
6. W. Huang, Q. Su, X. Wu, et al., Quantum key agreement against collective decoherence. Int. J. Theor. Phys. **53**(9), 2891–2901 (2014)
7. Y.F. He, W.P. Ma, Two-party quantum key agreement against collective noise. Quantum Inf. Process **15**(12), 5023–5035 (2016)
8. Y.F. He, W.P. Ma, Two quantum key agreement schemes immune to collective noise. Int. J. Theor. Phys. **56**(2), 328–338 (2017)
9. H. Gao, X.G. Chen, S.R. Qian, Two-party quantum key agreement schemes under collective noise channel. Quantum Inf. Process **17**(6), 140 (2018)
10. S.S. Wang, D.H. Jiang, G.B. Xu, et al., Quantum key agreement with bell states and cluster states under collective noise channels. Quantum Inf. Process **19**(6), 190 (2019)
11. Y.G. Yang, S. Gao, D. Li, et al., Two-party quantum key agreement over a collective noisy channel. Quantum Inf. Process **18**(3), 74 (2019)
12. Y.H. Zhou, M.F. Wang, W.M. Shi, et al., Two-party quantum key agreement against collective noisy channel. Quantum Inf. Process **19**(3), 100 (2020)
13. Y.H. Zhou, Y. Xu, Y.G. Yang, et al., Measurement-device-independent quantum key agreement against collective noisy channel. Int. J. Theor. Phys. **61**(7), 201 (2022)
14. J.H. Guo, Z. Yang, M.Q. Bai, et al., Quantum key agreement schemes with GHZ states under collective noise channels. Int. J. Theor. Phys. **61**(3), 63 (2022)
15. R.H. Shi, H. Zhong, Multi party quantum key agreement with bell states and bell measurements. Quantum Inf. Process **12**(2), 921–932 (2013)
16. B. Liu, F. Gao, W. Huang, et al., Multiparty quantum key agreement with single particles. Quantum Inf. Process **12**(4), 1797–1805 (2013)
17. Z. Sun, B. Wang, Q. Li, et al., Improvement on "multiparty quantum key agreement with single particles". Quantum Inf. Process **12**(11), 3411–3420 (2013)
18. Y.F. He, W.P. Ma, Two-party quantum key agreement based on four-particle GHZ states. Int. J. Quantum Inf. **14**(1), 1650007 (2016)
19. S.Q. Min, H.Y. Chen, L.H. Gong, Novel multi-party quantum key agreement scheme with G-like states and bell states. Int. J. Theor. Phys. **57**(6), 1811–1822 (2018)
20. T. Cai, M. Jiang, G. Gao, Multi-party quantum key agreement scheme with G-like states and bell states. Quantum Inf. Process **17**(5), 103 (2018)
21. Y.G. Yang, B.R. Li, S.Y. Kang, et al., New quantum key agreement schemes based on cluster states. Quantum Inf. Process **18**(3), 77 (2019)
22. H.N. Liu, X.Q. Liang, D.H. Jiang, et al., Multi-party quantum key agreement with four-qubit cluster states. Quantum Inf. Process **18**(8), 242 (2019)
23. C. Shukla, K. Thapliyal, A. Pathak, Semi-quantum communication: Schemes for key agreement, controlled secure direct communication and dialogue. Quantum Inf. Process **16**(12), 295 (2017)
24. W.J. Liu, Z.Y. Chen, S. Ji, et al., Multi-party semi-quantum key agreement with delegating quantum computation. Int. J. Theor. Phys. **56**(10), 3164–3174 (2017)

25. N.R. Zhou, K.N. Zhu, Y.Q. Wang, Three-party semi-quantum key agreement scheme. Int. J. Theor. Phys. **59**(3), 663–676 (2020)
26. H. Guo, Y.X. Li, J.H. Wei, et al., Semi-quantum key agreement based on maximally three-qubit entangled states, in *ICCIP 2021: 2021 the 7th International Conference on Communication and Information Processing*, (2021), pp. 132–136
27. H.H. Li, L.H. Gong, N.R. Zhou, New semi-quantum key agreement scheme based on high-dimensional single-particle states. Chinese Physics B **29**(11), 110304 (2020)
28. N.R. Zhou, Q. Liao, X.F. Zou, Multi-party semi-quantum key agreement scheme based on the four-qubit cluster states. Int. J. Theor. Phys. **61**(4), 114 (2022)
29. C. Liu, S. Cheng, H.H. Li, et al., New semi-quantum key agreement scheme based on the χ-type entanglement states. Int. J. Theor. Phys. **61**(3), 60 (2022)
30. Zhou N R, Xu Y, Yang Y G, et al. Semi-quantum key agreement scheme against dishonest third-party with delegating quantum measuring. PREPRINT (Version 2) available at Research Square, 2022
31. L.L. Yan, S.B. Zhang, Y. Chang, et al., Mutual semi-quantum key agreement scheme using bell states. Mod. Phys. Lett. A **34**(35), 1950294 (2019)
32. H. Gao, R. Zhou, Multi-party quantum key agreement scheme based on G-like states and $\chi+$ states. Int. J. Theor. Phys. **61**(2), 16 (2022)
33. Z. Zhao, T. Yang, Z.-B. Chen, J. Du, J.-W. Pan, Deterministic and highly efficient quantum cryptography with entangled photon pairs. arXiv:quant-ph/0211098 (2002)
34. M. Zhang, H. Li, Weak blind quantum signature scheme based on entanglement swapping. Photonics Res. **3**(6), 324 (2015)
35. F.G. Deng, G.L. Long, Y. Wang, L. Xiao, Increasing the efficiencies of random-choice-based quantum communication schemes with delayed measurement. Chin. Phys. Lett. **21**, 2097 (2004)
36. B. Liu, D. Xiao, H.Y. Jia, R.Z. Liu, Collusive attacks to "circle-type" multi-party quantum key agreement schemes. Quantum Inf. Process **15**, 2113–2124 (2016)
37. J. Lin, T. Hwang, New circular quantum secret sharing for remote agents. Quantum Inf. Process **12**, 685–697 (2013)
38. A. Cabello, Quantum key distribution in the Holevo limit. Phys. Rev. Lett. **85**(26), 5635–5638 (2000)
39. Q.Y. Cai, Eavesdropping on the two-way quantum communication schemes with invisible photons. Phys. Lett. A **351**(1), 23–25 (2006)
40. F.G. Deng, X.H. Li, H.Y. Zhou, Z. Zhang, Improving the security of multiparty quantum secret sharing against Trojan horse attack. Phys. Rev. A **72**(4), 044302 (2005)
41. M. Boyer, D. Kenigsberg, T. Mor, Quantum key distribution with classical bob. Phys. Rev. Lett. **99**(14), 140501 (2007)
42. G.L. Long, X.S. Liu, Theoretically efficient high-capacity quantum-key-distribution scheme. Phys. Rev. Lett. **65**(3), 032302 (2002)
43. Z.X. Man, Z.J. Zhang, L. Yong, Quantum dialogue revisited. Chin. Phys. Lett. **22**(1), 22–24 (2005)
44. T.Y. Ye, Robust quantum dialogue based on the entanglement swapping between any two logical bell states and the shared auxiliary logical bell state. Quantum Inf. Process **14**(4), 1469–1486 (2015)
45. F. Gao, S.J. Qin, F.Z. Guo, Cryptanalysis of quantum secure direct communication and authentication scheme via bell states. Chin. Phys. Lett. **28**(2), 20303 (2011)
46. L. Yan, S. Zhang, Y.S. Chang, Z. W. and Sun Y., Hu.: Semi-quantum key agreement and private comparison schemes using bell states. Int. J. Theor. Phys. **58**, 3852–3862 (2019)

Chapter 7
Quantum Private Query

Abstract Quantum privacy query is a quantum solution to the symmetric privacy information retrieval problem. Quantum privacy query scheme mainly uses the basic principles of quantum mechanics to protect the privacy of user Alice and database Bob so that the privacy query can be completed safely. Quantum privacy query is different from quantum key distribution. In the quantum key distribution scenario, the parties trust each other and do not attack each other. However, in quantum privacy query, the two parties do not trust each other. They both have private information that they do not want the other party to have, so they need to prevent the other party from stealing their own private information.

7.1 Introduction

With the rapid development of computer technology and network technology, the database is widely used in various industries. The complex, dynamic and open application environment makes the database face more and more security threats. Early research on database security focuses on ensuring data confidentiality, integrity, reliability, and availability. Some existing security measures, such as access control mechanism, backup and recovery policy, role-based management, etc., have played a certain protective role, but still face many problems in practical application. Among them, a more important threat is the privacy issue, privacy disclosure has become one of the main obstacles to information sharing.

For privacy protection, privacy information retrieval (PIR) is put forward first, which is a strategy adopted to ensure the privacy of users on the public platform [1–4]. When the user retrieves the database information, the relevant technology can be used to prevent the database owner from obtaining the relevant information of user query statement; therefore, the purpose of protecting user privacy is achieved. However, there is a lot of sensitive information in the database itself; thus, it has become an important issue to ensure the privacy of the database when providing services. How to effectively protect private information, especially how to combine the convenience of database application services with the security of private information is a meaningful and challenging topic. In the classical world, this kind of

R.-G. Zhou et al., *Design and Analysis of Secure Quantum Communication Schemes*,
https://doi.org/10.1007/978-3-031-78428-6_7

problem is called symmetric privacy information retrieval [5] (SPIR), which is a generalization of PIR (PIR only considers the privacy security of users). As we know, for most of the previous classical cryptographic schemes, their security is based on a mathematical assumption [6]. But with the introduction of quantum algorithms, the security of such schemes is facing severe threats. Fortunately, this weakness can be overcome by the security of quantum cryptography based on the laws of physics [7]. In view of the great success of quantum cryptography in the distribution of secret keys, researchers are also gradually trying to use quantum properties to design cryptographic schemes, looking for secure and practical quantum methods to achieve privacy query.

7.2 Principle and Research Progress of Quantum Private Query

Quantum privacy query (QPQ) scheme refers to the use of quantum mechanics to deal with such problems. It mainly uses the basic principles of quantum mechanics to protect the privacy of user Alice and database Bob, so that the privacy query can be completed safely. QPQ is different from quantum key distribution (QKD). In the quantum key distribution scheme, the communication parties trust each other and will not attack each other actively. However, in the QPQ scheme, the communicating parties do not trust each other. They both have private information that they do not want the other party to get, so they need to prevent the other party from stealing their own private information.

QPQ is not completely secure. This scheme is different from the quantum key distribution scheme. In the QKD scheme, the communication parties trust each other and will not attack each other actively. In the QPQ scheme, the communicating parties do not trust each other. They both have private information that they do not want the other party to get. Therefore, if you want to protect the privacy of both parties, the security requirements of QPQ are as follows:

(a) The database Bob deceives the query-taker Alice to obtain user Alice's query location. Fraud was detected by Alice with a nonzero rate;
(b) The query-taker Alice can randomly obtain a limited number of database elements in addition to the query elements. Alice's extraordinarily obtained entries are random and generally not what she needs. Bob doesn't usually attack at the risk of deceiving. If he is discovered, it will damage his credibility.

In 2008, Giovannetti et al. proposed the first QPQ scheme [8] (GLM scheme), in which the database is represented by an oracle operation. Compared with the classical SPIR schemes, G-scheme has an exponential decrease in computational and communication complexity [9]. Nevertheless, the scheme is not loss tolerant and the oracle dimension is very high when a large database is concerned. In order to solve this problem, Jacobi et al. proposed a significant QPQ scheme based on quantum key distribution [10] (J-scheme). In J-scheme, the database encrypted by an oblivious key is sent to the user. It is worth noting that the oblivious key means that the database owner knows all the key bits in the key sequence while the user only knows part of the key bits. Obviously, it is necessary to dilute the key bits known by the user to almost 1 bit in order to meet 1 out of N oblivious transfer. The method of dilution is also called post-processing algorithm. Most of the subsequent QPQ schemes are based on the construction of the oblivious key distribution [11–21] and post-processing algorithms [22–24] to achieve private query.

In 2012, Gao et al. implemented a generalized version of J-scheme using quantum states in $\{|0\rangle, |1\rangle, cos\theta|0\rangle + sin\ \theta|1\rangle, sin\theta|0\rangle - cos\ \theta|1\rangle\}$ [21]. By adjusting the value of a parameter θ, Gao et al.'s scheme shows higher communication efficiency and better user privacy security. However, considering quantum resources are not easy to prepare, Yang et al. designed a QPQ scheme using only one quantum state in 2015 [25]. Except for analyzing the communication efficiency and computational complexity, the security of database should also be taken into consideration when designing a QPQ scheme. The JM attack is the most effective attack method in QPQ scheme, which poses a significant threat to the security of database. Therefore, Wang et al. have presented a QPQ scheme that follows the procedure "receive → measure → re-prepare → send" to resist JM attack [26]. In real scenarios, QPQ schemes must consider the influence of noise on key. Therefore, Yang et al. presented a robust QPQ scheme based on the property that DF states can resist collective noise [27].

The abovementioned QPQ schemes focus on how to distribute oblivious key. In fact, effective post-processing cannot only dilute the user's conclusive key bits to nearly 1 bit, but also realize privacy queries with certain characteristics. In order to ensure that the conclusive key bits in the final key obtained by user are close to 1 bit, Jacobi proposed a kN → N post-processing algorithm in J-scheme, that is, the kN bits raw key is cut into k substrings of length N bits and add these strings bitwise. In order to reduce the communication complexity, Yang et al. proposed an effective and secure N → N post-processing algorithm [22], which only needs N-bit raw key to dilute the user's conclusive key bits to nearly one bit. By sacrificing communication complexity, Gao et al. presented gkN → N postprocessing method [23] to correct the errors in the final key. Recently, Liu et al. designed a subtle post-processing algorithm [24], which can resist JM attack by introducing a random vector into the post-processing process to eliminate the linear relationship between the raw key and the final key.

The above QPQ scheme is all for a single user to query the database. Recently, a multiuser QPQ scheme was proposed by Ye et al. [28]. However, Ye et al. only implemented the QPQ scheme [21] repeatedly according to the number of users, which greatly increased the communication complexity. In this paper, we first utilize the multiparticle symmetric W state [29–31] to make an oblivious key distribution among multiple users to complete the multiuser query in a particular situation, where only users who work together can query an item of common interest in the database.

7.3 Typical Quantum Private Query Scheme

7.3.1 QPQ Scheme Based on QKD

J-scheme is the first QPQ scheme [10] based on QKD. Suppose Bob has a database with N entries, and Alice wants to buy one of them. The main process for the J-scheme is as follows:

Step 1: Bob sends Alice a string of long random qubits with states $\{|0\rangle, |1\rangle, |+\rangle, |\text{-}\rangle\}$, where $|0\rangle$ and $|1\rangle$ represent 0 and $|+\rangle$ and $|\text{-}\rangle$ are encoded with 1.

Step 2: For each qubit that Alice receives, she randomly measures them using either the Z basis or the X basis.

Step 3: Alice announces which positions have been successfully measured, and then both participants should ignore all lost photons.

Step 4: For each qubit that Alice has successfully measured, Bob declares a pair of states, one of which is the quantum state he sent, and the other of which is randomly selected from the corresponding states of another base. For example, if Bob sends a status of $|+\rangle$, he will randomly declare $\{|+\rangle, |0\rangle\}$ or $\{|+\rangle, |1\rangle\}$.

Step 5: Alice will extrapolate her measurements in Step 2. Based on the state information published by Bob and the results measured by Alice using random basis, Alice can get the sent bit with a certain probability. Now, Alice and Bob share a raw key, K^r, where Bob knows all the keys and Alice only knows a few of them.

Step 6: Alice and Bob divide the original key into k sub-strings of length N bit (k is a security parameter, N is the number of entries in the database), then add each sub-string to get the final key K^f of length N bit. So far, Bob knows all the final keys, K^f, while Alice knows only about a 1-bit key.

Step 7: After Alice and Bob in Step 6, if Alice doesn't get any key bits, the scheme must be restarted.

Step 8: Bob encrypts the database, and Alice decrypts and gets the data she wants to buy. If the final key K^f is negotiated correctly between the participants, assume that Alice knows the key K_j^f of the j bit and wants to obtain the database's i data itemX_i. Alice announces a shift value, $s = j - i$, and Bob moves the final key by s bits and sends Alice N bits of encrypted data, $C_n = X_n \oplus K_{n+s}^f$; eventually, Alice uses the 1 bit key she owns to get her desired data X_i through $C_i = X_i \oplus K_j^f$.

7.3.2 QPQ Using W State

There is a problem in all proposed QPQ schemes. That is, if the querying party has more than one secret key number, then it is possible to perform a joint measurement on the database (this kind of attack has been proved in J-scheme [10] that the risk of a joint measurement attack is much greater than that of a single measurement). In order to solve this problem, in 2019, Li put forward a scheme based on GHZ state [32], the number of keys Alice can obtain is strictly controlled by a third party. However, the GHZ state is the largest entangled state, which is difficult to prepare, and the transmission is not stable in noisy information, and Li-scheme cannot resist the auxiliary system measurement attack, which brings great security threat to the database. Therefore, inspired by the work of Li, a QPQ scheme based on W state is now proposed [30]. This scheme not only retains the advantages of Li- scheme but also has better anti-attack ability.

7.3.2.1 Three-Qubit W State

The two basic single quantum measurement bases:

1. $\{|0\rangle, |1\rangle\}$ is a standard set of measurement bases, called B_z.
2. $\{|+\rangle, |-\rangle\}$ is also forms an orthonormal basis, called B_X.

Among them,

$$|+\rangle = \frac{1}{\sqrt{2}}(|0\rangle + |1\rangle), |-\rangle = \frac{1}{\sqrt{2}}(|0\rangle - |1\rangle) \tag{7.1}$$

From the above relation can be obtained

$$|0\rangle = \frac{1}{\sqrt{2}}(|+\rangle + |-\rangle), |1\rangle = \frac{1}{\sqrt{2}}(|+\rangle - |-\rangle) \tag{7.2}$$

It can be seen that the two sets of orthogonal bases $|B_z\rangle$ and $|B_X\rangle$ are nonorthogonal, that is, if one set of measurement bases is used to measure the particles in the other set of measurement basis states, the measurement results are uncertain.

For three-particle entangled state, the total is divided into two categories: Maximum entangled state (GHZ state) and nonmaximum entangled state (W state) [33, 34], although the degree of entanglement of the GHZ state is the largest, but its quantum entanglement is "fragile", as long as do to either particle trace operation or lost part, the remaining two body would collapse into separable state, and in a two-level system, the three body W state is an analogy GHZ state of quantum entanglement entangled state of "strong": for any particle trace is still remaining two body parts.

A simple proof is as follows. First of all, the general form of a GHZ is:$|GHZ\rangle = \frac{1}{\sqrt{2}}(|000\rangle + |111\rangle)$. If one of the particles in GHZ state is measured as 0, then the remaining two particles can be written as $|0\rangle \otimes |0\rangle$, which means that the remaining two particles are nonentangled states. Then if the form of a W state (the quantum circuit diagram for preparing W state is shown in Fig. 7.1 below) is $|W\rangle = \frac{1}{\sqrt{3}}(|001\rangle + |010\rangle + |100\rangle)$, Similarly, when one of the particles is

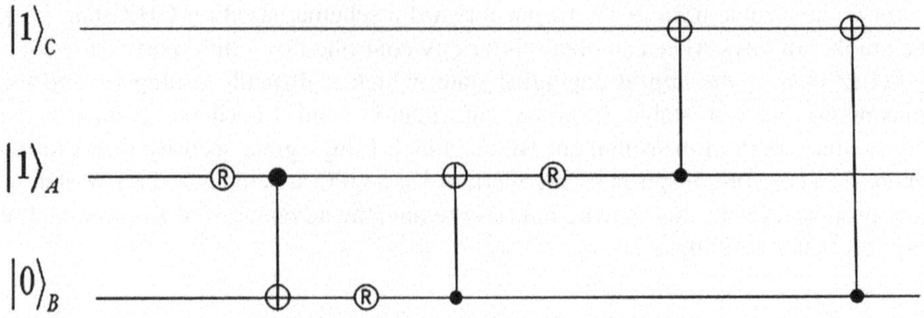

Fig. 7.1 Preparation of quantum circuits of three-particle W states

measured as 0, the remaining two particles ($|01\rangle + |10\rangle$) cannot be written as tensor product. Therefore, compared with choosing GHZ state, choosing W state is more practical.

The quantum physical principles that guarantee the security are as follows: Monogamy of Entanglement and No-Cloning. Among them, the essence of the entangled monogamy is that unlike the classical information, the information between them can be shared with each other. For the entangled particles, the information shared between them cannot be obtained by other parties in any way, this phenom;non is called monogamy of entanglement.

7.3.2.2 Specific Steps

The QPQ scheme is divided into three parts:

1. Preparation of W state and nonorthogonal state (Step 1).
2. Particle distribution and the first safety inspection (Step 2–3).
3. Measurement inquiry (Step 4–8).

The detail shows in Fig. 7.2,

Step 1: Charlie prepares some W states and nonorthogonal states, in which the W state is

$$|W\rangle = \frac{1}{\sqrt{3}}(|001\rangle + |010\rangle + |100\rangle)_{123} \tag{7.3}$$

The footer 123 represents particle 1, particle 2, and particle 3, respectively, in which the nonorthogonal state is $\{|0\rangle, |1\rangle, |+\rangle, |-\rangle\}$.

Step 2: Charlie keeps particle 1, distributes particle 2 to Alice and particle 3 to Bob, respectively. In order to ensure the safe transmission of particle sequence, a group of nonorthogonal particles $\{|0\rangle, |1\rangle, |+\rangle, |-\rangle\}$are randomly inserted into the distributed particle sequence, and each particle is randomly in one of four states.

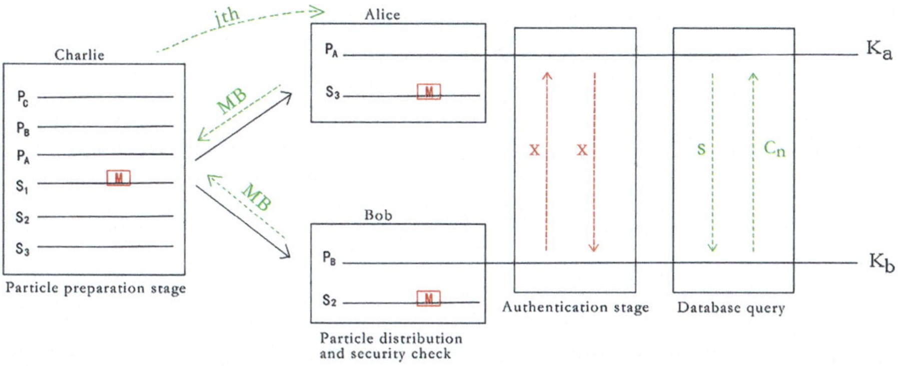

Fig. 7.2 The process of QPQ scheme is shown, where red represents the security detection process and green represents the operation performed by the query process, and solid lines represent quantum channels, and dashed lines represent classical channels

Step 3: After the distribution of particles, both parties participating in the inquiry process will publicly declare which positions of particles were not successfully received, and all participants will lose the particles that were not successfully received. It is worth noting that in this statement, they will not lie, because this Step does not reveal any private information, and lying will not get any benefits. Then in order to check whether the entangled particles are distributed safely, non-orthogonal particles are used for safety check. If the error rate is lower than the threshold, proceed to the next step; otherwise, return to Step 2 and start again.

Step 4: Next, Alice, Bob, and Charlie measure the particle sequence in their hands by randomly measuring the basis (X basis or Z basis), and record the selected measurement bases and measurement results.

Step 5: Alice and Bob tell Charlie over the classical channel which basis of measurement they have chosen. Charlie, according to the measurement base sequence published by Alice and Bob just now, randomly selects a W state, for instance, the j-th W state, which is measurement under Z basis by all parties. Then Charlie secretly tells Alice the information of this bit and her measurement results about this bit.

Step 6: Alice will infer Bob's measurement results based on the secret information Charlie told her. If Alice and Charlie have the same measurement result, Alice can infer that Bob's measurement result is $|1\rangle$; if their measurement result is different, Bob's measurement result is $|0\rangle$. Alice uses Bob's measurement of this state as her own key k_a, and Bob records all his measurement results as k_b.

Step 7: Through the above Steps, a raw secret key is created between Alice and Bob, where Bob knows the entire message of the secret key and Alice only knows one of them. Then Bob secretly sends a message X to Alice. Before Alice makes a privacy query, she must firstly send X to Bob to verify whether it is a legitimate user. If the authentication is successful, the database allows the query; if the authentication fails, the scheme will be terminated.

Table 7.1 The effects of Alice's and Charlie's measurements on Bob's state

<table>
<tr><td colspan="2"></td><td colspan="4">Charlie</td></tr>
<tr><td colspan="2">Bob</td><td>$|0\rangle$</td><td>$|1\rangle$</td><td>$|{+x}\rangle$</td><td>$|{-x}\rangle$</td></tr>
<tr><td rowspan="4">Alice</td><td>$|0\rangle$</td><td>$|1\rangle$</td><td>$|0\rangle$</td><td colspan="2" rowspan="2">When Charlie's measurement result is $\frac{1}{\sqrt{2}}(|{+x}\rangle + |-x\rangle)$, the result of Alice and bob is opposite</td></tr>
<tr><td>$|1\rangle$</td><td>$|0\rangle$</td><td>Nothingness</td></tr>
<tr><td>$|{+x}\rangle$</td><td colspan="2" rowspan="2">When Alice's measurement result is $\frac{1}{\sqrt{2}}(|{+x}\rangle + |-x\rangle)$, the result of Charlie and bob is opposite,</td><td>$|0\rangle$ or $|1\rangle$</td><td>$|1\rangle$</td></tr>
<tr><td>$|{-x}\rangle$</td><td>$|1\rangle$</td><td>$|0\rangle$ or $|1\rangle$</td></tr>
</table>

Step 8: In the query phase, suppose she knows the j-th bit k_j and wants the i-th bit of the database x_i. She then announces the number $s = j\text{-}i$ in order to allow Bob to encode the database by bitwise adding, shifted by s. So Bob announces N bits $c_n = x_n \oplus k_{n+s}$, where Alice can read $c_i = x_i \oplus k_j$ and thus obtain x_i.

Let's see how this works in more detail. The first thing to understand is that the effects of measurements by Alice and Charlie on the state of Bob's particle. For example, we can write

$$|W\rangle = \frac{1}{\sqrt{3}}\left[\frac{1}{2}(|+\rangle|+\rangle + |+\rangle|-\rangle + |-\rangle|+\rangle + |-\rangle|-\rangle)|1\rangle + (|+\rangle|+\rangle\text{-}|-\rangle|-\rangle)|0\rangle\right] \tag{7.4}$$

As can be seen from Eq. 7.4, the decomposition of $|W\rangle$ tell us what happens if both Alice and Charlie make measurements in the x direction. If they both get the same result, Bob's measurement results are uncertain; if they get different results, he will have the state $|1\rangle$. The following Table 7.1 summarizes the effects of Alice's and Charlie's measurements on Bob's state.

7.3.2.3 Features

Quantum secret keys are distributed in just two ways: based on single particles and based on entangled states. When using a single particle distribution for privacy query, multiple key numbers can be derived through query, and then joint measurement attack can be launched. According to Ref. [10], if Alice launches a joint measurement attack, the probability that the value of the final secret key bit can be correctly obtained is $p_{guess} = \frac{1}{2} + \frac{1}{2\sqrt{2^k}}$. To make it clear that joint measurements pose a significant risk to the database, let's assume that $k = 7$, analyze the differences between the number of bits that can be obtained by joint measurement attacks and normal conditions in different databases. From Fig. 7.3, it is clear that in the same situation, as the database gets larger and larger, the gap between the number of bits that can be obtained by using the joint measurement attack and the normal situation presents an exponential gap, which is very bad when faced with a large database.

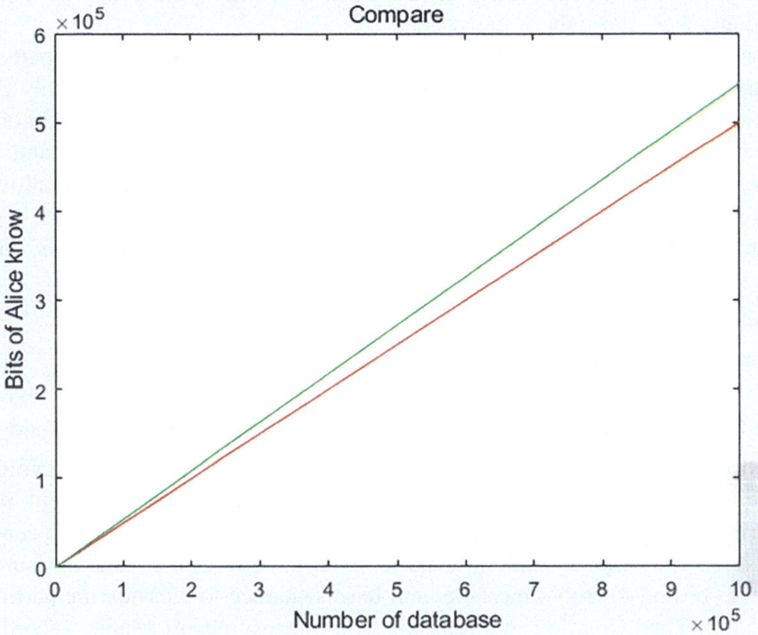

Fig. 7.3 Compare the bit value difference that Alice can get under two different conditions, where green means using joint measure attack and red means not using joint measure attack

It is also mentioned in Ref. [35] that EPR (entanglement state) can be used to conduct privacy query, and the proportion of Alice's raw key bits in the whole raw key up to *1/6*. So even if use entangled states, there is no perfect implement *1* out of *N*. Therefore, the proposed QPQ scheme proposes to distribute the secret key based on W state, using the three-particle W state, the inquirer can infer the measurement result of the inquirer only after receiving the secret information sent to her by the third party, so the number of key bits Alice can obtain is controlled by the third party. Thus, there is no JM attack. The security of the scheme has been greatly improved.

From the above analysis, it is necessary to put forward the QPQ scheme.

7.3.2.4 Security Analysis

To realize a secure quantum private query scheme, two security requirement, database security and user privacy, need to be satisfied simultaneously. The possible attacks in the query process are as follows: entanglement attack and auxiliary system attack can be used for the quantum state preparation side, quantum memory attack and joint measurement attack can be used for the receiver side. (Noting that in our scheme there is no joint measurement, because with the help of Charlie, only one secret key information is obtained at one time). In addition, there may be malicious external attackers throughout the system.

Although there are three participants in this scheme, Charlie, the third party, only plays an assisting role in the whole inquiry process and does not own any private information. Only Alice and Bob have private information. Therefore, without loss of generality, it is also necessary to analyze Charlie's attack on Alice and Bob's privacy. (One more thing that needs to be said in advance, Charlie is not allowed to collude with either Alice or Bob.)

Then all potential security threats are analyzed from these three aspects, and it is proved that the QPQ scheme can resist these attacks.

A. The Third Party's Attack

1. **Entangle-Measure Attack**
   Let's assume that Charlie's target is Bob. Specific as follows:Charlie prepares a fake entangled state in the form of $|0\rangle \otimes \frac{(|01\rangle+|10\rangle)}{\sqrt{2}}$ (Because Charlie knows the position of nonorthogonal particles in this sequence, she can easily avoid Step 3 of safety detection.). Then in order to obtain Bob's measurement results, Charlie will not measure his own particles immediately, but after Bob sends his measurement base sequence to Charlie, Charlie chooses the same measurement base according to Bob's measurement base sequence to measure the particles in her hand. Then Charlie can obtain the Bob's measurement results, according to the expansion equation under different measurement bases in the following:

$$\frac{1}{\sqrt{2}}(|01\rangle + |10\rangle) = \frac{1}{\sqrt{2}}(|+x\rangle|+x\rangle - |-x\rangle|-x\rangle) \tag{7.5}$$

   Then once Charlie gets Bob's measurement results, he can pretend to be a user to query to get the information in the database and make Bob use it to move his own measurement results, encrypt database X_n, and send E_n. At this time, although he successfully escaped the deceptions security check in Step 3, Charlie was still unable to obtain information about the database. Because the user needs to send message X to Bob before the query, once the authentication fails, the data of the database will be blocked and will not be open until the real user sends the correct message X. So even if launch entanglement attack, Charlie will not be able to get any information about the database.

   For Alice's privacy, the situation is similar to an attack on Bob, even though Charlie knows Alice's measurements, Charlie is still unable to obtain private information about Alice, because Charlie does not know s that Alice secretly sends to Bob.
2. **Entangled Auxiliary Particle Attack**
   In addition, Charlie can also launch entangled auxiliary particle attacks. The specific performance is as follows: Charlie entangled the prepared auxiliary particles into Bob's particle sequence through a local unitary operation and attempted to obtain the measurement result about Bob by measuring his own additional particle at a later time. Then Charlie can use auxiliary particles to

obtain measurement results about Alice and Bob because he knows where the particles used for safety detection are located, but Charlie still can't get the user's private information and data in the database because he doesn't know s and X.

What's more, compared with Ref. [32], this scheme has better performance against auxiliary measurement attacks.

In detail, in Ref. [32], when Charlie attaches an auxiliary system,

$$\bar{U}|0\rangle|E_{\mathrm{i}}\rangle = a|0\rangle|\delta_{00}\rangle + b|1\rangle|\delta_{01}\rangle \tag{7.6}$$

$$\bar{U}|1\rangle|E_{\mathrm{i}}\rangle = c|0\rangle|\delta_{10}\rangle + d|1\rangle|\delta_{11}\rangle \tag{7.7}$$

Then Charlie needs to meet the following conditions to accurately obtain information about Alice and Bob by measuring his own additional particles: $b = c = 0$. Although Charlie cannot obtain Alice's private information (because he does not know s), he can query the data in the database disguised as a legitimate user. However, this will not happen in this proposed QPQ scheme, because "second authentication" is added in Step 7, only legitimate users can carry out the database query. So even if Charlie can get Bob's measurements by measuring the auxiliary particles, he can't get the data from the database.

A. External Attack

Suppose that there is an external attacker Eve who can intercept particles prepared by Charlie and then Eve has been able to entangle ancilla particles $|E\rangle = \{|\mathrm{E}_1\rangle, |\mathrm{E}_2\rangle \ldots\}$ with the particle state that Alice, Bob, and Charlie are using. (Note that external attackers Eve are not aware of the existence of inspection particles) At some later time she can measure the ancilla to gain information about the measurement results of Alice, Bob, and Charlie.

Without loss of generality, we suppose Eve's attack on the qubit as follow:

$$\bar{U}|0\rangle|E_{\mathrm{i}}\rangle = a|0\rangle|\delta_{00}\rangle + b|1\rangle|\delta_{01}\rangle \tag{7.8}$$

$$\bar{U}|1\rangle|E_{\mathrm{i}}\rangle = c|0\rangle|\delta_{10}\rangle + d|1\rangle|\delta_{11}\rangle \tag{7.9}$$

$$\begin{aligned}\bar{U}|+\rangle|E_{\mathrm{i}}\rangle &= \frac{1}{\sqrt{2}}(a|0\rangle|\delta_{00}\rangle + b|1\rangle|\delta_{01}\rangle + c|0\rangle|\delta_{10}\rangle + d|1\rangle|\delta_{11}\rangle) \\ &= \frac{1}{2}[|+\rangle(a|\delta_{00}\rangle + b|\delta_{01}\rangle + c|\delta_{10}\rangle + d|\delta_{11}\rangle) + |-\rangle(a|\delta_{00}\rangle \\ &\quad - b|\delta_{01}\rangle + c|\delta_{10}\rangle - d|\delta_{11}\rangle)]\end{aligned} \tag{7.10}$$

$$\begin{aligned}\bar{U}|-\rangle|E_{\mathrm{i}}\rangle &= \frac{1}{\sqrt{2}}(a|0\rangle|\delta_{00}\rangle + b|1\rangle|\delta_{01}\rangle - c|0\rangle|\delta_{10}\rangle - d|1\rangle|\delta_{11}\rangle) \\ &= \frac{1}{2}[|+\rangle(a|\delta_{00}\rangle + b|\delta_{01}\rangle - c|\delta_{10}\rangle - d|\delta_{11}\rangle) + |-\rangle(a|\delta_{00}\rangle \\ &\quad - b|\delta_{01}\rangle - c|\delta_{10}\rangle + d|\delta_{11}\rangle)],\end{aligned} \tag{7.11}$$

where a, b. c, $d \in [0, 1]$, $|E_i\rangle$is an additional particle of Charlie. And $|a|^2 + |b|^2 = 1$, $|c|^2 + |d|^2 = 1$, $|\delta_{00}\rangle$, $|\delta_{01}\rangle$, $|\delta_{10}\rangle$, $|\delta_{11}\rangle$ is a quantum state that Eve needs to distinguish when measuring. If Eve wants to successfully avoid Step 3 of safety inspection, she must meet the following conditions:

$$\begin{cases} b = c = 0 \\ a|\delta_{00}\rangle - b|\delta_{01}\rangle + c|\delta_{10}\rangle - d|\delta_{11}\rangle = 0 \\ a|\delta_{00}\rangle + b|\delta_{01}\rangle - c|\delta_{10}\rangle - d|\delta_{11}\rangle = 0 \end{cases} \tag{7.12}$$

It is easy to see that this kind of attack is also ineffective. It can be obtained from the above conditions: $a|\delta_{00}\rangle = d|\delta_{11}\rangle$. This equation means that Eve can't distinguish $|\delta_{00}\rangle$and $|\delta_{11}\rangle$ if she wants to avoid security detection successfully. Therefore, this kind of attack will still be found with a probability. Obviously, $\mathrm{p}_0 = 1 - a^2$, $\mathrm{p}_1 = 1 - b^2$, $\mathrm{p}_+ = \frac{1}{4}\langle\phi_-|\phi_-\rangle$, $\mathrm{p}_- = \frac{1}{4}\langle\phi_+|\phi_+\rangle$ In which $\phi_- = a|\delta_{00}\rangle - b|\delta_{01}\rangle + c|\delta_{10}\rangle - d|\delta_{11}\rangle$, $\phi_+ = a|\delta_{00}\rangle + b|\delta_{01}\rangle - c|\delta_{10}\rangle - d|\delta_{11}\rangle$. To sum up, the probability of being discovered is: $P_{found} = p_0 + p_1 + p_+ + p_-$.

B. Inside Participant's Attack

Since the third party does not carry any private information, Alice and Bob will only attack each other.

First, analyze it from Bob's point of view, Bob will attack Alice to get the user's search address, that is, such as the things she plans to buy, the stocks she plans to buy and so on, which may reveal some important private information of the inquirer. In principle, these are private information that should be protected. Then in order to get user privacy, Bob must know which key bits Alice knows or doesn't know, so that he can successfully infer Alice's private information with high probability. In addition, Bob must guess the original key bit value correctly when Alice knows the key bit; otherwise. he may provide Alice with wrong database information, and then be found to be "dishonest". Therefore, if Bob wants to infer user privacy without being discovered by Alice, he must ensure that he can clearly know which raw key bits Alice knows, and can correctly know their specific values, or which secret keys Bob can know are unknown to Alice.

In the QPQ scheme, Bob is not the producer or distributor of particles, so entanglement measurement and auxiliary system attack cannot be implemented. Not only that, Bob never knows Alice's measurement basis during the whole scheme process, so he cannot carry out quantum memory attacks.

Then analyze from Alice's point of view, different from Bob, Alice can perform quantum memory attacks. When Alice launched an attack, she did not directly measure the particles in her hand but waited for Bob to announce the measurement basis and use the same basis as Bob to measure. Nevertheless, without the assistance of the third-party, Alice cannot infer Bob's measurement results from her own measurements.

7.3.2.5 Discussion

Here we mainly make two points of discussion:

1. First, Ref. [32] proposed a scheme with multiple third parties. The quantum state in that scheme is the GHZ state. However, there are two problems:
    1. GHZ is the maximum entangled state, and its performance is not stable and robustness is not good in the transmission process.
    2. The privacy query scheme proposed in Ref. [32] cannot resist the auxiliary system measurement attack, which brings great security threat to the database.

    Therefore, inspired by that scheme, this QPQ scheme proposes to conduct privacy query in W state. Although the quantum W state does not belong to the maximum entangled state between three particles like the GHZ state, it is more robust than the GHZ state, and W state can better maintain the quantum entanglement property in the case of particle loss. That is to say, for the three particles in the state of W, even if any one of them is lost, the remaining two particles still remain entangled. Moreover, some research show that the state of W is the most "strongest" three-body two-level entangled state, while the GHZ state is the most "fragile" three-body two-level entangled state [36–39]. Therefore, using quantum entanglement and nonlocality of W state to query quantum privacy not only retains the advantages of the original privacy query scheme based on GHZ state but also has better practical significance. Moreover, the privacy scheme proposed has two security verifiers, which fully guarantees the privacy of the querying party and the security of the queried party.
2. Second, to estimate the efficiency of the private query, define the length of the database as *N*. In Table 7.2, the efficiency of the proposed QPQ scheme is compared with G-scheme [8], J-scheme [10], Y-scheme [40]. From the table, in this scheme, Alice can only get one secret key information with the help of a third party, so there is no need for complicated post-processing. What's more, in such cases, joint measurements cannot be made.

Table 7.2 Comparison between proposed QPQ scheme and other QPQ schemes

	G-scheme [8]	J-scheme [10]	Y-scheme [40]	Our scheme
Quantum state	N-qubit register	Single qubit	Nonorthogonal quantum state	W state
Qubit efficiency	1 or 0	1	1/2	1
For the raw key sequence, Alice can know the proportion	1/N	1/4	$sin^2\theta/2$	1/N
Can the scheme resist malicious intrusion attacks by third parties?	No	No	No	Yes
Can the scheme resist joint measurement attack?	Yes	No	No	Yes

7.3.2.6 Summary

From the above discussion, the proposed QPQ scheme can realize the function of protecting privacy and security between both parties involved in the query.

In general, this section has presented a novel QPQ scheme with W state, the scheme is like an improved version of Ref. [32]. This QPQ scheme well retains the advantages of using GHZ state for privacy query. Because of the entangled state, with the help of a third party, the key bit information obtained by the query can be limited to one bit, so there is no need for a classical postprocessing process, so it is naturally impossible to perform joint measurement attack (this attack is considered as a powerful attack). Moreover, compared with Ref [32], this QPQ scheme has better performance. The first selected quantum state is W state. Compared with the GHZ state, the performance of W state in transmission is more stable. And two security checks are set in the query, which can effectively resist external attacks and malicious attacks from third parties, so the security is relatively higher. In addition, the proposed QPQ scheme only considers the query of a single bit. In the process of practical application, it may be meaningful for the user to obtain multiple bits in one query, the QPQ scheme can be achieved with the help of a third party. Of course, we also hope to achieve a higher breakthrough in the following research.

7.3.3 MQPQ Scheme

In this section, we first utilize the multiparticle symmetric W state to make an oblivious key distribution among multiple users to complete the multiuser query in a particular situation [41], where only users who work together can they query an item of common interest in the database. Different from Ye's QPQ scheme [28], the scheme we proposed only carries out key distribution once to realize multiuser common query instead of repeating key distribution among different users, which significantly improves communication efficiency. In addition, with the increase in the number of users, the more sensitive the users are to the cheating of the database. Moreover, the security parameter k decreases as the number of users increases, which makes the scheme practical.

To simplify the process, the following QPQ scheme starts with two users. As shown in Fig. 7.4, Alice1 and Alice2 are two users who will work together to gain the same entry d_j in Bob's database $D = \{d_1, d_2, d_3 \ldots d_N\}$. As a supplement, users (Alice1 and Alice2) are sometimes referred to as Alice. Here we introduce a semi-honest third-party Trent to assist Bob in key distribution.

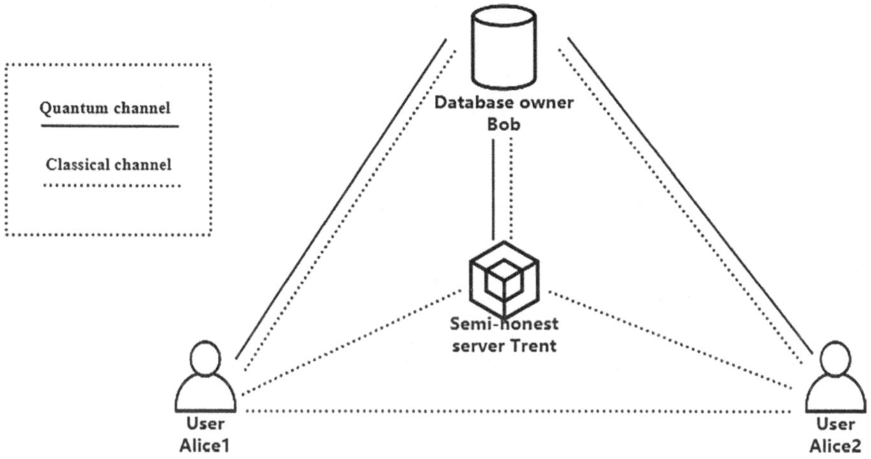

Fig. 7.4 QPQ scheme model with two users

7.3.3.1 Four-Qubit Symmetric W State

The four-qubit symmetric W state $|W_4\rangle = \frac{1}{2}(|1000\rangle + |0100\rangle + |0010\rangle + |0001\rangle)_{1234}$ can be written in $Z-Z-X-X$, $Z-X-Z-X$, $X-Z-X-Z$, $Z-X-X-Z$, $X-Z-Z-X$, and $X-X-Z-Z$ bases, the formula is represented as follows:

$$
\begin{aligned}
|W\rangle_{1234} =& \frac{1}{2}|00\rangle_{12}(|{+}{+}\rangle - |{-}{-}\rangle)_{34} + \frac{1}{4}(|10\rangle + |01\rangle)_{12}(|{+}{+}\rangle + |{+}{-}\rangle + |{-}{+}\rangle + |{-}{-}\rangle)_{34} \\
=& \frac{1}{2}(|0{+}0{+}\rangle - |0{-}0{-}\rangle)_{1234} + \frac{1}{4}(|1{+}0{+}\rangle + |1{+}0{-}\rangle + |1{-}0{+}\rangle + |1{-}0{-}\rangle)_{1234} \\
&+ \frac{1}{4}(|0{+}1{+}\rangle + |0{+}1{-}\rangle + |0{-}1{+}\rangle + |0{-}1{-}\rangle)_{1234} \\
=& \frac{1}{2}(|{+}0{+}0\rangle - |{-}0{-}0\rangle)_{1234} + \frac{1}{4}(|{+}1{+}0\rangle + |{+}1{-}0\rangle + |{-}1{+}0\rangle + |{-}1{-}0\rangle)_{1234} \\
&+ \frac{1}{4}(|{+}0{+}1\rangle + |{+}0{-}1\rangle + |{-}0{+}1\rangle + |{-}0{-}1\rangle)_{1234} \\
=& \frac{1}{2}(|0{+}{+}0\rangle - |0{-}{-}0\rangle)_{1234} + \frac{1}{4}(|1{+}{+}0\rangle + |1{+}{-}0\rangle + |1{-}{+}0\rangle + |1{-}{-}0\rangle)_{1234} \\
&+ \frac{1}{4}(|0{+}{+}1\rangle + |0{+}{-}1\rangle + |0{-}{+}1\rangle + |0{-}{-}1\rangle)_{1234} \\
=& \frac{1}{2}(|{+}00{+}\rangle - |{-}00{-}\rangle)_{1234} + \frac{1}{4}(|{+}10{+}\rangle + |{+}10{-}\rangle + |{-}10{+}\rangle + |{-}10{-}\rangle)_{1234} \\
&+ \frac{1}{4}(|{+}01{+}\rangle + |{+}01{-}\rangle + |{-}01{+}\rangle + |{-}01{-}\rangle)_{1234} \\
=& \frac{1}{2}(|{+}{+}\rangle - |{-}{-}\rangle)_{12}|00\rangle_{34} + \frac{1}{4}(|{+}{+}\rangle + |{+}{-}\rangle + |{-}{+}\rangle + |{-}{-}\rangle)_{12}(|10\rangle + |01\rangle)_{34}
\end{aligned}
\tag{7.13}
$$

From Eq. 7.13, once the four-qubit symmetric W state is measured with the six sequences of single-qubit measurements $\{Z_1Z_2X_3X_4, Z_1X_2Z_3X_4, X_1Z_2X_3Z_4, Z_1X_2X_3Z_4, X_1Z_2Z_3X_4, X_1X_2Z_3Z_4\}$, mutual relations among the measurement results can be established. To illustrate these relations concisely, it is defined that $Z+$ denotes the measurement result $|0\rangle$ if Z basis is used, and $X+$ denotes the measurement result $|+\rangle$ if X basis is used. The mutual relations among the measurement results of the above measurement base sequences are exhibited in Table 7.3. It is obvious that if the two qubits in the four-qubit symmetric W state are $Z+$, then the other two qubits must be collapsed into $(X+, X+)$ or $(X-, X-)$. On the contrary, if the two qubits in the four-qubit W state are different, that is, one is $Z+$ and the other is $Z-$, then the other two qubits may be collapsed into one of the four different combinations which are described as $(X+, X+)$, $(X+, X-)$, $(X-, X+)$ and $(X-, X-)$.

It is worth noting that there are also other combinations of bases, such as $X-X-X-X$, $Z-X-X-X$, $Z-Z-Z-X$ and so on, which are not very helpful. This is because with the increase of the dimension of symmetric W state, like $|W_n\rangle = \frac{1}{\sqrt{n}}(|100...0\rangle + |010...0\rangle + |001...0\rangle + \ldots + |000...1\rangle)_{123...n}$, the measurement outcomes of other measurement bases like $X_1X_2X_3...X_n$, $Z_1X_2X_3...X_n$, $Z_1Z_2Z_3...X_n$ and so on do not have stable mutual relations. Fortunately, for the n-qubit W symmetric state ($n \geq 4$), if we choose a measurement base sequence composed of $(n-2)$ Z bases and two X bases, like $X_1X_2Z_3Z_4...Z_n$, the correlations between measurement results similar to those in Table 7.3 can be obtained. That is, if the $n-2$ qubits of n-qubit symmetric W state measured by Z basis are both $Z+$, then the results of the remaining two qubits measured by the X basis are all $X+$ or $X-$; otherwise, the qubits measured by X basis may be in $(X+, X+)$,$(X+, X-)$,$(X-, X+)$ and $(X-, X-)$.

7.3.3.2 The Construction of MQPQ Scheme

The QPQ scheme includes two parts: (1) cooperative query among users; (2) extending two users to multiple users. As a supplement to the scheme, Bob and Alice use the measurement results of the measurement base sequence $X_BX_TZ_{A1}Z_{A1}$ as the key bits in the scheme because other similar measurement base sequences in Table 7.3 may pose risks to the security of the QPQ scheme. For example, $X_BZ_TX_{A1}Z_{A2}$ enables Trent to determine the conclusiveness of the user's key bit with a probability of *0.25*. Furthermore, it can cause dishonest cooperation among users.

7.3.3.3 Encoding Operation

Imagine that there are four participants in a QPQ scheme: Bob (database holder), Trent (third party), Alice1 (database inquirer), and Alice2 (database inquirer), and they share some four-qubit symmetric W states prepared and sent by Bob. For

Table 7.3 Relationship of measurement outcomes on measurement base sequences

Case	(q_1)	(q_2)	(q_3)	(q_4)
$Z_1Z_2X_3X_4$	Z+	Z+	X+	X+
	Z+	Z+	X−	X−
	Z+	Z−	X+ or X−	X+ or X−
	Z−	Z+	X+ or X−	X+ or X−
$Z_1X_2Z_3X_4$	Z+	X+	Z+	X+
	Z+	X−	Z+	X−
	Z+	X+ or X−	Z−	X+ or X−
	Z−	X+ or X−	Z+	X+ or X−
$X_1Z_2X_3Z_4$	X+	Z+	X+	Z+
	X−	Z+	X−	Z+
	X+ or X−	Z+	X+ or X−	Z−
	X+ or X−	Z−	X+ or X−	Z+
$Z_1X_2X_3Z_4$	Z+	X+	X+	Z+
	Z+	X−	X−	Z+
	Z+	X+ or X−	X+ or X−	Z−
	Z−	X+ or X−	X+ or X−	Z+
$X_1Z_2Z_3X_4$	X+	Z+	Z+	X+
	X−	Z+	Z+	X−
	X+ or X−	Z+	Z−	X+ or X−
	X+ or X−	Z−	Z+	X+ or X−
$X_1X_2Z_3Z_4$	X+	X+	Z+	Z+
	X−	X−	Z+	Z+
	X+ or X−	X+ or X−	Z+	Z−
	X+ or X−	X+ or X−	Z−	Z+

convenience, q_1, q_2, q_3, q_4 are used to represent the qubit in a four-qubit symmetric W state held by Bob, Trent, Alice1, and Alice2, respectively. If Bob chooses to use *Z* basis to measure q_1 and use the measurement result as a key bit, other participants in the QPQ scheme can obtain part of Bob's key bits directly. To fully illustrate this point, suppose that all participants choose to use *Z* basis to measure their own particles.

As shown in Table 7.4, when Bob's measurement result is *Z*+, the participant whose measurement result is *Z*− can infer Bob's measurement result, resulting in the disclosure of the key. For this reason, Bob chooses to use X basis for measurement and encodes the measurement result as a key bit. The coding rules are as follows:

$$X+ \quad \leftrightarrow \quad 0, \; X- \quad \leftrightarrow \quad 1 \tag{7.14}$$

7.3.3.4 Cooperative Query Among Users

Step 1: Bob firstly prepares a sequence of four-qubit symmetric W states $|W_4\rangle = \frac{1}{2}(|1000\rangle + |0100\rangle + |0010\rangle + |0001\rangle)_{1234}$. Then he sends the second,

Table 7.4 Relationships of measurement outcomes on measurement base sequence $Z_1Z_2Z_3Z_4$

Case	q_1	q_2	q_3	q_4
$Z_1Z_2Z_3Z_4$	Z+	Z−	Z−	Z−
	Z−	Z+	Z−	Z−
	Z−	Z−	Z+	Z−
	Z−	Z−	Z−	Z+

third, and fourth qubits in each state to Trent, Alice 1, and Alice 2, respectively. Finally, each participant gets a qubit sequence.

Step 2: After the distribution of particles, all participants announce the positions of particles that were not successfully received or detected in their sequences and discard them. For example, the qubit sequence of Alice1 is $\left\{|W_4\rangle_3^1, |W_4\rangle_3^2, |W_4\rangle_3^3 \ldots |W_4\rangle_3^N \ldots\right\}$, where the superscript N represents the N-th four-qubit symmetric W state sent by Bob, and the subscript *3* indicates that Bob only sends the third particle of each W state to Alice1 at a time. If Alice1 does not receive $|W_4\rangle_3^i$, Bob, Alice2 and Trent will discard the i-th particle in their own qubit sequence. After discarding all the lost particles, Bob's qubit sequence is labeled S_B, Trent‘s is S_T, Alice1 and Alice2 are S_{A1} and S_{A2}, respectively. So far, the users cannot get any benefit by cheating because these entangled particles have not yet been used to carry any information. Therefore, the scheme is completely loss tolerant.

Step 3: (Security detection 1) Bob may send false states instead of the W states for obtaining the common query privacy of Alice1 and Alice2, so it is necessary to check Bob's honesty before proceeding to the next step.

1. The semi-honest Trent selects some particles in S_T as detection particles and publishes their positions in the sequence.
2. Participants randomly measure each particle using basis $Z = \{|0\rangle, |1\rangle\}$ or basis $X = \{|+\rangle, |-\rangle\}$, and then all participants publish their measurement results at the same time.
3. Check the measurement results of the detection particles especially those obtained by the measurement base sequences containing two X bases, like$Z_BZ_TX_{A1}X_{A2}$, $Z_BX_TZ_{A1}X_{A2}$, $X_BZ_TX_{A1}Z_{A2}$, $Z_BX_TX_{A1}Z_{A2}$, $X_BZ_TZ_{A1}X_{A2}$ and $X_BX_TZ_{A1}Z_{A2}$, and calculate the error rate of these measurement results. If the error rate is higher than the preset threshold, the scheme is terminated.

Step 4: Discard those detection particles, and the qubit sequences of the participants are denoted as S'_B, S'_T, S'_{A1}, and S'_{A2}. Bob chooses X basis to measure the particles in S'_B in order to prevent his measurement results from leaking and asks Trent to use X basis as well. As for all users in the MQPQ scheme, Bob asks them to use Z basis to measure their qubit sequences.

Step 5: (Security detection 2) For the dishonest Alice1 and Alice2, they can choose other combinations of bases, such as $Z_{A1}Z_{A2}$, $Z_{A1}X_{A2}$, $X_{A1}Z_{A2}$, and $X_{A1}X_{A2}$ to obtain more Bob's measurement results. Obviously, it is necessary to detect the users' honesty before obtaining the key.

1. Bob selects some measurement outcomes and announces the positions of these measurement outcomes in the qubit sequence. Next, Bob asks Alice1, Alice2, and Trent to announce the measurement results of these particles.
2. From Table 7.3, Bob can judge whether Alice is honest or not, because dishonest Alice will provide wrong measurement results, such as $|+-00\rangle_{BTA_1A_2}$, $|-+00\rangle_{BTA_1A_2}$. Then Bob calculates the error rate. If the error rate is higher than the threshold, the scheme will be terminated.

Similarly, if Alice1 and Alice2 are honest users, Trent uses the Z basis to measure particles in its hands and will obtain the conclusiveness of Alice measurements with a probability of 0.25. To avoid detection, Trent can only randomly announce X+ or X− when Trent and Bob both announce the use of X basis, which can lead to incorrect measurement results such as $|+-00\rangle_{BTA1A2}$, $|-+00\rangle_{BTA1A2}$.

Step 6: The sequence S''_B, S''_T, S''_{A1} and S''_{A2} represent new sequences after discarding the detection particles in the previous Step. Bob records the measurement results in S''_B as the raw key K_r according to the encoding operation in Subsect. 7.3.3.3, and Trent secretly announces the measurement results of all particles in the S''_T to the users.

Step 7: All users secretly exchange their measurement results through the classical channels. The specific method is that Alice1 and Alice2 follow such a rule: they encode the measurement results Z+ as *0* and Z− as *1*. According to their common query location *j* in the database, Alice1 shifts *j* bits of sequence S''_{A1} to the right to generate ciphertext S'''_{A1} and sends it to Alice2. Alice2 shifts *j* bits of the sequence S''_{A2} to the left to generate ciphertext S'''_{A2} and sends it to Alice1.

$$\begin{aligned} S''_{A1} = \{0_1\ 1_2\ 0_3\ 0_4\ 1_5\ 0_6\ 0_7\ 0_8\} &\xrightarrow{\text{Shift 5 bits to the right}} S'''_{A1} = \{0_4\ 1_5\ 0_6\ 0_7\ 0_8\ 0_1\ 1_2\ 0_3\} \\ S''_{A2} = \{0_1\ 0_2\ 0_3\ 1_4\ 0_5\ 0_6\ 0_7\ 1_8\} &\xrightarrow{\text{Shift 5 bits to the left}} S'''_{A2} = \{0_6\ 0_7\ 1_8\ 0_1\ 0_2\ 0_3\ 1_4\ 0_5\} \end{aligned} \tag{7.15}$$

Equation (7.15) is an example of *8*-bit sequence S''_{A1} shifting *5* bits to the right and *8*-bit sequence S''_{A2} shifting *5* bits to the left. When both sides receive the ciphertext (S'''_{A1} or S'''_{A2}) sent by the other party and decrypt it, the measurement results of other users can be achieved. Through (7.13) and Trent's announcement, Alice1 and Alice2 can obtain elements in K_r with a probability of $p_c = \frac{1}{2}$, which means only when the measurement results of Alice1 and Alice2 are *Z+* can they obtain the conclusive key bits.

Step 8: Enough qubits should be transmitted so that the raw key K_r of $k \times N$ bits is shared between Bob and Alice (*k* is a security parameter). The raw key is cut into *k* substrings in the way that each substring has *N* bits. These substrings are added bitwise to obtain the final key K_f, and the number of the conclusive key bits in K_f obtained by Alice1 and Alice2 is roughly reduced to one bit.

Step 9: In order to maintain the security of the database and the interests of each user, Bob requires all users to encrypt their common query location by using the first conclusive key bit in their shared K_f sequence. For example, the i-th bit K_i in K_f is the first conclusive key bit shared between users, and they want to query the j-th element d_j in the database. Then Alice1 and Alice2 send $s1 = i - j$ and $s2 = i - j$ to Bob, respectively. Bob compares whether $s1$ is equal to $s2$. If it is equal, he shifts K_f by $i - j$ bits and encrypts the database with the shifted K'_f. If it is not equal, the scheme will be terminated or restarted. It is apparent that K_i is employed to encrypt d_j, and users can recover d_j from the database when they receive the encrypted database from Bob.

7.3.3.5 Extending Two Users to Multiuser

Suppose that there are n ($n \geq 3$) users such as Alice1, Alice2, ..., and Alicen cooperating with each other to query the same target d_j in the database, then the resource state becomes the $(n + 2)$—qubit symmetric W state $|W_{n+2}\rangle = \frac{1}{\sqrt{n+2}}(|100...0\rangle + |010...0\rangle + \ldots + |0...001\rangle)_{123...(n+2)}$. Bob prepares such quantum states and retains particle *1* of each state. Particle *2* is sent to Trent, and the remaining particles are sent to Alice1, Alice2, ..., Alicen in turn. In the multiuser cooperation phase, users with odd tags such as Alice1, Alice3, and so on need to move their sequences to the left by j bits, while users with even labels need to move their sequences to the right by the same length. The other steps in the MQPQ scheme can be implemented completely according to the QPQ scheme in the case of two users. Furthermore, the probability that the user can obtain a conclusive key bit in K_r is $p_c = \frac{2}{n+2.}$

7.3.3.6 Security Analysis

As described in Ref [18], dishonest participants are more powerful than external eavesdroppers in the process of multiparty communication. Therefore, the main attacks in the communication process can be attributed to attacks initiated by dishonest participants in the scheme. It is mentioned that we only analyze the security of the QPQ scheme in the case of two users. The security analysis of multiuser can refer to the situations of two users.

Some security assumptions about the proposed MQPQ scheme are as follows:

1. Trent is a semi-honest third party, which means honest but curious. Specifically, Trent is inclined to execute the scheme honestly and not collude with any party, but he still remains curious about the privacy of users and database in the meantime.
2. In the whole communication system, there is only one quantum channel between Bob and other participants, and all participants can access the public classical channel.

3. Similar to other QPQ schemes, the proposed MQPQ scheme supposes that Bob wants to obtain Alice's query location j while Alice wants to get more entries than d_j in a query.

A. User Privacy

According to the security assumption, Bob and the semi-honest Trent are the main threats to Alice's query privacy. The different attacks from them can be described as follows:

1. **Fake State Attack**
   Bob may use single particle states instead of the four-qubit W symmetric states for key distribution, which cannot only obtain Alice's measurement results but also saves the cost of preparing entangled states. Because the positions of detection particles with two X bases are completely random in Step 3, the false single-particle sequences sent to other participants can only be randomly in $\{|0\rangle, |1\rangle, |+\rangle, |-\rangle\}$. At this time, there are *16* possible results for all measurement base sequences containing two X bases, but only *10* of them approved. If the number of particles detected by the measurement base sequences containing two X bases is t, then the probability that Bob cannot pass the safety detection is $1-\left(\frac{5}{8}\right)^t$. When t is large enough, the probability of dishonest Bob being detected tends to *1*.
2. **Entangle-Measure Attack from Bob**
   In this attack mode, Bob will attack each qubit sent to Alice1, Alice2 and Trent in order to obtain Alice's query privacy. Specifically, Bob can entangle the auxiliary particles $E = \{|E_1\rangle, |E_2\rangle, |E_3\rangle, \ldots, |E_i\rangle, \ldots\}$ with each traveling particles and hopes to perform a unitary operation $\hat{U}$ to pass the detection. Note that $\hat{U}$ is expressed as follows:

$$\hat{U}\cdot|0\rangle|E_i\rangle = \lambda_1|0\rangle|e_{00}\rangle + \lambda_2|1\rangle|e_{01}\rangle \tag{7.16}$$

$$\hat{U}\cdot|1\rangle|E_i\rangle = \lambda_3|0\rangle|e_{10}\rangle + \lambda_4|1\rangle|e_{11}\rangle \tag{7.17}$$

$$\begin{aligned}\hat{U}\cdot|+\rangle|E_i\rangle = \frac{1}{2}(&|+\rangle(\lambda_1|e_{00}\rangle + \lambda_2|e_{01}\rangle + \lambda_3|e_{10}\rangle + \lambda_4|e_{11}\rangle)\\ &+|-\rangle(\lambda_1|e_{00}\rangle - \lambda_2|e_{01}\rangle + \lambda_3|e_{10}\rangle - \lambda_4|e_{11}\rangle))\end{aligned} \tag{7.18}$$

$$\begin{aligned}\hat{U}\cdot|-\rangle|E_i\rangle = \frac{1}{2}(&|+\rangle(\lambda_1|e_{00}\rangle + \lambda_2|e_{01}\rangle - \lambda_3|e_{10}\rangle - \lambda_4|e_{11}\rangle)\\ &+|-\rangle(\lambda_1|e_{00}\rangle - \lambda_2|e_{01}\rangle - \lambda_3|e_{10}\rangle + \lambda_4|e_{11}\rangle)),\end{aligned} \tag{7.19}$$

   where $|\lambda_1|^2 + |\lambda_2|^2 = |\lambda_3|^2 + |\lambda_4|^2 = 1$, $\{|e_{00}\rangle, |e_{01}\rangle, |e_{10}\rangle, |e_{11}\rangle\}$ are the four states can be distinguished by Bob. When Bob uses this attack, the four-qubit W symmetric state can be written in $Z-Z-Z-Z$ bases as follows:

$$
\begin{aligned}
|W_4\rangle = \frac{1}{2}\Big[& \lambda_1^3|1000\rangle|e_{00}\rangle|e_{00}\rangle|e_{00}\rangle + \lambda_1^2\lambda_2|1001\rangle|e_{00}\rangle|e_{00}\rangle|e_{01}\rangle \\
& + \lambda_1^2\lambda_2|1010\rangle|e_{00}\rangle|e_{01}\rangle|e_{00}\rangle + \lambda_1\lambda_2^2|1011\rangle|e_{00}\rangle|e_{01}\rangle|e_{01}\rangle \\
& + \lambda_1^2\lambda_2|1100\rangle|e_{01}\rangle|e_{00}\rangle|e_{00}\rangle + \lambda_1\lambda_2^2|1101\rangle|e_{01}\rangle|e_{00}\rangle|e_{01}\rangle \\
& + \lambda_1\lambda_2^2|1110\rangle|e_{01}\rangle|e_{01}\rangle|e_{00}\rangle + \lambda_2^3|1111\rangle|e_{01}\rangle|e_{01}\rangle|e_{01}\rangle \\
& + |0000\rangle\Big(\lambda_1^2\lambda_3|e_{10}\rangle|e_{00}\rangle|e_{00}\rangle + \lambda_1^2\lambda_3|e_{00}\rangle|e_{10}\rangle|e_{00}\rangle \\
& + \lambda_1^2\lambda_3|e_{00}\rangle|e_{00}\rangle|e_{10}\rangle\Big) + |0100\rangle\Big(\lambda_1^2\lambda_4|e_{11}\rangle|e_{00}\rangle|e_{00}\rangle \\
& + \lambda_1\lambda_2\lambda_3|e_{01}\rangle|e_{10}\rangle|e_{00}\rangle + \lambda_1\lambda_2\lambda_3|e_{01}\rangle|e_{00}\rangle|e_{10}\rangle\Big) \\
& + |0010\rangle\Big(\lambda_1\lambda_2\lambda_3|e_{10}\rangle|e_{01}\rangle|e_{00}\rangle + \lambda_1^2\lambda_4|e_{00}\rangle|e_{11}\rangle|e_{00}\rangle \\
& + \lambda_1\lambda_2\lambda_3|e_{00}\rangle|e_{01}\rangle|e_{11}\rangle\Big) + |0110\rangle\Big(\lambda_1\lambda_2\lambda_4|e_{11}\rangle|e_{01}\rangle|e_{00}\rangle \\
& + \lambda_1\lambda_2\lambda_4|e_{01}\rangle|e_{11}\rangle|e_{00}\rangle + \lambda_2^2\lambda_3|e_{01}\rangle|e_{01}\rangle|e_{10}\rangle\Big) \\
& + |0001\rangle\Big(\lambda_1\lambda_2\lambda_3|e_{10}\rangle|e_{00}\rangle|e_{01}\rangle + \lambda_1\lambda_2\lambda_3|e_{00}\rangle|e_{10}\rangle|e_{01}\rangle \\
& + \lambda_1^2\lambda_4|e_{00}\rangle|e_{00}\rangle|e_{11}\rangle\Big) + |0011\rangle\Big(\lambda_2^2\lambda_3|e_{10}\rangle|e_{01}\rangle|e_{01}\rangle \\
& + \lambda_1\lambda_2\lambda_4|e_{00}\rangle|e_{11}\rangle|e_{01}\rangle + \lambda_1\lambda_2\lambda_4|e_{00}\rangle|e_{01}\rangle|e_{11}\rangle\Big) \\
& + |0101\rangle\Big(\lambda_1\lambda_2\lambda_4|e_{11}\rangle|e_{00}\rangle|e_{01}\rangle + \lambda_1\lambda_2\lambda_4|e_{01}\rangle|e_{00}\rangle|e_{11}\rangle \\
& + \lambda_2^2\lambda_3|e_{01}\rangle|e_{10}\rangle|e_{01}\rangle\Big) + |0111\rangle\Big(\lambda_2^2\lambda_4|e_{11}\rangle|e_{01}\rangle|e_{01}\rangle + \lambda_2^2\lambda_4|e_{01}\rangle|e_{11}\rangle|e_{01}\rangle \\
& + \lambda_2^2\lambda_4|e_{01}\rangle|e_{01}\rangle|e_{11}\rangle\Big)\Big]
\end{aligned} \tag{7.20}
$$

Under this attack, the four-qubit W state can also be expressed in the base sequences containing two X bases as follows:

$$
\begin{aligned}
|W_4\rangle = \frac{1}{8}\Bigg[& |{+}{+}\rangle_{ij}(\overrightarrow{C_1}\overrightarrow{C_1} - \overrightarrow{D_1}\overrightarrow{D_1}) + |{+}{-}\rangle_{ij}(\overrightarrow{C_1}\overrightarrow{C_2} - \overrightarrow{D_1}\overrightarrow{D_2}) \\
& + |{-}{+}\rangle_{ij}(\overrightarrow{C_2}\overrightarrow{C_1} - \overrightarrow{D_2}\overrightarrow{D_1})|{-}{-}\rangle_{ij}(\overrightarrow{C_2}\overrightarrow{C_2} - \overrightarrow{D_2}\overrightarrow{D_2})\Bigg] \\
& \otimes|00\rangle_{kp} + \frac{1}{16}[|{+}{+}\rangle_{ij}(\overrightarrow{C_1}\overrightarrow{C_1} + \overrightarrow{C_1}\overrightarrow{D_1} + \overrightarrow{D_1}\overrightarrow{C_1} + \overrightarrow{D_1}\overrightarrow{D_1}) \\
& + |{+}{-}\rangle_{ij}(\overrightarrow{C_1}\overrightarrow{C_2} + \overrightarrow{C_1}\overrightarrow{D_2} + \overrightarrow{D_1}\overrightarrow{C_2} + \overrightarrow{D_1}\overrightarrow{D_2}) \\
& + |{-}{+}\rangle_{ij}(\overrightarrow{C_2}\overrightarrow{C_1} + \overrightarrow{D_2}\overrightarrow{C_1} + \overrightarrow{C_2}\overrightarrow{D_1} + \overrightarrow{D_2}\overrightarrow{D_1}) \\
& + |{-}{-}\rangle_{ij}(\overrightarrow{C_2}\overrightarrow{C_2} + \overrightarrow{C_2}\overrightarrow{D_2} + \overrightarrow{D_2}\overrightarrow{C_2} + \overrightarrow{D_2}\overrightarrow{D_2})] \otimes |10\rangle_{kp} \\
& + \frac{1}{16}[|{+}{+}\rangle_{ij}(\overrightarrow{C_1}\overrightarrow{C_1} + \overrightarrow{C_1}\overrightarrow{D_1} + \overrightarrow{D_1}\overrightarrow{C_1} + \overrightarrow{D_1}\overrightarrow{D_1}) \\
& + |{+}{-}\rangle_{ij}(\overrightarrow{C_1}\overrightarrow{C_2} + \overrightarrow{C_1}\overrightarrow{D_2} + \overrightarrow{D_1}\overrightarrow{C_2} + \overrightarrow{D_1}\overrightarrow{D_2}) \\
& + |{-}{+}\rangle_{ij}(\overrightarrow{C_2}\overrightarrow{C_1} + \overrightarrow{D_2}\overrightarrow{C_1} + \overrightarrow{C_2}\overrightarrow{D_1} + \overrightarrow{D_2}\overrightarrow{D_1}) \\
& + |{-}{-}\rangle_{ij}(\overrightarrow{C_2}\overrightarrow{C_2} + \overrightarrow{C_2}\overrightarrow{D_2} + \overrightarrow{D_2}\overrightarrow{C_2} + \overrightarrow{D_2}\overrightarrow{D_2})] \otimes |01\rangle_{kp},
\end{aligned}
\tag{7.21}
$$

where $\overrightarrow{C_{1(2)}} = \lambda_1|e_{00}\rangle \pm \lambda_2|e_{01}\rangle + \lambda_3|e_{10}\rangle \pm \lambda_4|e_{11}\rangle$ and $\overrightarrow{D_{1(2)}} = \lambda_1|e_{00}\rangle \pm \lambda_2|e_{01}\rangle - \lambda_3|e_{10}\rangle \mp \lambda_4|e_{11}\rangle$. Table 7.5 shows the specific values of *(i,j,k,p)*. In order not to be detected by the Step 3 of the scheme, Bob must set the parameters of $\hat{U}$ appropriately. According to the original form of four-qubit W symmetric state, Bob sets the condition that ${\lambda_2}^3 = 0$ and $\lambda_1^2\lambda_3|e_{10}\rangle|e_{00}\rangle|e_{00}\rangle + \lambda_1^2\lambda_3|e_{00}\rangle|e_{10}\rangle|e_{00}\rangle + \lambda_1^2\lambda_3|e_{00}\rangle|e_{00}\rangle|e_{10}\rangle = \vec{0}$, where $\vec{0}$ denotes a zero vector. It is obvious that $\lambda_1 = 1$ due to $|\lambda_1|^2 + |\lambda_2|^2 = 1$ and $\lambda_2 = 0$. To make $\lambda_1^2\lambda_3|e_{10}\rangle \times |e_{00}\rangle|e_{00}\rangle + \lambda_1^2\lambda_3|e_{00}\rangle|e_{10}\rangle|e_{00}\rangle + \lambda_1^2\lambda_3|e_{00}\rangle|e_{00}\rangle|e_{10}\rangle$ equal to $\vec{0}$, λ_3 must be set to *0*. It can be seen that $\lambda_4 = 1$ because $|\lambda_3|^2 + |\lambda_4|^2 = 1$. So far, we can only extract information from the valid ancillas $|e_{00}\rangle$ and $|e_{11}\rangle$.

In addition, $\overrightarrow{C_1}\,\overrightarrow{C_2} - \overrightarrow{D_1}\,\overrightarrow{D_2}$ and $\overrightarrow{C_2}\,\overrightarrow{C_1} - \overrightarrow{D_2}\,\overrightarrow{D_1}$ must be set to $\vec{0}$. According to (Eq. 7.13), we can calculate that $|e_{00}\rangle|e_{11}\rangle = |e_{11}\rangle|e_{00}\rangle$, then the formula $|e_{00}\rangle = |e_{11}\rangle$ can be derived, which means that $|e_{00}\rangle$ and $|e_{11}\rangle$ are indistinguishable. In conclusion, on the condition that $|e_{00}\rangle$ and $|e_{11}\rangle$ are indistinguishable, although Bob's attack will not be detected by Alice, he will be incapable to get any information about Alice and Trent. If Bob violates the parameter setting of $\hat{U}$, then Bob's attack will be detected by Alice.

Table 7.5 The specific values of *i, j, k* and *p* under different base sequences

MBS	$Z_1Z_2X_3X_4$	$Z_1X_2Z_3X_4$	$X_1Z_2X_3Z_4$	$Z_1X_2X_3Z_4$	$X_1Z_2Z_3X_4$	$X_1X_2Z_3Z_4$
(i, j, k, p)	*(3, 4, 1, 2)*	*(2, 4, 1, 3)*	*(1, 3, 2, 4)*	*(2, 3, 1, 4)*	*(1, 4, 2, 3)*	*(1, 2, 3, 4)*

3. **Eavesdropping the Classical Channel Provided by Trent**
   Now consider a special case. If a semi-honest Trent just wants to complete his task and does not provide secure channels for announcing information to Alice, then Bob can infer which positions in K_r are inconclusive key bits for Alice by eavesdropping the measurement results published by Trent. Specifically, when the measurement results of Bob and Trent are different, that is, when one is $X+$ the other is $X-$, Bob can determine that the corresponding key bit (marked as K_B) is an inconclusive key bit in K_r for Alice. If there is at least one K_B in the k raw key bits contributing to one final key bit in K_f, Bob can infer that the final key bit is an inconclusive key bit for Alice. According to (Eq. 7.13), the probability that the measurement result is $X + X-$ or $X - X+$ is 0.25. Only when K_B is not included in the k raw key bits, Bob cannot determine which final key bits are conclusive to Alice and which are not. In order to clarify the attack effect, suppose that Bob's database size is $N = 10^4$, then Alice's conclusive key bits in K_f are $n_a = N(0.5)^k$ and Bob's indistinguishable key bits in K_f are $n_b = N(0.75)^k$. Fig. 7.5 shows that when the security parameter is $k = 13$, Alice's conclusive key bits in the final key sequence K_f can be diluted to almost one bit, while there are about *237* bits in the K_f that cannot be distinguished by Bob. Bob has no better way to guess which of the *237* key bits may be Alice's conclusive key bit randomly, so it is not beneficial for Bob to guess Alice's privacy by accessing the public announcement of Trent.
4. **Intercept-Resend Attack**

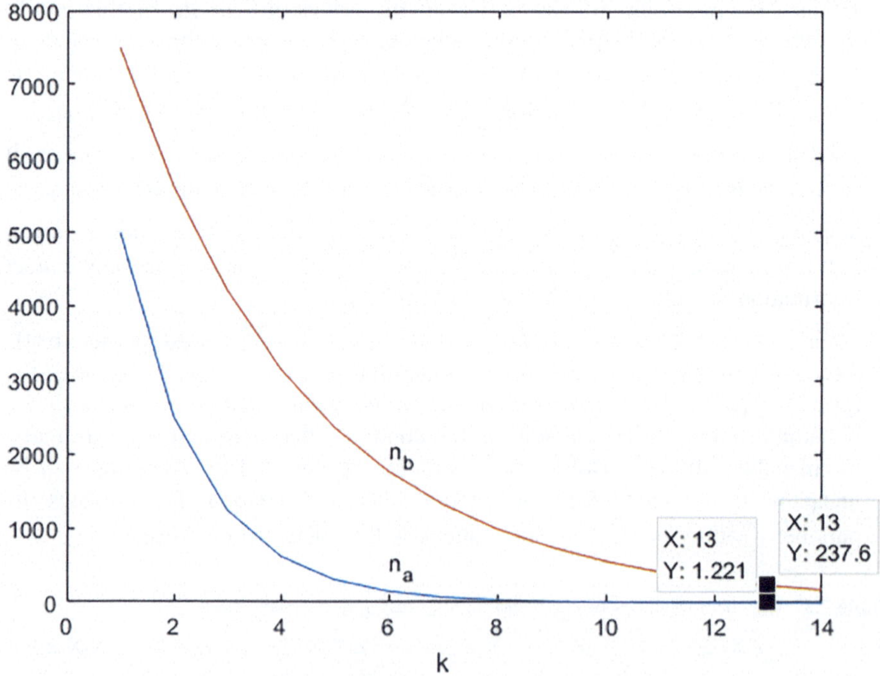

Fig. 7.5 The effect of the unsafe classical channels offered by Trent when $N = 10{,}000$

In order to get Alice's query privacy, Trent can intercept the particles Bob sent to Alice1 and Alice2, then use Z basis to measure them, and finally make the corresponding measurement results sent to users. What can be discovered is that the four-qubit W state will collapse to $|0010\rangle_{BTA1A2}$ or $|0001\rangle_{BTA1A2}$ with a probability of $\frac{1}{4}$, and to $(|10\rangle + |01\rangle)_{BT}|00\rangle_{A1A2}$ with a probability of $\frac{1}{2}$. However, Step 3 of the MQPQ scheme can detect not only Bob's fake state attacks but also dishonest Trent. For the first two cases, only when all Z bases in the measurement base sequences containing two X bases are used to measure particles in $|0\rangle$ can errors be caused with a probability of $\frac{1}{2}$. In the last case, except for $X_BX_TZ_{A1}Z_{A2}$ and $Z_BZ_TX_{A1}X_{A2}$ bases, other measurement sequences with two X bases will bring wrong measurement results with a probability of $\frac{1}{4}$. After the above analysis, Eve's attack will bring wrong results with a probability of $\frac{5}{24}$. As the number of detection particles with two X base increases, dishonest Trent will be detected.

5. **Entangle-Measure Attack from Trent**
   Trent can also use the same attack as Bob, that is, entangle the additional particles with the particles transmitted to Alice, hoping to obtain the measurement results of Alice1 and Alice2 by measuring the additional particles. According to (Eqs. 7.16, 7.17, 7.18, and 7.19), the four-qubit W state written in $Z - Z - Z - Z$ bases is

$$
\begin{aligned}
|W_4\rangle = \frac{1}{2}\Big[&\lambda_1^2|1000\rangle|e_{00}\rangle|e_{00}\rangle + \lambda_1\lambda_2|1001\rangle|e_{00}\rangle|e_{01}\rangle + \lambda_1\lambda_2|1010\rangle|e_{01}\rangle|e_{00}\rangle \\
&+ \lambda_2^2|1011\rangle|e_{01}\rangle|e_{01}\rangle + \lambda_1^2|0100\rangle|e_{00}\rangle|e_{00}\rangle + \lambda_1\lambda_2|0101\rangle|e_{00}\rangle|e_{01}\rangle \\
&+ \lambda_1\lambda_2|0110\rangle|e_{01}\rangle|e_{00}\rangle + \lambda_2^2|0111\rangle|e_{01}\rangle|e_{01}\rangle + \lambda_1\lambda_3|0000\rangle|e_{11}\rangle|e_{00}\rangle \\
&+ \lambda_2\lambda_3|0001\rangle|e_{10}\rangle|e_{01}\rangle + \lambda_4\lambda_1|0010\rangle|e_{11}\rangle|e_{00}\rangle + \lambda_4\lambda_2|0011\rangle|e_{11}\rangle|e_{01}\rangle \\
&+ \lambda_1\lambda_3|0000\rangle|e_{00}\rangle|e_{10}\rangle + \lambda_1\lambda_4|0001\rangle|e_{00}\rangle|e_{11}\rangle + \lambda_2\lambda_3|0010\rangle|e_{01}\rangle|e_{10}\rangle \\
&+ \lambda_4\lambda_2|0011\rangle|e_{01}\rangle|e_{11}\rangle\Big]
\end{aligned}
\tag{7.22}
$$

According to the W state of the original form, Trent sets the qualification that $\lambda_2^2 = 0$ and $\lambda_1\lambda_3 = 0$. Because the coefficients λ_1, λ_2, λ_3, λ_4 meet the conditions that $|\lambda_1|^2 + |\lambda_2|^2 = 1$, $|\lambda_3|^2 + |\lambda_4|^2 = 1$,$\lambda_1 = 1$, $\lambda_4 = 1$.Thus the four-qubit symmetric W state under the circumstance of Trent's attack can be expressed in the same form as Eq. 7.22 with the bases containing two X bases. From the above Bob's attack results, if Trent wants to obtain Alice's measurement through this attack, Trent will definitely introduce the wrong measurement results, which will be detected by Alice.

B. Database Security

1. **The Joint Measurement Attack Implemented by Users**
   For users, the joint measurement attack is the most effective way to obtain more conclusive bits in K_f, and it will not be detected Ref [18] pointed out that dishonest users can implement JM attacks only if they hold the states at the end of the scheme and know which states contribute to one final key bit. However, in the MQPQ scheme, only n-qubit W symmetric states (n depends on the number of users) are used as carrier states, so there is no need to use joint USD measurement $p_{\mathrm{JM}} = 1 - tr\sqrt{\rho_0^{\frac{1}{2}}\rho_1\rho_0^{\frac{1}{2}}}$ to distinguish the two quantum states ρ_0 and ρ_1. Moreover, even if users know those carrier states contributing to one final bit, they only get some particles in each carrier state, so Alice cannot execute a JM attack.
2. **The Attack from the Semi-Honest Trent**
   Although the semi-honest Trent will execute the scheme honestly and will not collude with other participants, he also wants to obtain some information in K_f through his own qubit sequence and the public information. It is noted that Trent can obtain the key information only after achieving Alice's measurement results, the reason of which is that Bob will not publish any information during the key generation phase. The analysis in the previous part has shown that other attack methods of Trent will be discovered by Alice, and no information can be obtained from the public channels between users, since the communication between users is encrypted by their common query location in the database. Therefore, Trent has no better way but to guess the key information at random according to Table 7.3.

7.3.3.7 Discussion

In addition to the advantages of extending the QPQ scheme of two users to multiple users, the MQPQ scheme also has the advantages of other QPQ schemes, such as loss tolerance and cheat sensitivity. It is imperative that as the number of users increases, our scheme also presents some peculiar advantages.

A. More Cheat Sensitive

When Bob uses the fake state attack in Sect. 7.3.3.6, the security of the QPQ scheme can be improved with the increasing number of users in the MQPQ scheme, making the scheme more cheat-sensitive. In general, when n-qubit symmetric W state is used as the resource state of the scheme, the number of users is $(n-2)$. For Bob, because the positions of detection particles with two X bases are completely random, the false single-particle sequences sent to other participants can only be randomly in $\{|0\rangle, |1\rangle, |+\rangle, |-\rangle\}$. At this time, there are 2^n possible results for all measurement base sequences containing two X bases, but only $4n-6$ of them are approved. Assuming that t is the number of particles detected using two X bases, then the probability that Bob cannot pass the detection is $p_f = 1 - \left(\frac{4n-6}{2^n}\right)^t$. As can be seen from Fig. 7.6, if the number n remains unchanged, dishonest Bob will eventually be detected as t increases. On condition that the threshold p_f is set to *0.95*, it can be seen from

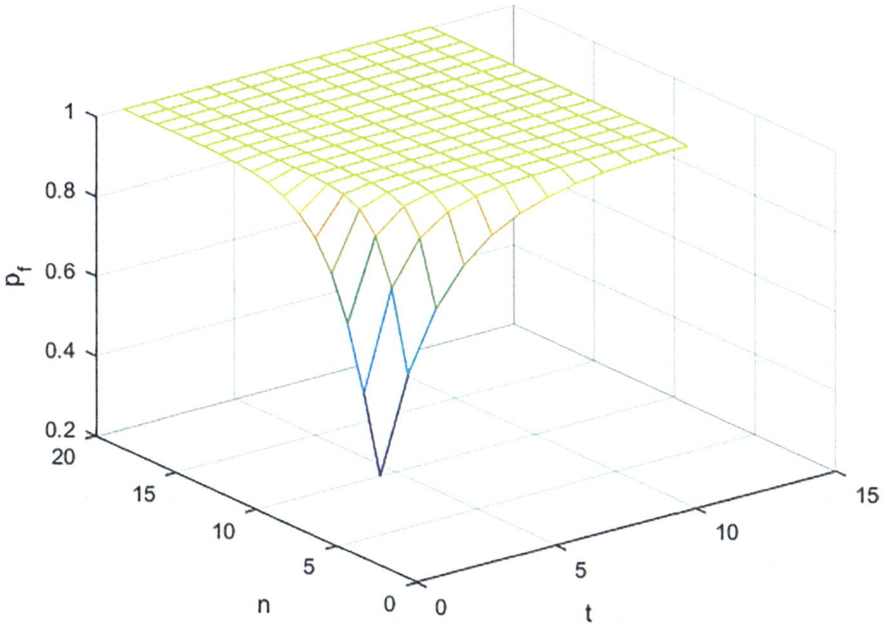

Fig. 7.6 The relations between the detection probability P_f of dishonest Bob and the number of detection particles with two X basis under the circumstance of different amounts of the users denoted by n

Table 7.6 Example of the number of detection particles with *2* X basis with different number of users denoted by U_n, when $p_f = 0.95$

U_n	2	3	4	5	6	7	8	...	18
t	7	4	3	2	2	2	1	1	1
p_f	0.9627	0.9634	0.9778	0.9705	0.9897	0.9966	0.9668	...	0.9999

Table 7.6 that as the number of users increases, fewer detection particles with two X bas*es* are demanded to detect Bob's attack. When the scheme involves eight or more users, only one detection particle measured by a measurement base sequence containing two X bases can detect dishonest Bob with an efficiency higher than *0.95*. Therefore, the scheme is more cheat-sensitive with the increase of users.

B. More Practical

We find that with the increase of the number of users, the higher dimensional symmetric W state will be used as the resource state of the MQPQ scheme. If the value of the safety parameter k enlarges with the number of users increasing, the MQPQ scheme will become impractical because the higher dimensional symmetric W state is more difficult to prepare. However, the probability of a conclusive key bit in K_r got by users will decrease, which means that when the size of the database to be queried remains unchanged, the more users there are, a smaller k is. When the n-qubit symmetric W state is used, the probability of users cooperating with each other

Table 7.7 Example of possible choice of k and the values of p_0 and n when the number of users increases with database size $N = 10{,}000$

n-qubit W state	4	10	20	30	44
U_n	2	8	18	28	42
k	13	5	3	3	2
$-n$	1.221	3.2	10	2.963	20.66
P_0	0.2950	0.0407	4.5173×10–5	0.0516	1.0416×10–9

to obtain a conclusive key bit in K_r is $p_c = \frac{2}{n}$. Based on the post-processing algorithm of $kN \rightarrow N$ used in the scheme, the average number of the conclusive key bits obtained by users in the final key K_f is $-n = Np_c^k$, and the probability that users do not know any key bit in K_f is $p_0 = \left(1 - p_c^k\right)^N$, which means that the MQPQ scheme needs to be restarted. To fully illustrate the advantage of the MQPQ scheme, we choose $N = 10^4$ as an example to observe the selection of the value of k with the increase of the number of users. It can be seen from Table 7.7 that as the number of users increases, the value of safety factor k decreases. In addition, Table 7.7 only enumerates the case of the lower limit of the corresponding number of users with $k = 2$. With the increase of the number of users, the theoretical value of k can be taken as *1*, but the corresponding resource state is too difficult to prepare, so it is not discussed here.

7.3.3.8 Summary

In summary, a novel multiuser quantum private query scheme using multiparticle symmetric W state is proposed. Compared with existing QPQ schemes, the proposed MQPQ scheme has the following differences:

1. The MQPQ scheme can be well extended to multiuser scenarios by utilizing the property of multiparticle symmetric W state in local measurement with X basis and Z basis.
2. In security analysis, both the database security and the user privacy security can be well protected, and with the increase in the number of users, users are more sensitive to the fake single-particle attack of the database holder.
3. As the number of users increases, the smaller security factor k is required to obtain the final key K_f, which means that the scheme is more practical because the higher dimensional W state is difficult to prepare.

Although there are several improvements of the proposed MQPQ scheme comparing with previous QPQ schemes, the query efficiency of our MQPQ scheme is not ideal which means that multiuser can only query the same items in one query. Therefore, the future research will focus on improving the efficiency of the scheme and making it more perfect.

7.3.4 QPQ Based on Six-Qubit DF State

Many QPQ schemes are based on the ideal environment, that is, the case without quantum channel noise. In practice, quantum channel noise may cause Alice to get incorrect query results, which makes Alice misunderstand the database holder. Considering the actual noisy channel, Chan proposed an error-tolerant QKD-based QPQ scheme in which an error correction method for the oblivious key is used in QKD-based QPQ schemes [19]. In addition, Gao put forward $gkN \to N$ post-processing algorithm on the basis of $kN \to N$ post-processing with linear error correction code, which has significant effect on error correction [23]. However, they all correct the classical information after the oblivious key distribution, but there are few robust oblivious key distribution with certain anti-noise ability. Fortunately, not all quantum states are equally fragile when they interact with the environment. Decoherence-free states are the quantum states that remain unchanged no matter how strong the collective noise is as they are transmitted through the quantum channel. This is because the collective errors that occur on different qubits can cancel each other out when a particularly relevant symmetry arise. Wang proposed a flexible QPQ scheme based on four-particle decoherence-free state, in which the specific security requirements are achieved by adjusting parameter θ [42]. At the same time, Li presented a QPQ scheme which can resist collective noise by using four-particle decoherence-free state [43]. However, the above two schemes need to build two projective measurement bases, which makes their schemes difficult to implement under current technologies. In order to reduce the impact of noise, Wei recently proposed a $k_1k_2N \to N$ post-processing method, which determines whether the current query content is valid by determining the error rate of k_2 middle keys [44]. In addition, Gao also proposed a feasible QPQ scheme to deal with the error caused by noise [45]. Unfortunately, these two post-processing methods deal with the impact of noise by increasing the communication complexity. In order to deal with the error of oblivious key distribution caused by quantum channel noise, a new quantum privacy query (QPQ) scheme with six-particle decoherence-free states is proposed [46].

7.3.4.1 Decoherence-Free Space

It has been found that decoherence is usually caused by the quantum channel noise, which is caused by the coupling of a quantum system to its environment [47]. The existence of noise in quantum channel usually makes the quantum states fragile and easily destroyed during transmission, which leads to errors. An external attacker may use the error to conceal his attack, making the whole quantum communication system unsafe and even resulting in information leakage.

As mentioned in [48], a special symmetry occurs when the coupling between each qubit and the environment is not distinguished. This particular symmetry, also known as collective noise, is a kind of decorrelation. Collective noise occurs when particles pass through some media such as optical fiber. If the time delay between the transmitted particles is small enough, the collective noise acting on these particles can be described by a unitary operation $U(t)$, where t denotes the time of transmission and indicates a temporal dependence. The influence of collective noise on an N-qubit state is shown in Eq. 7.23:

$$\rho_N \Rightarrow [U(t)]^{\otimes N} \rho_N \left[U(t)^{\dagger}\right]^{\otimes N} \tag{7.23}$$

In 2004, Bourennane proved that decoherence-free space (DFS) can remain invariant under collective decoherence no matter how strong the interaction is [49]. The invariance of N-qubit DF state can be described by the following Eq. 7.24:

$$\rho_N = [U(t)]^{\otimes N} \rho_N \left[U(t)^{\dagger}\right]^{\otimes N} \tag{7.24}$$

In Bourennane's experiment, two four-qubit DF state is selected to transmit in optical fiber to verify the invariance property of DFS under collective noise. One of the two states is $|\bar{0}\rangle_{abcd} = |\phi^-\rangle_{ab} \otimes |\phi^-\rangle_{cd}$, where $|\phi^-\rangle_{ab} = \frac{1}{\sqrt{2}}(|01\rangle - |10\rangle)$ is the only DF state in the case of two particles, and the other four-qubit DF state is $|\bar{1}\rangle = \frac{1}{2\sqrt{3}}(2|0011\rangle - |0101\rangle - |0110\rangle - |1001\rangle - |1010\rangle + 2|1100\rangle)$, which is prepared by spontaneous parametric down-conversion. Cabello then proposed DF state preparation methods for six and eight qubits. Compared with the DF state of four particles, the DF state of six particles have a correlation in the single base measurement sequences [50–54].

The four six-qubit DF states are described as follows:

$$\begin{aligned}
|0'\rangle &= |\psi^-\rangle_{12} \otimes |\bar{1}\rangle \\
&= \frac{1}{2\sqrt{6}}(2|010011\rangle\text{-}|010101\rangle\text{-}|010110\rangle\text{-}|011001\rangle \\
&\quad - |011010\rangle + 2|011100\rangle\text{-}2|100011\rangle + |100101\rangle \\
&\quad +|100110\rangle + |101001\rangle + |101010\rangle\text{-}2|101100\rangle)_{123456}
\end{aligned} \tag{7.25}$$

$$\begin{aligned}
|1'\rangle &= P_{1\leftrightarrow 3, 2\leftrightarrow 4}|0'\rangle \\
&= |\psi'\rangle_{34} \otimes |\bar{1}\rangle_{1256} \\
&= \frac{1}{2\sqrt{6}}(2|000111\rangle - |010101\rangle - |010110\rangle - |100101\rangle \\
&\quad - |100110\rangle + 2|110100\rangle - 2|001011\rangle + |011001\rangle \\
&\quad +|011010\rangle + |101001\rangle + |101010\rangle - 2|111000\rangle)_{123456}
\end{aligned} \tag{7.26}$$

$$\begin{aligned}
|+'\rangle &= P_{1\leftrightarrow 3}|0'\rangle \\
&= |\psi'\rangle_{32} \otimes |\overline{1}\rangle_{1456} \\
&= \frac{1}{2\sqrt{6}}(2|010011\rangle - |010101\rangle - |010110\rangle - |110001\rangle \\
&\quad - |110010\rangle + 2|110100\rangle - 2|001011\rangle + |001101\rangle \\
&\quad + |001110\rangle + |101001\rangle + |101010\rangle - 2|101100\rangle)_{123456}
\end{aligned} \tag{7.27}$$

$$\begin{aligned}
|-'\rangle &= P_{2\leftrightarrow 4}|0'\rangle \\
&= |\psi'\rangle_{14} \otimes |\overline{1}\rangle_{3256} \\
&= \frac{1}{2\sqrt{6}}(2|000111\rangle - |010101\rangle - |010110\rangle - |001101\rangle \\
&\quad - |001110\rangle + 2|011100\rangle - 2|100011\rangle + |110001\rangle \\
&\quad + |110010\rangle + |101001\rangle + |101010\rangle - 2|111000\rangle)_{123456,}
\end{aligned} \tag{7.28}$$

where $P_{i \leftrightarrow j}$ intends permuting the ith qubit and the jth qubit in the quantum state. These four states satisfy

$$\begin{aligned}
\langle 0'|1'\rangle = \langle +'|-'\rangle = 0 \\
|\langle 0'|+'\rangle| = |\langle 0'|-'\rangle| = |\langle 1'|+'\rangle| = |\langle 1'|-'\rangle| = \frac{1}{2}
\end{aligned} \tag{7.29}$$

7.3.4.2 Specific Steps

In this subsection, Bob is defined as the owner of the database, and Alice is defined as the query party of the database. Bob's database size is N bit, and r and k are the number of N bits keys, respectively. In particular, k is the security parameter of the scheme, that is, kN-bit keys are needed to dilute Alice's final key to nearly *1* bit.

Step 1: Bob randomly generates a binary sequence of length $r \times N$ ($r > k$)and marks it as $S_k = \{b_1, b_2, b_3 \ldots, b_{r \times N}\}$. In addition, Bob needs to prepare enough quantum carrier states in $|0'\rangle$.

Step 2: Bob prepares the quantum carrier sequence S_q according to the elements in S_k. The specific operation is as follows:

1. If b_i is *0*, Bob randomly performs $1 \leftrightarrow 3$ and $2 \leftrightarrow 4$ operations or nothing on the i-th $|0'\rangle$ state.
2. If b_i is *1*, Bob randomly performs $1 \leftrightarrow 3$ operation or $2 \leftrightarrow 4$ operation on the i-th $|0'\rangle$ state.

Bob sends the prepared quantum carrier sequence $S_q = \{q_1, q_2, q_3, \ldots, q_{r \times N}\}$ to Alice.

Step 3: Alice measures each received six-qubit DF state randomly in one of the two sequences of single-qubit measurements $\{Z_1Z_2Z_3Z_4X_5X_6, X_1X_2Z_3Z_4Z_5Z_6\}$. Then

Alice tells Bob which six-qubit DF state in the sequence S_q have not been successfully received, they disgard the lost DFS. Sequences S_k and S_q changed accordingly, and the changed S_k and S_q are recorded as binary sequence S_{ck} and the new quantum carrier sequence S_{cq}, respectively.

Step 4: Bob randomly selects some quantum states in S_{cq} as detection particles and asks Alice to announce the corresponding measurement results. Bob calculates the probability of the results in $\{|0110\pm\pm\rangle, |1001\pm\pm\rangle, |1100\pm\pm\rangle, |0011\pm\pm\rangle, |\pm\pm\mp\mp 00\rangle, |\pm\pm\mp\mp 11\rangle, |\pm\mp\mp\pm 00\rangle, |\pm\mp\mp\pm 11\rangle\}$. If it is higher than a preset threshold, the scheme is terminated. Otherwise, the elements in the corresponding positions of the detection particles in S_{cq} and S_{ck} are discarded, and the scheme continues to be executed. The changed S_{ck} and S_{cq} are recorded as S_{fk} and S_{fq}, respectively.

Step 5: For each DF state that Alice has successfully measured in sequence S_{fq}, Bob announces two states. For example, if $q_i = |0'\rangle$ or $|+'\rangle$, Bob will announce $\{|0'\rangle, |+'\rangle\}$. Otherwise, Bob will announce $\{|1'\rangle, |-'\rangle\}$.

Step 6: Alice interprets the DF state of S_{fq} based on her measurements. Alice can obtain the operation of Bob in Step 2 with a certain probability. Tables 7.8 and 7.9 show that Alice obtains the conclusive measurement results with a probability of $\frac{1}{2}$. For example, if $b_i = 0$ and Bob sends $|0'\rangle$, Alice uses the measurement sequence $Z_1Z_2Z_3Z_4X_5X_6$ to get a measurement result of $\{|1000\pm\pm\rangle, |1000\pm\mp\rangle, |0110\pm\pm\rangle, |1001\pm\pm\rangle, |0111\pm\pm\rangle, |0111\pm\mp\rangle\}$, then she can determine $b_i = 0$.

Step 7: Bob randomly selects kN bits from S_{fk} as raw key K_r and announces the location of the kN bits in the S_{fk}. Then K_r is cut into k substrings. These substrings are added bitwise to get the final key K_f, where k is also called the security factor. If the probability of Alice getting a conclusive key bit in K_r is p_c, then the probability of Alice getting a conclusive key bit in K_f is p_c^k. The appropriate

Table 7.8 The case of distinguishing $|\mathbf{0}'\rangle$ from $|+'\rangle$ by $Z_1Z_2Z_3Z_4X_5X_6$ and $X_1X_2Z_3Z_4Z_5Z_6$

	$\|0'\rangle$		$\|+'\rangle$	
Alice's MBS	Alice's MR	Alice's conclusiveness	Alice's MR	Alice's conclusiveness
$Z_1Z_2Z_3$ $Z_4X_5X_6$	$\|0100\pm\pm\rangle\|0100\pm\mp\rangle$	?	$\|0100\pm\pm\rangle\|0100\pm\mp\rangle$	?
	$\|0101\pm\pm\rangle\|1010\pm\pm\rangle$	?	$\|0101\pm\pm\rangle\|1010\pm\pm\rangle$	?
	$\|1011\pm\pm\rangle\|1011\pm\mp\rangle$	?	$\|1011\pm\pm\rangle\|1011\pm\mp\rangle$	?
	$\|1000\pm\pm\rangle\|1000\pm\mp\rangle$	0	$\|0010\pm\pm\rangle\|0010\pm\mp\rangle$	1
	$\|0110\pm\pm\rangle\|1001\pm\pm\rangle$	0	$\|1100\pm\pm\rangle\|0011\pm\pm\rangle$	1
	$\|0111\pm\pm\rangle\|0111\pm\mp\rangle$	0	$\|1101\pm\pm\rangle\|1101\pm\mp\rangle$	1
$X_1X_2X_3$ $X_4Z_5Z_6$	$\|\pm\mp\pm\pm 11\rangle\|\pm\mp\pm\mp 11\rangle$	?	$\|\pm\mp\pm\pm 11\rangle\|\pm\mp\pm\mp 11\rangle$	?
	$\|\pm\mp\pm\pm 01\rangle\|\pm\mp\pm\pm 10\rangle$	?	$\|\pm\mp\pm\pm 01\rangle\|\pm\mp\pm\pm 10\rangle$	?
	$\|\pm\mp\pm\pm 00\rangle\|\pm\mp\pm\mp 00\rangle$	?	$\|\pm\mp\pm\pm 00\rangle\|\pm\mp\pm\mp 00\rangle$	?
	$\|\pm\mp\mp\mp 11\rangle\|\mp\pm\pm\mp 11\rangle$	0	$\|\pm\pm\mp\mp 11\rangle\|\pm\pm\mp\pm 11\rangle$	1
	$\|\pm\mp\mp\mp 01\rangle\|\pm\mp\mp\mp 10\rangle$	0	$\|\pm\pm\mp\pm 01\rangle\|\pm\pm\mp\pm 10\rangle$	1
	$\|\pm\mp\mp\mp 00\rangle\|\pm\mp\mp\pm 00\rangle$	0	$\|\pm\pm\mp\mp 00\rangle\|\pm\pm\mp\pm 00\rangle$	1

Table 7.9 The case of distinguishing $|\mathbf{1}'\rangle$ from $|-'\rangle$ by $Z_1Z_2Z_3Z_4X_5X_6$ and $X_1X_2Z_3Z_4Z_5Z_6$

	$\|1'\rangle$		$\|-'\rangle$	
Alice's MBS	Alice's MR	Alice's conclusiveness	Alice's MR	Alice's conclusiveness
$Z_1Z_2Z_3$	$\|0001\pm\pm\rangle\|0001\pm\mp\rangle$	?	$\|0001\pm\pm\rangle\|0001\pm\mp\rangle$	?
$Z_4X_5X_6$	$\|0101\pm\pm\rangle\|1010\pm\pm\rangle$	?	$\|0101\pm\pm\rangle\|1010\pm\pm\rangle$	?
	$\|1110\pm\pm\rangle\|1110\pm\mp\rangle$	?	$\|1110\pm\pm\rangle\|1110\pm\mp\rangle$	?
	$\|0010\pm\pm\rangle\|0010\pm\mp\rangle$	0	$\|0111\pm\pm\rangle\|0111\pm\mp\rangle$	1
	$\|0110\pm\pm\rangle\|1001\pm\pm\rangle$	0	$\|0011\pm\pm\rangle\|1100\pm\pm\rangle$	1
	$\|1101\pm\pm\rangle\|1101\pm\mp\rangle$	0	$\|1000\pm\pm\rangle\|1000\pm\mp\rangle$	1
$X_1X_2X_3$	$\|\pm\mp\pm\mp 11\rangle\|\pm\pm\pm\mp 11\rangle$	?	$\|\pm\mp\pm\mp 11\rangle\|\pm\pm\pm\mp 11\rangle$	?
$X_4Z_5Z_6$	$\|\pm\pm\pm\mp 01\rangle\|\pm\pm\pm\mp 10\rangle$	?	$\|\pm\pm\pm\mp 01\rangle\|\pm\pm\pm\mp 10\rangle$	?
	$\|\pm\mp\pm\mp 00\rangle\|\pm\pm\pm\mp 00\rangle$	?	$\|\pm\mp\pm\mp 00\rangle\|\pm\pm\pm\mp 00\rangle$	?
	$\|\pm\pm\mp\pm 11\rangle\|\pm\mp\mp\pm 11\rangle$	0	$\|\pm\mp\mp\mp 11\rangle\|\pm\pm\mp\mp 11\rangle$	1
	$\|\pm\pm\mp\pm 01\rangle\|\pm\pm\mp\pm 10\rangle$	0	$\|\pm\mp\mp\mp 01\rangle\|\pm\mp\mp\mp 10\rangle$	1
	$\|\pm\pm\mp\pm 00\rangle\|\pm\mp\mp\pm 00\rangle$	0	$\|\pm\mp\mp\mp 00\rangle\|\pm\pm\mp\mp 00\rangle$	1

value of k is to make Alice's conclusive key bits $n_c = Np_c^k$ in K_f diluted to roughly *1* bit. If Alice doesn't get any bits in K_f after $kN \rightarrow N$, the scheme has to be executed again.

Step 8: Suppose Alice knows the j-th key bit in K_f and wants to query the i-th element d_i. She then publishes the value $s = j - i$. Then Bob shifts K_f by s and uses the shifted key to encrypt his database. Because the ciphertext of the i-th bit in the encrypted database is $C_i = K_f^j \oplus d_i$, Alice can correctly get d_i using K_f.

7.3.4.3 Security Analysis

This subsection will analyze the security of the DFQPQ scheme. Because quantum privacy query scheme is based on the scheme between two untrusted communicators, dishonest participants bring more harm than external eavesdroppers [55], so we do not analyze external attacks here.

A. Database Security

Dishonest Alice may attack the database in some ways, hoping to get more elements in the database. Because the database is encrypted by the final key, Alice must obtain more key bits in the raw key K_r or more final key bits in the K_f to achieve this goal.

1. **Loss Cheating Attack**
   Getting more key bits in the raw key K_r is the indirect way to get more key bits in the final key K_f. In order to get more key bits in the K_f, Alice may claim that the DF state whose measurement results are not conclusive is missing. For example, in Step 2 of the DFQPQ scheme, Alice can use the measurement sequence $Z_1Z_2Z_3Z_4X_5X_6$ to measure the received quantum states and to judge whether the measurement results are conclusive based on the measurements of the first

four particles. However, only $|0110\rangle_{1234}$, $|1001\rangle_{1234}$, $|1100\rangle_{1234,}$ and $|0011\rangle_{1234}$ are valid in Alice's measurement results without Bob announcing any messages, that is, Alice is able to obtain a raw key bit with a probability of 1/6. Measurement results $|1000\rangle_{1234}$, $|0010\rangle_{1234}$, $|0111\rangle_{1234}$ and $|1101\rangle_{1234}$ are invalid for Alice at this time because Bob has not published any information yet. For example, Alice's measurements is $|1000\rangle_{1234}$, and Alice is not possible to determine whether Bob sent $|0^{'}\rangle$ or $|-^{'}\rangle$ so Alice could not guess the corresponding raw key bits. However, in Step 3 of the DFQPQ scheme, Bob can detect the attack launched by Alice, because the measurement results announced by dishonest Alice are *100%* in $\{|0110\pm\pm\rangle, |1001\pm\pm\rangle, |1100\pm\pm\rangle$, $|0011\pm\pm\rangle$, $|\pm\pm\mp\mp 00\rangle, |\pm\pm\mp\mp 11\rangle, |\pm\mp\mp\pm 00\rangle, |\pm\mp\mp\pm 11\rangle\}$ instead of *16.7%*.

2. **Joint Measurement Attack**
   In order to determine which DF state Bob sent, Alice uses a delay attack in Step 2. Alice first stores the received DF state and then performs a more effective measurement on them after Bob's announcement in Step 5 of the DFQPQ scheme. First, we consider the measurement of a single DF state. In this case, the most effective measurement is optimal unambiguous state discrimination (USD) measurement. Because there are two groups of DF state in the scheme that need to be distinguished, $\{|0^{'}\rangle, |+^{'}\rangle\}$ and $\{|1^{'}\rangle, |-^{'}\rangle\}$, the success probability of this USD measurement is bounded by $1 - F(\rho_0, \rho_1)$, where $F(\rho_0, \rho_1)$ is the fidelity between two six-qubit DF state to be discriminated. Through a simple calculation, the probability that using USD measurement to distinguish the transmitted DF state is $p^{USD} = 1 - \langle 0'|+'\rangle = 1 - \langle 1'|-'\rangle = \frac{1}{2}$, which is the same as the probability of using the two sequences of single qubit measurement $\{Z_1Z_2Z_3Z_4X_5X_6, X_1X_2Z_3Z_4Z_5Z_6\}$. Obviously, this attack can't bring Alice more conclusive key bits in the raw key sequence S_k.

   The abovementioned USD measurement is an effective way to distinguish a single quantum carrier. However, the joint measurement (JM) attack can get the parity of a final key bit corresponding to k DF state with a certain probability $p_{JM} = 1 - tr\sqrt{\rho_0^{\frac{1}{2}}\rho_1\rho_0^{\frac{1}{2}}}$, where ρ_0 and ρ_1 correspond to k mixed states of two DF state, respectively. For example, when $k = 2$, the message corresponding to two raw key bits announced by Bob is $\{|0^{'}\rangle, |+^{'}\rangle\}$, then ρ_0 is $\frac{1}{2} \times (|+'+'\rangle\langle +'+'| + |0'0'\rangle\langle 0'0'|)$ and ρ_1 is $\frac{1}{2}(|+'0'\rangle\langle +'0'| + |0'+'\rangle\langle 0'+'|)$. However, Fig. 7.7 shows that JM attack in this scheme decreases sharply with the increase of k, that is, Alice cannot distinguish nonorthogonal quantum states well, thus preventing Alice from obtaining a large number of additional keys.

3. **Other Measurement Base Sequences Attack**
   It is worth noting that in addition to the single-qubit measurement sequences $\{Z_1Z_2Z_3Z_4X_5X_6, X_1X_2Z_3Z_4Z_5Z_6\}$, there are other measurement basis combinations, such as $Z_1Z_2Z_3Z_4Z_5Z_6$, $X_1Z_2Z_3Z_4Z_5Z_6$, $X_1X_2Z_3Z_4Z_5Z_6$ and so on. However, the merging of these measurement bases does not give Alice an advantage. When Bob randomly sends $|0^{'}\rangle$ or $|+^{'}\rangle$, Alice uses other measurement base sequences to determine that the probability of $|0^{'}\rangle$ sent by Bob is p_{A1}, and the probability of

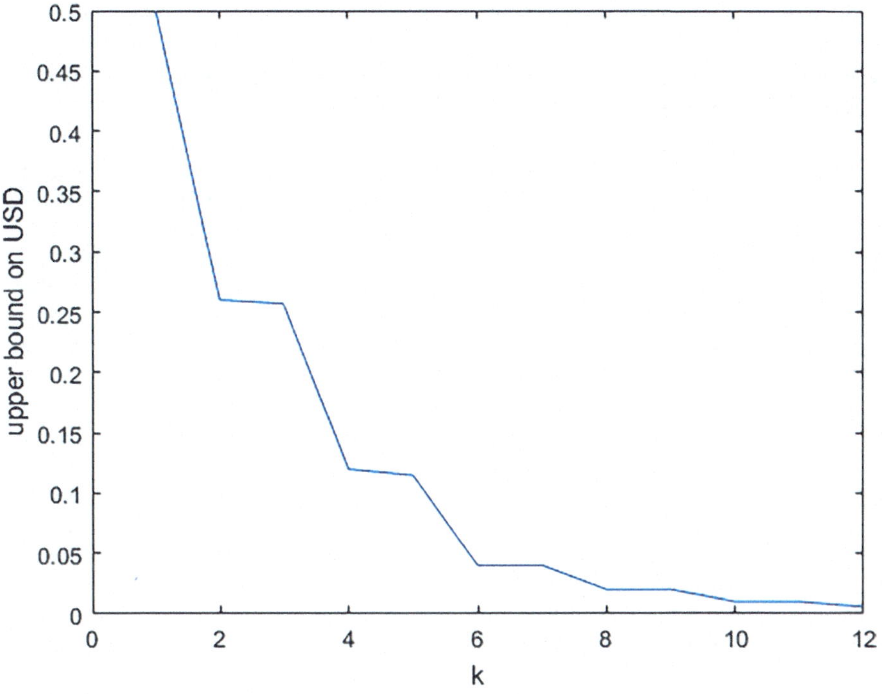

Fig. 7.7 The probability of JM attack success decreases sharply with the increase of k

$|+'\rangle$ is p_{A2}, then $P_c = \frac{1}{2}(p_{A1} + p_{A2})$. The following Tables 7.10 and 7.11 shows the values of P_c that can be determined by other measurement base sequences.

B. User Privacy

In this subsection, dishonest Bob may use false state attack or entanglement attack to obtain Alice's query location, so the DFQPQ scheme must ensure that Alice is sensitive to deception from Bob.

1. **Fake State Attack**
   It is well known that the entangled state is more difficult to prepare than the single particle quantum state. In order to reduce the implementation cost of the whole scheme, Bob may choose the single particle quantum states rather than the six-qubit DF state to implement the DFQPQ scheme. Since Alice randomly selects the six single-qubit measurement base from the two sequences single-qubit sequence $\{Z_1Z_2Z_3Z_4X_5X_6, X_1X_2Z_3Z_4Z_5Z_6\}$ for each received DF state, the fake single-particle sequence sent by Bob can only be random in $\{|0\rangle, |1\rangle, |+\rangle, |-\rangle\}$. According to Tables 7.9 and 7.10, for each six single-qubit measurements, there are *64* possible results, but only *38* of them are approved. Therefore,

Table 7.10 The probability of distinguishing $|0'\rangle$ from $|+'\rangle$ using other measurement base sequences

$X_1X_2X_3$ $X_4X_5X_6$	$Z_1X_2X_3$ $X_4X_5X_6$	$X_1Z_2X_3$ $X_4X_5X_6$	$X_1X_2Z_3$ $X_4X_5X_6$	$X_1X_2X_3$ $Z_4X_5X_6$	$X_1X_2X_3$ $X_4Z_5X_6$
0.5	0.5	0.5	0.5	0.5	0.5
$X_1X_2X_3$ $X_4X_5Z_6$	$Z_1Z_2X_3$ $X_4X_5X_6$	$Z_1X_2Z_3$ $X_4X_5X_6$	$Z_1X_2X_3$ $Z_4X_5X_6$	$Z_1X_2X_3$ $X_4Z_5X_6$	$Z_1X_2X_3$ $X_4X_5Z_6$
0.5	0.3125	0	0.458	0.292	0.317
$X_1Z_2Z_3$ $X_4X_5X_6$	$X_1Z_2X_3$ $Z_4X_5X_6$	$X_1Z_2X_3$ $X_4Z_5X_6$	$X_1Z_2X_3$ $X_4X_5Z_6$	$X_1X_2Z_3$ $Z_4X_5X_6$	$X_1X_2Z_3$ $X_4Z_5X_6$
0.3125	0.375	0.083	0.089	0.396	0.292
$X_1X_2Z_3$ $X_4X_5Z_6$	$X_1X_2X_3$ $Z_4Z_5X_6$	$X_1X_2X_3$ $Z_4X_5Z_6$	$X_1X_2X_3$ $X_4Z_5Z_6$	$Z_1Z_2Z_3$ $X_4X_5X_6$	$Z_1X_2Z_3$ $Z_4X_5X_6$
0.292	0.5	0.5	0.5	0.5	0.5
$Z_1X_2X_3$ $Z_4Z_5X_6$	$Z_1X_2X_3$ $X_4Z_5Z_6$	$Z_1Z_2X_3$ $Z_4X_5X_6$	$Z_1X_2Z_3$ $X_4Z_5X_6$	$Z_1X_2X_3$ $Z_4X_5Z_6$	$Z_1Z_2X_3$ $X_4Z_5X_6$
0.271	0.417	0.417	0	0.25	0.323
$Z_1X_2Z_3$ $X_4X_5Z_6$	$Z_1Z_2X_3$ $X_4X_5Z_6$	$X_1Z_2Z_3$ $Z_4X_5X_6$	$X_1Z_2Z_3$ $X_4Z_5X_6$	$X_1Z_2Z_3$ $X_4X_5Z_6$	$X_1Z_2X_3$ $Z_4Z_5X_6$
0	0.323	0.432	0.188	0.292	0
$X_1Z_2X_3$ $Z_4X_5Z_6$	$X_1Z_2X_3$ $X_4Z_5Z_6$	$X_1X_2Z_3$ $Z_4Z_5X_6$	$X_1X_2Z_3$ $Z_4X_5Z_6$	$X_1X_2Z_3$ $X_4Z_5Z_6$	$X_1X_2X_3$ $Z_4Z_5Z_6$
0	0.167	0.271	0.229	0.354	0.5
$X_1X_2Z_3$ $Z_4Z_5Z_6$	$X_1Z_2X_3$ $Z_4Z_5Z_6$	$X_1Z_2Z_3$ $X_4Z_5Z_6$	$X_1Z_2Z_3$ $Z_4X_5Z_6$	$X_1Z_2Z_3$ $Z_4Z_5X_6$	$Z_1X_2X_3$ $Z_4Z_5Z_6$
0.375	0	0.458	0.323	0.292	0.375
$Z_1X_2Z_3$ $X_4Z_5Z_6$	$Z_1X_2Z_3$ $Z_4X_5Z_6$	$Z_1X_2Z_3$ $Z_4Z_5X_6$	$Z_1Z_2X_3$ $X_4Z_5Z_6$	$Z_1Z_2X_3$ $Z_4X_5Z_6$	$Z_1Z_2X_3$ $Z_4Z_5X_6$
0.33	0.146	0.083	0.458	0.292	0.292
$Z_1Z_2Z_3$ $X_4X_5Z_6$	$Z_1Z_2Z_3$ $X_4Z_5X_6$	$Z_1Z_2Z_3$ $Z_4X_5X_6$	$X_1Z_2Z_3$ $Z_4Z_5Z_6$	$Z_1X_2Z_3$ $Z_4Z_5Z_6$	$Z_1Z_2X_3$ $Z_4Z_5Z_6$
0.5	0.5	0.5	0.5	0.5	0.5
$Z_1Z_2Z_3$ $X_4Z_5Z_6$	$Z_1Z_2Z_3$ $Z_4X_5Z_6$	$Z_1Z_2Z_3$ $Z_4Z_5X_6$	$Z_1Z_2Z_3$ $Z_4Z_5Z_6$	Null	Null
0.5	0.5	0.5	0.5	Null	Null

Alice has *59.375%* probability to detect the behavior that Bob sends single particle state sequence instead of DF state.

2. **Entangled-Measure Attack**
   In Chang's scheme of quantum secure direct communication using six-particle DF state, there is a similar description of this attack [40]. Generally speaking, Bob will prepare some auxiliary particles $|e\rangle$ and implement unitary transformation U to entangle these auxiliary particles with quantum carriers. Bob's attack entangles each single photon $|0\rangle$ and $|1\rangle$ in the DF state with auxiliary particles as follows:

Table 7.11 The probability of distinguishing $|\mathbf{1}'\rangle$ from $|-'\rangle$ using other measurement base sequences

$X_1X_2X_3$ $X_4X_5X_6$	$Z_1X_2X_3$ $X_4X_5X_6$	$X_1Z_2X_3$ $X_4X_5X_6$	$X_1X_2Z_3$ $X_4X_5X_6$	$X_1X_2X_3$ $Z_4X_5X_6$	$X_1X_2X_3$ $X_4Z_5X_6$
0.5	0.5	0.5	0.5	0.5	0.5
$X_1X_2X_3$ $X_4X_5Z_6$	$Z_1Z_2X_3$ $X_4X_5X_6$	$Z_1X_2Z_3$ $X_4X_5X_6$	$Z_1X_2X_3$ $Z_4X_5X_6$	$Z_1X_2X_3$ $X_4Z_5X_6$	$Z_1X_2X_3$ $X_4X_5Z_6$
0.5	0.458	0	0.375	0.292	0.292
$X_1Z_2Z_3$ $X_4X_5X_6$	$X_1Z_2X_3$ $Z_4X_5X_6$	$X_1Z_2X_3$ $X_4Z_5X_6$	$X_1Z_2X_3$ $X_4X_5Z_6$	$X_1X_2Z_3$ $Z_4X_5X_6$	$X_1X_2Z_3$ $X_4Z_5X_6$
0.458	0.083	0.5	0.5	0.25	0.292
$X_1X_2Z_3$ $X_4X_5Z_6$	$X_1X_2X_3$ $Z_4Z_5X_6$	$X_1X_2X_3$ $Z_4X_5Z_6$	$X_1X_2X_3$ $X_4Z_5Z_6$	$Z_1Z_2Z_3$ $X_4X_5X_6$	$Z_1X_2Z_3$ $Z_4X_5X_6$
0.292	0.083	0.083	0.5	0.042	0.5
$Z_1X_2X_3$ $Z_4Z_5X_6$	$Z_1X_2X_3$ $X_4Z_5Z_6$	$X_1Z_2X_3$ $Z_4X_5Z_6$	$X_1Z_2X_3$ $X_4Z_5Z_6$	$Z_1Z_2X_3$ $Z_4X_5X_6$	$Z_1X_2Z_3$ $X_4Z_5X_6$
0.271	0.458	0	0.5	0.417	0
$Z_1X_2X_3$ $Z_4X_5Z_6$	$X_1Z_2X_3$ $Z_4Z_5X_6$	$Z_1Z_2X_3$ $X_4Z_5X_6$	$Z_1X_2Z_3$ $X_4X_5Z_6$	$X_1Z_2Z_3$ $X_4X_5Z_6$	$X_1X_2X_3$ $Z_4Z_5Z_6$
0.271	0	0.271	0	0.271	0.17
$Z_1Z_2X_3$ $X_4X_5Z_6$	$X_1Z_2Z_3$ $X_4Z_5X_6$	$X_1X_2Z_3$ $X_4Z_5Z_6$	$X_1Z_2Z_3$ $Z_4X_5X_6$	$X_1X_2Z_3$ $Z_4X_5Z_6$	$X_1X_2Z_3$ $Z_4Z_5X_6$
0.271	0.271	0.458	0.292	0.271	0.271
$X_1X_2Z_3$ $Z_4Z_5Z_6$	$X_1Z_2X_3$ $Z_4Z_5Z_6$	$X_1Z_2Z_3$ $X_4Z_5Z_6$	$X_1Z_2Z_3$ $Z_4X_5Z_6$	$X_1Z_2Z_3$ $Z_4Z_5X_6$	$Z_1X_2X_3$ $Z_4Z_5Z_6$
0.458	0	0.375	0.292	0.292	0.458
$Z_1X_2Z_3$ $X_4Z_5Z_6$	$Z_1X_2Z_3$ $Z_4X_5Z_6$	$Z_1X_2Z_3$ $Z_4Z_5X_6$	$Z_1Z_2X_3$ $X_4Z_5Z_6$	$Z_1Z_2X_3$ $Z_4X_5Z_6$	$Z_1Z_2X_3$ $Z_4Z_5X_6$
0.33	0.5	0.5	0.375	0.292	0.292
$Z_1Z_2Z_3$ $X_4X_5Z_6$	$Z_1Z_2Z_3$ $X_4Z_5X_6$	$Z_1Z_2Z_3$ $Z_4X_5X_6$	$X_1Z_2Z_3$ $Z_4Z_5Z_6$	$Z_1X_2Z_3$ $Z_4Z_5Z_6$	$Z_1Z_2X_3$ $Z_4Z_5Z_6$
0.083	0.083	0.5	0.5	0.5	0.5
$Z_1Z_2Z_3$ $X_4Z_5Z_6$	$Z_1Z_2Z_3$ $Z_4X_5Z_6$	$Z_1Z_2Z_3$ $Z_4Z_5X_6$	$Z_1Z_2Z_3$ $Z_4Z_5Z_6$	Null	Null
0.5	0.5	0.5	0.5	Null	Null

$$U \otimes |0e\rangle = |\varphi_1\rangle = a|0e_{00}\rangle + b|1e_{01}\rangle \tag{7.30}$$

$$U \otimes |1e\rangle = |\varphi_2\rangle = c|0e_{10}\rangle + d|1e_{11}\rangle \tag{7.31}$$

$$\begin{aligned} U \otimes |+e\rangle = |\varphi_3\rangle = \frac{1}{2}(&|+\rangle(a|e_{00}\rangle + b|e_{01}\rangle + c|e_{10}\rangle + d|e_{11}\rangle) \\ &+|-\rangle(a|e_{00}\rangle - b|e_{01}\rangle + c|e_{10}\rangle - d|e_{11}\rangle)) \end{aligned} \tag{7.32}$$

$$\begin{aligned} U \otimes |-e\rangle = |\varphi_4\rangle = \frac{1}{2}(&|+\rangle(a|e_{00}\rangle + b|e_{01}\rangle - c|e_{10}\rangle - d|e_{11}\rangle) \\ &+|-\rangle(a|e_{00}\rangle - b|e_{01}\rangle - c|e_{10}\rangle + d|e_{11}\rangle)) \end{aligned} \tag{7.33}$$

Here $|a|^2 + |b|^2 = 1$ and $|c|^2 + |d|^2 = 1$. In addition, $|0'\rangle$ can be expressed as Eq. 7.34 in the measurement sequence $Z - Z - Z - Z - X - X$.

$$\begin{aligned}|0'\rangle = \frac{1}{2\sqrt{6}}(&|0100++\rangle - |0100+-\rangle - |0100-+\rangle + |0100--\rangle\\ &- |0101++\rangle + |0100--\rangle - |0110++\rangle + |0110--\rangle + |1001++\rangle\\ &- |1001--\rangle + |1010++\rangle - |1010--\rangle + |0111++\rangle + |0111+-\rangle\\ &+ |0111-+\rangle + |0111--\rangle - |1000++\rangle + |1000+-\rangle + |1000-+\rangle\\ &- |1000--\rangle - |1011++\rangle - |1011+-\rangle - |1011-+\rangle - |1011--\rangle)\end{aligned} \tag{7.34}$$

Next, we analyze the influence of Bob's entanglement attack on Alice with DF state $|0^{'}\rangle$. Due to Bob's attack, the combined system $|0^{'}\rangle$ changes to Eq. 7.35:

$$\begin{aligned}|\psi\rangle_{Bob} = U\otimes U\otimes U\otimes U\otimes U\otimes U\otimes \Big[\frac{1}{2\sqrt{6}}(&|0e1e0e0e+e+e\rangle - |0e1e0e0e+e-e\rangle\\ &- |0e1e0e0e-e+e\rangle - |0e1e0e0e-e-e\rangle - |0e1e0e1e+e+e\rangle\\ &+ |0e1e0e1e-e-e\rangle - |0e1e1e0e+e+e\rangle + |0e1e1e0e-e-e\rangle\\ &+ |1e0e0e1e+e+e\rangle - |1e0e0e1e-e-e\rangle + |1e0e1e0e+e+e\rangle\\ &- |1e0e1e0e-e-e\rangle + |0e1e1e1e+e+e\rangle + |0e1e1e1e+e-e\rangle\\ &+ |0e1e1e1e-e+e\rangle + |0e1e1e1e-e-e\rangle - |1e0e0e0e+e+e\rangle\\ &+ |1e0e0e0e+e-e\rangle + |1e0e0e0e-e+e\rangle - |1e0e0e0e-e-e\rangle\\ &- |1e0e1e1e+e+e\rangle - |1e0e1e1e+e-e\rangle - |1e0e1e1e-e+e\rangle\\ &- |1e0e1e1e-e-e\rangle)\Big]\end{aligned} \tag{7.35}$$

Here,

$$\begin{aligned}&U\otimes U\otimes U\otimes U\otimes U\otimes U\otimes |0e1e0e0e+e+e\rangle\\ &= |\varphi_1\rangle\otimes|\varphi_2\rangle\otimes|\varphi_1\rangle\otimes|\varphi_1\rangle\otimes|\varphi_3\rangle\otimes|\varphi_3\rangle\\ &= \frac{1}{4}\Big(|p_1\rangle + |p_2\rangle + |p_3\rangle + |p_4\rangle + |p_5\rangle + |p_6\rangle + |p_7\rangle + |p_8\rangle + |p_9\rangle + |p_{10}\rangle\\ &\quad + |p_{11}\rangle + |p_{12}\rangle + |p_{13}\rangle + |p_{14}\rangle + |p_{15}\rangle + |p_{16}\rangle\Big)\end{aligned} \tag{7.36}$$

$$
\begin{aligned}
|p_1\rangle = {} & a^5c|0e_{00}0e_{10}0e_{00}0e_{00}+e_{00}+e_{00}\rangle + a^4bc|0e_{00}0e_{10}0e_{00}0e_{00}+e_{00}+e_{01}\rangle - a^4c^2|0e_{00}0e_{10}0e_{00}0e_{00}+e_{00}+e_{10}\rangle \\
& - a^4cd|0e_{00}0e_{10}0e_{00}0e_{00}+e_{00}+e_{11}\rangle + a^5c|0e_{00}0e_{10}0e_{00}0e_{00}+e_{00}-e_{00}\rangle - a^4bc|0e_{00}0e_{10}0e_{00}0e_{00}+e_{00}-e_{01}\rangle \\
& - a^4c^2|0e_{00}0e_{10}0e_{00}0e_{00}+e_{00}-e_{10}\rangle + a^4cd|0e_{00}0e_{10}0e_{00}0e_{00}+e_{00}-e_{11}\rangle + a^4bc|0e_{00}0e_{10}0e_{00}0e_{00}+e_{01}+e_{00}\rangle \\
& + a^3b^2c|0e_{00}0e_{10}0e_{00}0e_{00}+e_{01}+e_{01}\rangle - a^3bc^2|0e_{00}0e_{10}0e_{00}0e_{00}+e_{01}+e_{10}\rangle - a^3bcd|0e_{00}0e_{10}0e_{00}0e_{00}+e_{01}+e_{11}\rangle \\
& + a^4bc|0e_{00}0e_{10}0e_{00}0e_{00}+e_{01}-e_{00}\rangle - a^3b^2c|0e_{00}0e_{10}0e_{00}0e_{00}+e_{01}-e_{01}\rangle - a^3bc^2|0e_{00}0e_{10}0e_{00}0e_{00}+e_{01}-e_{10}\rangle \\
& + a^3bcd|0e_{00}0e_{10}0e_{00}0e_{00}+e_{01}-e_{11}\rangle + a^4c^2|0e_{00}0e_{10}0e_{00}0e_{00}+e_{10}+e_{00}\rangle + a^3bc^2|0e_{00}0e_{10}0e_{00}0e_{00}+e_{10}+e_{01}\rangle \\
& - a^3c^3|0e_{00}0e_{10}0e_{00}0e_{00}+e_{10}+e_{10}\rangle - a^4c^2|0e_{00}0e_{10}0e_{00}0e_{00}+e_{10}+e_{11}\rangle + a^4c^2|0e_{00}0e_{10}0e_{00}0e_{00}+e_{10}-e_{00}\rangle \\
& - a^3bc^2|0e_{00}0e_{10}0e_{00}0e_{00}+e_{10}-e_{01}\rangle - a^3c^3|0e_{00}0e_{10}0e_{00}0e_{00}+e_{10}-e_{10}\rangle + a^4c^2|0e_{00}0e_{10}0e_{00}0e_{00}+e_{10}-e_{11}\rangle \\
& + a^4cd|0e_{00}0e_{10}0e_{00}0e_{00}+e_{11}+e_{00}\rangle + a^3bcd|0e_{00}0e_{10}0e_{00}0e_{00}+e_{11}+e_{01}\rangle - a^3c^2d|0e_{00}0e_{10}0e_{00}0e_{00}+e_{11}+e_{10}\rangle \\
& - a^3cd^2|0e_{00}0e_{10}0e_{00}0e_{00}+e_{11}+e_{11}\rangle + a^4cd|0e_{00}0e_{10}0e_{00}0e_{00}+e_{11}-e_{00}\rangle - a^3bcd|0e_{00}0e_{10}0e_{00}0e_{00}+e_{11}-e_{01}\rangle \\
& - a^3c^2d|0e_{00}0e_{10}0e_{00}0e_{00}+e_{11}-e_{10}\rangle + a^3cd^2|0e_{00}0e_{10}0e_{00}0e_{00}+e_{11}-e_{11}\rangle + a^5c|0e_{00}0e_{10}0e_{00}0e_{00}-e_{00}+e_{00}\rangle \\
& + a^4bc|0e_{00}0e_{10}0e_{00}0e_{00}-e_{00}+e_{01}\rangle - a^4c^2|0e_{00}0e_{10}0e_{00}0e_{00}-e_{00}+e_{10}\rangle - a^4cd|0e_{00}0e_{10}0e_{00}0e_{00}-e_{00}+e_{11}\rangle \\
& + a^5c|0e_{00}0e_{10}0e_{00}0e_{00}-e_{00}-e_{00}\rangle - a^4bc|0e_{00}0e_{10}0e_{00}0e_{00}-e_{00}-e_{01}\rangle - a^4c^2|0e_{00}0e_{10}0e_{00}0e_{00}-e_{00}-e_{10}\rangle \\
& + a^4cd|0e_{00}0e_{10}0e_{00}0e_{00}-e_{00}-e_{11}\rangle - a^4bc|0e_{00}0e_{10}0e_{00}0e_{00}-e_{01}+e_{00}\rangle - a^3b^2c|0e_{00}0e_{10}0e_{00}0e_{00}-e_{01}+e_{01}\rangle \\
& + a^3bc^2|0e_{00}0e_{10}0e_{00}0e_{00}-e_{01}+e_{10}\rangle + a^3bcd|0e_{00}0e_{10}0e_{00}0e_{00}-e_{01}+e_{11}\rangle - a^4bc|0e_{00}0e_{10}0e_{00}0e_{00}-e_{01}-e_{00}\rangle \\
& + a^3b^2c|0e_{00}0e_{10}0e_{00}0e_{00}-e_{01}-e_{01}\rangle + a^3bc^2|0e_{00}0e_{10}0e_{00}0e_{00}-e_{01}-e_{10}\rangle - a^3bcd|0e_{00}0e_{10}0e_{00}0e_{00}-e_{01}-e_{11}\rangle \\
& + a^4c^2|0e_{00}0e_{10}0e_{00}0e_{00}-e_{10}+e_{00}\rangle + a^3bc^2|0e_{00}0e_{10}0e_{00}0e_{00}-e_{10}+e_{01}\rangle - a^3c^3|0e_{00}0e_{10}0e_{00}0e_{00}-e_{10}+e_{10}\rangle \\
& - a^4c^2|0e_{00}0e_{10}0e_{00}0e_{00}-e_{10}+e_{11}\rangle + a^4c^2|0e_{00}0e_{10}0e_{00}0e_{00}-e_{10}-e_{00}\rangle - a^3bc^2|0e_{00}0e_{10}0e_{00}0e_{00}-e_{10}-e_{01}\rangle \\
& - a^3c^3|0e_{00}0e_{10}0e_{00}0e_{00}-e_{10}-e_{10}\rangle + a^4c^2|0e_{00}0e_{10}0e_{00}0e_{00}-e_{10}-e_{11}\rangle - a^4cd|0e_{00}0e_{10}0e_{00}0e_{00}-e_{11}+e_{00}\rangle \\
& - a^3bcd|0e_{00}0e_{10}0e_{00}0e_{00}-e_{11}+e_{01}\rangle + a^3c^2d|0e_{00}0e_{10}0e_{00}0e_{00}-e_{11}+e_{10}\rangle + a^3cd^2|0e_{00}0e_{10}0e_{00}0e_{00}-e_{11}+e_{11}\rangle \\
& - a^4cd|0e_{00}0e_{10}0e_{00}0e_{00}-e_{11}-e_{00}\rangle + a^3bcd|0e_{00}0e_{10}0e_{00}0e_{00}-e_{11}-e_{01}\rangle + a^3c^2d|0e_{00}0e_{10}0e_{00}0e_{00}-e_{11}-e_{10}\rangle \\
& - a^3cd^2|0e_{00}0e_{10}0e_{00}0e_{00}-e_{11}-e_{11}\rangle
\end{aligned}
\tag{7.37}
$$

The specific expansion of $|p_1\rangle$ in Eq. 7.36 is shown in Eq. 7.37, and the other *15* items can be calculated similarly. Each of $|p_1\rangle$, $|p_2\rangle$, $|p_3\rangle$, …, $|p_{15}\rangle$, $|p_{16}\rangle$ in Eq. 7.7. contains 2^6 components. However, only the quantum states $|0e_{ij}1e_{ij}0e_{ij}0e_{ij}+e_{ij}+e_{ij}\rangle$, $|0e_{ij}1e_{ij}0e_{ij}0e_{ij}-e_{ij}-e_{ij}\rangle, \ldots, |1e_{ij}1e_{ij}0e_{ij}1e_{ij}-e_{ij}+e_{ij}\rangle$ in Table 7.8 are valid, where i, j belongs to $\{0, 1\}$. Through calculation, the probability of Alice using $Z_1Z_2Z_3Z_4X_5X_6$ measurement base sequence to get correct measurement results is

$$
\begin{aligned}
p(|\psi\rangle_{Bob}) = \frac{1}{1536}\Big(& 24|4a^3c^2d|^2 + 28|4a^3cd^2|^2 + 16|4a^3d^3|^2 + 88|4a^2bc^2d|^2 \\
& + 28|4a^2bd^3|^2 + 88|4a^2bcd^2|^2 + 80|4ab^2c^2d|^2 + 24|4ab^2d^3|^2 \\
& + 24|4a^2bc^3|^2 + 28|4ab^2c^3|^2 + 80|4ab^2cd^2|^2 + 16|4b^3c^3|^2 \\
& + 28|4b^3c^2d|^2 + 24|4b^3cd^2|^2 + 7|2b^2cd^2|^2 + 42|2a^3c^2d|^2 \\
& + 140|2a^2bc^2d|^2 + 55|2a^3cd^2|^2 + 149|2a^2bcd^2|^2 + 6|2a^4cd|^2 \\
& + 21|2a^3bcd|^2 + 13|2a^3d^3|^2 + 7|2a^2cd^3|^2 + 9|2a^3bd^2|^2 \\
& + 4|2a^4d^2|^2 + 52|2a^2bd^3|^2 + 162|2ab^2c^2d|^2 + 20|2a^2b^2cd|^2 \\
& + 139|2ab^2cd^2|^2 + 40|2ab^2d^3|^2 + 8|2abcd^3|^2 + 5|2a^2b^2d^2|^2 \\
& + 42|2a^2bc^3|^2 + 54|2b^2c^3|^2 + 6|2a^3bc^2|^2 + 10|2a^2b^2c^2|^2 \\
& + 6|2abc^3d|^2 + 6|2abc^2d^2|^2 + 7|2a^2d^4|^2 + 8|2b^2c^3d|^2 \\
& + 8|2b^2c^2d^2|^2 + 57|2b^3c^2d|^2 + 49|2b^3cd^2|^2 + 5|2ab^3cd|^2 \\
& + 8|2b^3c^3|^2 + 2|2ab^3c^2|^2 + 8|2abd^4|^2\Big).
\end{aligned} \tag{7.38}
$$

Among which $|a|^2 + |b|^2 = 1$ and $|c|^2 + |d|^2 = 1$, $0 \le |a| \le 1$and $0 \le |c| \le 1$. According to Eq. 7.38, the probability of Bob being found is $p_e = 1 - p(|\psi\rangle_{Bob})$. Fig. 7.8 shows the change of p_e with $|c|$ and $|a|$, and the lowest value of p_e is between 40% and 50%. Especially when $|c|$ and $|a|$ are both *0* or *1*, the probability of entanglement attack is *1*. The above analysis is about the probability of finding Bob's entanglement attack when the six-particle DF state is $|0'\rangle$ and the measurement base sequence that Alice uses are $Z_1Z_2Z_3Z_4X_1X_2$. In other cases, Bob's attack can be detected according to a similar process.

C. Other Noise

Although it has been proved that DF state can resist collective decorrelation noise, there are other noises in quantum channel. Here, we extend the post-processing of the scheme to $gkN \rightarrow N$, which can correct any error caused by noises to a certain extent. Assuming that [3, 4, 7] Hamming code is used in $gkN \rightarrow N$ post-processing and $N = 10^3$, the probability of a conclusive middle key bit obtained by dishonest Alice is as Eq. 7.39. As you can see from Table 7.12, when $g = 9$, nine middle keys are required to make the dishonest Alice get a *1* bit from the final key.

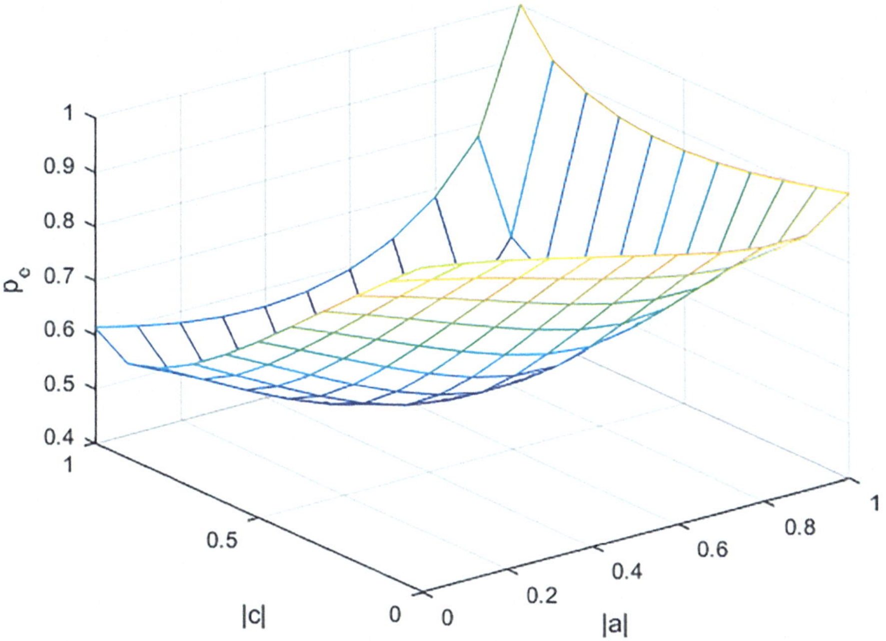

Fig. 7.8 When Bob sends the DF state $|\mathbf{0}'\rangle$, Alice uses the measurement basis sequence $\boldsymbol{Z_1Z_2Z_3Z_4X_1X_2}$ to detect the change trend of the probability $\boldsymbol{p_e}$ of entangle-measure attack with $|\boldsymbol{a}|$ and $|\boldsymbol{c}|$

Table 7.12 Here $n = 10^3$, $k = 7$, the number of final key bits that Alice can get and the number of corresponding middle keys

g	1	2	3	4	5	6	7	8	9
n_A	500	250	125	62	31	15	7	3	1

$$p_{\overline{c}} = \sum_{i=4,5,6,7} \binom{i}{7} \left(\frac{1}{2}\right)^i \left(\frac{1}{2}\right)^{7-i} \tag{7.39}$$

Assuming that the probability that Alice and Bob have a different raw key bit is e, that is, the error rate caused by noise, then the error probability of a middle key bit and a final key bit formed by the bitwise addition of middle keys are respectively described in Eqs. 7.40 and 7.41. Figure 7.9 shows that p_e increases with the increase of e in the $(0,\ 0.3)$ interval, and p_e tends to 0.5 when $e > 0.3$. Obviously, when $e < 0.3$, the error correction effect of $gkN \rightarrow N$ is good.

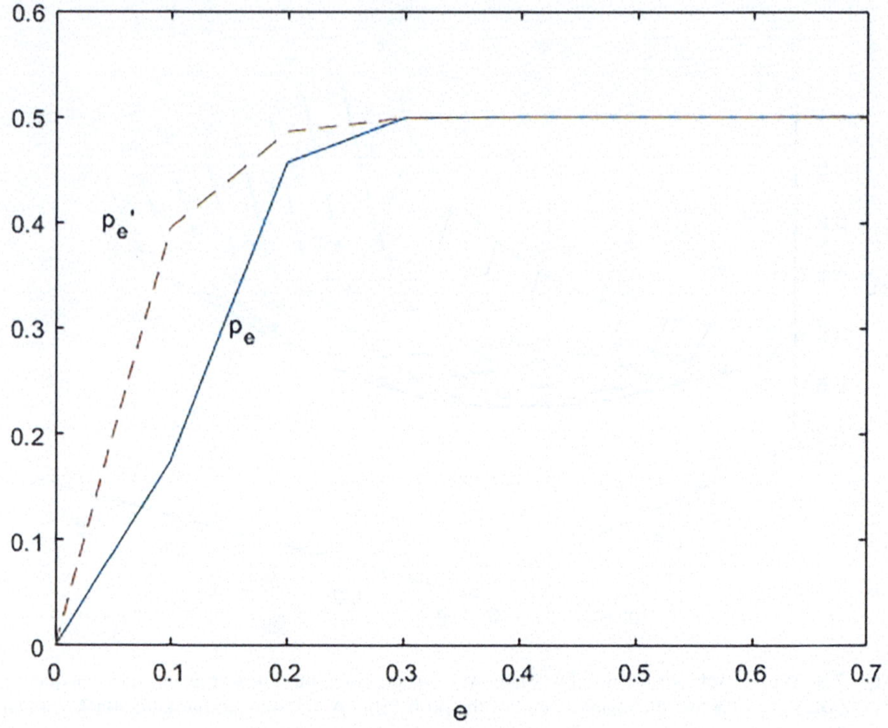

Fig. 7.9 The error rate of the final key. Here $\boldsymbol{e}$ is the error rate of the oblivious key distribution caused by noise, and $\boldsymbol{p_e}$ and $\boldsymbol{p'_e}$, respectively, represent the final key error rate of this scheme and J-scheme after $\boldsymbol{gkN \rightarrow N}$ processing

$$p_{em} = \sum_{i=3,5,\ldots,i \le k} \binom{k}{i} (e)^i (1-e)^{k-i} \tag{7.40}$$

$$p_e = \sum_{i=1,3,5,\ldots,i \le g} \binom{g}{i} (p_{em})^i (1-p_{em})^{g-i} \tag{7.41}$$

On the other hand, the number of middle keys in $gkN \rightarrow N$ post-processing algorithm is determined by the probability of a conclusive raw key bit in the scheme. In most of the proposed schemes [13, 24, 25, 30, 41, 43, 45, 47, 48], the probability of obtaining a conclusive raw key bit by Alice is 1/4, so we select J-scheme for comparison. It can be seen from Fig. 7.9 that the error correction performance of the scheme in the *(0, 0.3)* interval is significantly better than that of J-scheme using $gkN \rightarrow N$.

7.3.4.4 Summary

This section proposes a new robust quantum private query scheme with six-qubit decoherence-free state which can be used to deal with the collective noise problem in the quantum channel. Two sequences of single-qubit measurement is used, which makes the whole DFQPQ scheme easy to be implement by exploiting current technology [27]. Of course, there are other noises besides collective noise in the channel. In order to deal with other transmission errors, we extend the post-processing method to $gkN \rightarrow N$, which is a significant error correction method at the cost of communication complexity. Therefore, in future work, we will focus on more efficient error correction methods for the oblivious key distribution.

7.4 Brief Summary

In this chapter, we give a brief introduction to quantum privacy queries. In addition, we design three quantum privacy query schemes based on three different multiparticle quantum states, which are quantum private query using W state, multiuser quantum private query using symmetric multiparticle W state and a robust quantum private query based on six-qubit DF state. Compared with the existing quantum privacy query schemes, the proposed three schemes have the advantages of high efficiency and simple implementation.

References

1. Y. Li, R.G. Zhou, R. Xu, J. Luo, W. Hu, A quantum deep convolutional neural network for image recognition. Quantum Sci. Technol. **5**(4), 044003 (2020)
2. Y. Li, R. Zhou, R. Xu, J. Luo, S. Jiang, A quantum mechanics-based framework for EEG signal feature extraction and classification. IEEE Trans. Emerg. Top. Comput. **10**(1), 211–222 (2020)
3. B. Chor et al., J. ACM **45**, 965 (1998); C. Cachin, S. Micali, M. Stadler, in *Advances in Cryptology-EUROCRYP99* (Springer-Verlag, Berlin, 1999); C. Gentry and Z. Ramzan, in *Proceedings of the 32nd ICALP* (Springer-Verlag, Berlin, 2005), p. 803; S. Yekhanin, Technical Report No. ECCC TR06–127, 2006
4. M. Naor, B. Pinkas, Oblivious transfer with adaptive queries. Lect. Notes Comput. Sci **791**, 573–590 (1999)
5. S. Wiesner, ACM SIGACT News, (Livermore, 1983), Vol. 15, p. 78; M.O. Rabin, Harvard Aiken Computational Laboratory, Technical Report No. TR-81, 1981; A. Jakoby, M. Liskiewicz, and A. Madry, arXiv:quantph/0605150v1
6. Ambainis, A., *Proceedings of the 24th ICALP, Lecture Notes in Computer Science* (Springer-Verlag, Berlin, 1997), Vol. 1256, p. 401
7. Brassard, G., Crepeau, C., Jozsa, R., Langlois, D., *Proceedings of 1993 IEEE 34th Annual Foundations of Computer Science – A Quantum Bit Commitment Scheme Provably Unbreakable by Both Parties* (1993), pp. 362–371
8. V. Giovannetti, S. Lloyd, L. Maccone, Quantum private queries. Phys. Rev. Lett. **100**(23), 230502 (2008)

9. V. Giovannetti, S. Lloyd, L. Maccone, Quantum private queries: Security analysis. IEEE Trans. Inf. Theory **56**(7), 3465–3477 (2010)
10. M. Jakobi, C. Simon, N. Gisin, et al., Practical private database queries based on a quantum-key-distribution scheme. Phys. Rev. A **83**(2), 22301–22301 (2011)
11. J.L. Zhang, F.Z. Guo, F. Gao, et al., Private database queries based on counterfactual quantum key distribution. Phys. Rev. A **88**(2), 195–201 (2013)
12. Y.G. Yang, S.J. Sun, J. Tian, et al., Secure quantum private query with real-time security check. Optik **125**(19), 5538–5541 (2014)
13. C.Y. Wei, F. Gao, Q.Y. Wen, et al., Practical quantum private query of blocks based on unbalanced-state Bennett-Brassard-1984 quantum-key-distribution scheme. Sci. Rep. **4**, 7537 (2014)
14. F. Yu, D. Qiu, Coding-based quantum private database query using entanglement. Quantum Inf. Comput. **14**(1–2), 91–106 (2014)
15. B. Liu, F. Gao, W. Huang, et al., QKD-based quantum private query without a failure probability. Sci.China Phys. Mech. Astron. **58**(10), 100301 (2015)
16. L. Hong, M.A. Orgun, J. Pieprzyk, et al., Controllable quantum private queries using an entangled Fibonacci-sequence spiral source. Phys. Lett. A **379**(40), 2561 (2015)
17. Y.G. Yang, B.R. Li, D. Li, et al., New quantum key agreement schemes based on bell states. Quantum Inf. Process **18**(10), 1–17 (2019)
18. C.Y. Wei, T.Y. Wang, F. Gao, Practical quantum private query with better performance in resisting joint-measurement attack. Phys. Rev. A **93**(4), 042318 (2016)
19. P. Chan, I. Lucio-Martinez, X. Mo, et al., Performing private database queries in a real-world environment using a quantum scheme. Rep **4**(23), 5233 (2014)
20. M. Rao, M. Jakobi, Towards communication-efficient quantum oblivious key distribution. Phys. Rev. A **87**(1), 012331 (2012)
21. F. Gao, B. Liu, Q.Y. Wen, et al., Flexible quantum private queries based on quantum key distribution. Opt. Express **20**(16), 17411–17420 (2012)
22. Y.G. Yang, Z.C. Liu, X.B. Chen, et al., Novel classical post-processing for quantum key distribution-based quantum private query. Quantum Inf. Process **15**(9), 1–8 (2016)
23. G. Fei, B. Liu, H. Wei, et al., Post-processing of the oblivious key in quantum private queries. IEEE J. Sel. Top. Quantum Electron. **21**(3), 98–108 (2014)
24. Y.G. Yang, Z.C. Liu, J. Li, et al., Quantum private query with perfect user privacy against a joint-measurement attack. Phys. Lett. A **380**(48), 4033–4038 (2016)
25. Y.G. Yang, M.O. Zhang, R. Yang, Private database queries using one quantum state. Quantum Inf. Process **14**(3), 1017–1024 (2015)
26. Y. Wang, F.Z. Guo, L. Liu, et al., A new scheme for quantum private query against joint-measurement attack. Int. J. Theor. Phys. **58**, 1828 (2019)
27. Y.G. Yang, Z.C. Liu, X.B. Chen, et al., Robust QKD-based private database queries based on alternative sequences of single-qubit measurements. Sci. China: Phys. Mech. Astron. **60**(12), 120311 (2017)
28. T.Y. Ye, H.K. Li, J.L. Hu, Multi-user quantum private query scheme. Int. J. Theor. Phys. **16**, 253 (2020)
29. H.J. Cao, H.S. Song, Quantum secure direct communication scheme using a W state and teleportation. Chin. Phys. Lett. **23**(2), 290–292 (2006)
30. R.G. Zhou, Y. Hua, Quantum private query using W state. Int. J. Theor. Phys. **23**, 1–12 (2020)
31. X.B. Chen, Q.Y. Wen, F.Z. Guo, et al., Controlled quantum secure direct communication with W state. Int. J. Quantum Inf. **6**(04), 899 (2008)
32. N. Li, J. Li, X. Chen, Y. Yang, Quantum wireless network private query with multiple third parties. IEEE Access **7**, 33964–33969 (2019)
33. D.M. Greenberger, M.A. Horne, A. Shimony, et al., Bell's theorem without inequalities. Am. J. Phys. **58**(12), 1131–1143 (1990)
34. W. Dür, G. Vidal, J.I. Cirac, Three qubits can be entangled in two inequivalent ways. Phys. Rev. A **62**, 062314 (2000)

35. C. Yan, X. Jinxin, X. Gao, Z. Shibin, Y. Lili, Quantum private query scheme based on EPR pairs. Chin. J. Electron. **27**(2), 256–262 (2018)
36. V. Cofiman, J. Kundu, W.K. Wootters, Analysis of radiatively stable entanglement in a system of two dipole interacting three-level atoms. Phys. Rev. A **61**, 052306 (2000)
37. R. Lohmayer, A. Osterloh, J. Siewert, A. Uhlmann, Entangled three-qubit states without concurrence and three-tangle. Phy. Rev. Lett. **97**, 260502 (2006)
38. G. Svetlichny, Distinguishing three-body from two-body nonseparability by a bell-type inequality. Phys. Rev. D. **35**, 3066–3071 (1987)
39. S. Ghoes, N. Sinclair, S. Debnath, et al., Tripartite entanglement versus tripartite nonlocality in three-qubit Greenberger-Horne-Zeilinger-class states. Phys. Rev. Lett. **102**, 250404 (2009)
40. Y.-G. Yang, S.-J. Sun, P. Xu, J. Tian, Flexible scheme for quantum private query based on B92 scheme. Quantum Inf. Process **13**(3), 805–813 (2014)
41. H.P. Wang, R.G. Zhou, Multi-user quantum private query using symmetric multi-particle w state. Int. J. Theor. Phys. **61**(3), 71 (2022)
42. T.Y. Wang, S.Y. Wang, J.F. Ma, Robust quantum private queries. Int. J. Theor. Phys. **55**, 3309–3317 (2016)
43. N. Li, J. Li, X. Chen, et al., Quantum private query with perfect performance universally applicable against collective-noise. IEEE Access **7**, 29313 (2019)
44. C.Y. Wei, X.Q. Cai, T.Y. Wang, et al., Error tolerance bound in QKD-based quantum private query. IEEE J Sel Areas Commun **38**(3), 517–527 (2020)
45. F. Gao, S.J. Qin, W. Huang, et al., Quantum private query: A new kind of practical quantum cryptographic scheme. Sci. China Phy. Mech. Astron. **62**, 1–12 (2019)
46. H.P. Wang, R.G. Zhou, A robust quantum private query based on six-qubit with decoherence. Optik **253**, 168507 (2022)
47. M. Bourennane, M. Eibl, S. Gaertner, et al., Decoherence-free quantum information processing with four-photon entangled states. Phys. Rev. Lett. **92**(10), 107901 (2004)
48. D.A. Lidar, K. Birgitta Whaley, Decoherence-free subspaces and subsystems, in *Irreversible quantum dynamics*, (Springer, Berlin/Heidelberg, 2003), pp. 83–120
49. A. Cabello, Six-qubit permutation-based decoherence-free orthogonal basis. Phys. Rev. A **75**(2), 020301 (2007)
50. J.C. Boileau, D. Gottesman, R. Laflamme, et al., Robust polarization-based quantum key distribution over a collective-noise channel. Phys. Rev. Lett. **92**(1), 017901 (2004)
51. T.Y. Chen, J. Zhang, J.C. Boileau, et al., Experimental quantum communication without a shared reference frame. Phys. Rev. Lett. **96**(15), 150504–150504 (2006)
52. P.G. Kwiat, A.J. Berglund, J.B. Altepeter, et al., Experimental verification of decoherence-free subspaces. Science **290**(5491), 498–501 (2000)
53. A. Cabello, N-particle N-level singlet states: Some properties and applications. Phys. Rev. Lett. **89**(10), 100402 (2002)
54. A. Cabello, Bell's theorem without inequalities and without alignments. Phys. Rev. Lett. **91**(23), 230403 (2003)
55. A. Karlsson, M. Koashi, N. Imoto, Quantum entanglement for secret sharing and secret splitting. Phys. Rev. A **59**(1), 162–168 (1999)

Chapter 8
Quantum Network Coding

Abstract With the development of quantum communication and quantum computation, the concept of a quantum communication network is proposed. However, there is a throughput bottleneck when quantum information is transmitted in the common channel of quantum communication networks; therefore, it has become a hot research topic to solve this problem by using effective quantum network coding. Network coding is the idea of using the intermediate node of the network to encode multiple data. With the introduction of network coding technology, multiple data can be transmitted in the same network, which can greatly improve the network throughput.

8.1 Introduction

Quantum computer processing scale is often larger compared to the classical computer, which also brings challenges to the classical network transmission. Network coding is gradually introduced into the quantum field because of its powerful throughput and robustness, forming the field of quantum communication network (QNC), which gradually attracted people's attention. Different from classical network coding, QNC has many new features. For example, the quantum noncloning theorem guarantees the security of QNC, which is one of the reasons why QNC has become a research hotspot. However, it is almost impossible to achieve multicast or data replication and forwarding, which also determines that QNC is different from classical network coding.

The concept of network coding was proposed by Ahlswede in 2000 [1]. They redefined the intermediate node of the network and put forward the idea of using the intermediate node of the network to encode multiple data on the basis of the classical network. In the classical network, the intermediate node only has the function of store and forward, while the encoding operation only occurs in the receiver and sender. The data flow is often forwarded one by one, so when the amount of data reaches a certain scale, the throughput will reach a bottleneck and cause network

R.-G. Zhou et al., *Design and Analysis of Secure Quantum Communication Schemes*,
https://doi.org/10.1007/978-3-031-78428-6_8

congestion. However, multiple data can be transmitted in the same network with the introduction of network coding technology. It is a new feature that nodes are forwarded at the same time, which will greatly improve the throughput of the network.

8.2 Principle and Research Progress of Quantum Network Coding

The network coding redefined the intermediate node of the network and put forward the idea of using the intermediate node of the network to encode multiple data on the basis of the classical network. In the classic network, the intermediate node only has the function of store and forward, while the encoding operation only occurs in the receiver and sender. The data flow is often forwarded one by one, so when the amount of data reaches a certain scale, the throughput will reach a bottleneck and cause network congestion. However, multiple data can be transmitted in the same network with the introduction of network coding technology. It is a new feature that nodes are forwarded at the same time, which will greatly improve the throughput of the network.

Classical network coding allows intermediate nodes to encode the received data to maximize communication efficiency. However, with the development of quantum algorithm, its security is seriously threatened. The current QNC schemes have many limitations. Because of the complexity of real-world scenarios, it requires the designer to consider the scope of application of the scheme. The traditional routing network only has the ability to make the intermediate nodes copy the information and then transmit it backward, so when the network nodes increase, that is, when multiple information arrive at the intermediate node at the same time, it will lead to communication or routing conflicts between bottleneck nodes, resulting in limited network throughput. As a new communication technology. QNC solves this problem by encoding, compressing, and forwarding the received information to downstream nodes.

Network coding is proposed to solve the bottleneck problem in the process of transmission. The main idea is to allow intermediate nodes in the process of information transmission by using linear or nonlinear operations to perform the coding operation. By using the encoding operation, the transmission efficiency and the transmission security are improved, and the resource consumption is reduced compared to the routing transmission scheme. In classical network coding, copy operation is essential and can be realized in the process of classical information transmission. But because of quantum nocloning theorem, such an operation is not possible in a quantum context. To solve this problem, researchers have proposed a number of solutions in three areas: free classical communication (QNC-FCC) (allowing free classical communication between any pair of network nodes, thus achieving perfect quantum communication), quantum teleportation (QNC-QT) (the use of quantum teleportation to transmit quantum information), and approximate cloning (QNC-AC) (the use of quantum coding operations to transmit quantum information).

QNC-FCC allows free classical communication between nodes, thus achieving perfect quantum communication. This direction is a hot research direction of QNC, which has played a huge role in the development of QNC. In 2009, Kobayashi et al. proposed a QNC scheme [2] that allows transmission of classical bit information in a butterfly network without any entanglement pair assistance. It is possible to realize efficient transmission of quantum bits by encoding quantum information. In 2010, Kobayashi et al. proposed a similar QNC scheme based on the encoding scheme over a finite field in classical network coding [3], which was the first encoding scheme to be implemented without the use of teleportation, and subsequently, they also extended the scheme to the case of finite rings. In 2010, Ma et al., in the scientific work of Hayashi, proposed a scheme to achieve M-qubit cross-transmission by having two senders share nonmaximally entangled states in advance [4]. Li et al. optimized the classical auxiliary QNC and designed a new structure that separates the classical communication from the classical network solution, which improves the communication efficiency and reduces the communication cost [5]. Then Li et al. solved the quantum multicast problem in directed acyclic networks by using globally entangled 2D and 3D cluster states [6]. In 2019, Li et al. focused on the security and efficiency of QNC and proposed a relatively secure QNC scheme [7, 8]. At the same time, the transmission efficiency of QNC scheme is considered, which promotes the development of QNC.

QNC-QT can transmit quantum information by quantum teleportation, which makes a great contribution to the development of QNC in the direction of quantum teleportation. In 2006, Hayashi et al. achieved the cross-probability transmission of two arbitrary quantum states with the help of the butterfly network model [9], which is an important symbol of the possible implementation of QNC. In 2009, Leung et al. implemented QNC for various network models [10], which is no longer limited to the traditional butterfly network. In 2012, Satoh et al. introduced repeaters into QNC and proved that QNC can improve the transmission rate of quantum relay grids [11]. In 2016, Shang et al. made full use of the channel property of quantum teleportation to ensure the safe transmission of quantum states, and eavesdroppers could be detected by quantum channel verification [12].

QNC-AC mainly allows the use of quantum coding operations to achieve quantum information transmission. Such a transmission is based on the difficulty of achieving perfect transmission; approximate transmission may be imperfect transmission. In 2007, Hayashi introduced quantum teleportation, in which the maximum entangled states are shared between two senders in advance, thus achieving cross-transmission of quantum states in a butterfly-shaped network and achieving perfect QNC [13]. In 2014, Shang et al. improved the butterfly network and proposed QNC based on controlled teleportation [14], which can control the decoding process of unicast stream by introducing a third-party controller. This helps to extend these schemes to generic networks with unicast or multicast. Then Shang et al. put forward a QNC scheme based on continuous variables, which improves the executability of QNC [15].

To sum up, QNC research is still in the initial stage, and the overall framework is not perfect. Both in theory and application, there is a huge space for development, especially in the approximate cloning research. Therefore, it is necessary to study the design and improvement of QNC scheme based on approximate cloning, which can provide a good basis for the development of quantum information.

8.3 Typical Quantum Network Coding Scheme

8.3.1 QNC Scheme

In 2000, Ahlswede, Cai, Li, and Yeung proposed the idea of network coding to solve the bottleneck problem in the transmission of classical communication networks [1]. The main idea is to allow intermediate nodes to perform encoding operations by using linear or nonlinear operations in the transmission of information. By using the encoding operation, the transmission efficiency and the transmission security are improved and the resource consumption is reduced compared with the routing transmission scheme.

The development of classical network coding has been relatively mature. Many scholars begin to try to apply classical network coding scheme to quantum networks and put forward many important schemes. Among them, Hayashi et al. are the first researchers who try to encode the butterfly quantum network, a special topological model [9]. They confirmed that for each channel is a unit capacity butterfly quantum network, without the aid of auxiliary coding channel, cannot achieve error-free quantum bit transmission. On this basis, they also give a scheme to prove that in the butterfly quantum network model, the upper limit of the fidelity of the information obtained by directly encoding and transmitting the quantum bit information is 1/2. In order to show that the side channel in the butterfly quantum network can affect the coding fidelity, the following theorem is given:

Theorem If the two side-path channels in butterfly topology are removed, it is impossible to have any coding scheme that makes the coding fidelity more than 1/2.

In 2007, Hayashi [13] showed that error-free cross-transmission of quantum bits in butterfly quantum networks can be achieved by means of maximum entanglement pairs and classical bit strings. In order to realize the perfect QNC, the maximum entangled state is shared between two senders in advance and the quantum teleportation is introduced as an auxiliary coding channel. The coding process of the scheme is described in detail, and the schematic diagram of the scheme is shown in Fig. 8.1.

Suppose two senders A_1 and A_2 share two pairs of maximally entangled states in advance, with the first pair represented by nodes $A_{1,\ 1}$ and $A_{2,\ 1}$ and the second pair represented by nodes $A_{1,\ 2}$ and$A_{2,\ 2}$. Two senders A_1 and A_2 need to send the state$|\psi_1\rangle$, $|\psi_2\rangle$, respectively, encoding steps are as follows:

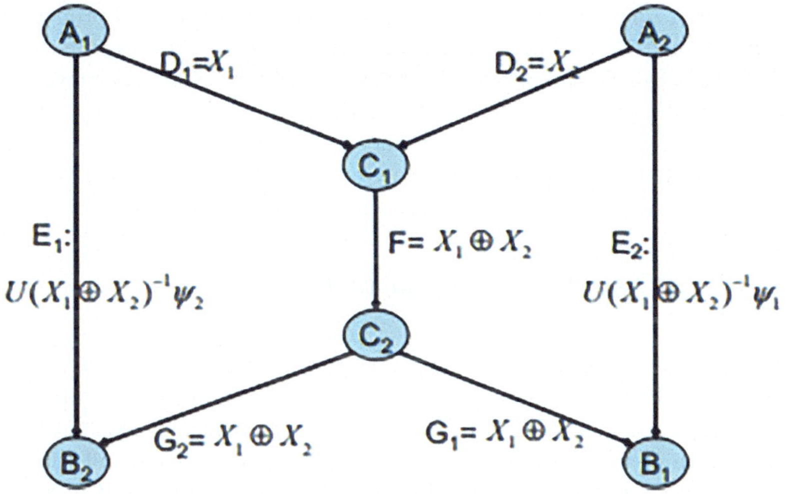

Fig. 8.1 Principle diagram of butterfly QNC based on maximum entangled state

Step 1: The sender A_i directly product the entangled state of the node and the state to be sent to form the joint system $|\psi_i\rangle \otimes A_{i,\ i}$ and measure it with Bell basis $\{\Phi^+, \Phi^-, \Psi^+, \Psi^-\}$. For the four kinds of measurement results, the precoded information is X_i by using (0,0), (1,0), (0,1), (1,1) corresponding to Φ^+, Φ^-, Ψ^+, Ψ^-, respectively. In this case, the state on the remaining node$A_{i,\ i \oplus 1}$ changes to$U(X_{i \oplus 1})^{-1}|\psi_{i \oplus 1}\rangle$, where $U(X)$ is a recovery unitary operation with result X for stealth transmission.

Step 2: The sender A_i makes a comparison with the local remaining node$A_{i,\ i \oplus 1}$, a performs unitary operations $U(X_{i \oplus 1})^{-1}$, when the remaining node $A_{i,\ i \oplus 1}$, the state on a changes to $U(X_i)^{-1}U(X_{i \oplus 1})^{-1}|\psi_{i \oplus 1}\rangle = c(X_i, X_{i \oplus 1})U(X_1 \oplus X_2)^{-1}|\psi_{i \oplus 1}\rangle$, where $|c(X_i, X_{i \oplus 1})| = 1$. Then sender A_i sends system $A_{i,\ i \oplus 1}$ to $B_{i \oplus 1}$ via channel E_i and classic information X_i via channel D_i.

Step 3: The intermediate node C_1 sends the classical information $X_1 \oplus X_2$ to node C_2 via channel F and node C_2 copies the classical information $X_1 \oplus X_2$ and forwards it to receivers B_1 and B_2 via channels G_1 and G_2, respectively.

Step 4: The receiver B_i performs unitary operation $U(X_1 \oplus X_2)$ on the received state $U(X_1 \oplus X_2)^{-1}|\psi_i\rangle$ and finally recovers the initial state$U(X_1 \oplus X_2)U(X_1 \oplus X_2)^{-1}|\psi_i\rangle = |\psi_i\rangle$.

In the proposed scheme, two pairs of maximally entangled states and classical bit information are shared by the sender and transmitted by an intermediate coded channel to realize the cross-transmission of quantum states in the butterfly quantum network. This scheme is also the first scheme to use auxiliary quantum states to complete QNC, and it is a model in the field of QNC.

But under the nonlaboratory condition, the quantum state of any quantum system will be disturbed and influenced by various external factors and will gradually transform into a mixed state. They are generally nonmaximal entangled states.

And the current scientific and technological conditions still cannot properly solve this problem. In 2010, MA et al. proposed a QNC scheme based on nonmaximally entangled states [4]. The premise of this scheme is that the classical channel and the quantum channel coexist in the butterfly quantum network, and the nonmaximally entangled state can be used as an auxiliary coding channel to realize the cross-transmission of the single-qubit state.

8.3.2 CQNC Based on Quantum Walk

In 2007, the prior entanglement QNC scheme based on the butterfly network (here it is represented as PE) was proposed by Hayashi [13]. Then Shang [14] improved this QNC scheme by adding a third party and proposed a controllable QNC scheme (here it is represented as CT). Inspired by Hayashi, our CQNC scheme still adopted the classical butterfly network. And inspired by Shang [14], we added a third-party controller to enhance the security of the network. According to the needs of this CQNC scheme, we simplified the third-party controller to one, as shown in Fig. 8.2. All links adopted the classical communication, except for two lines of A1 (A2)→B1 (B2), which used quantum communication.

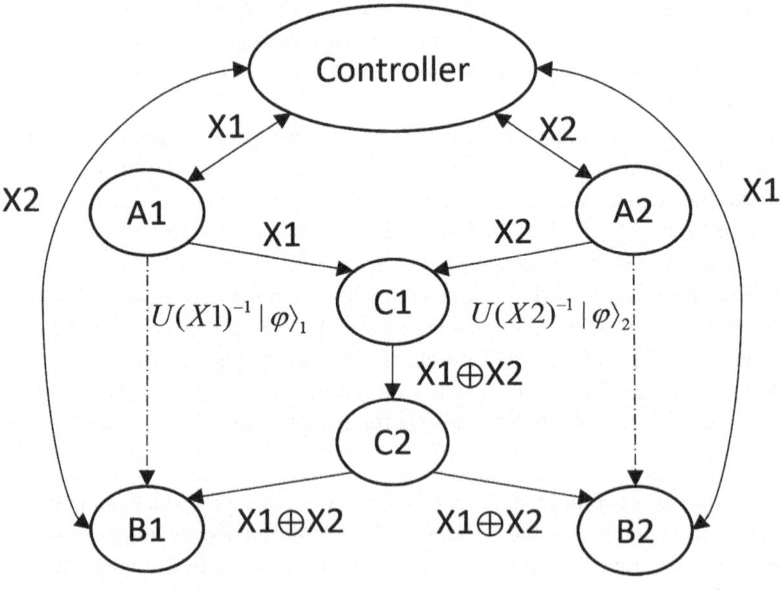

Fig. 8.2 CQNC scheme based on quantum walk

The main points of the scheme [16] are as follows:

1. The CQNC scheme uses quantum walk to realize the quantum entangled state in the transmission process, which can realize the process of quantum teleportation. Therefore, it does not need to prepare the entangled state in advance. For the sender and receiver, it only needs to prepare a single particle.
2. The CQNC scheme also tries to extend the quantum network coding scheme to the case of N pair nodes of n sender and n receiver, which is more suitable for the requirements of the current network structure. At the same time, we can further expand the load balancing.
3. Simplifying the quantum network coding scheme including the third party. Without affecting the communication, the CQNC scheme reduced the third-party controller to one, forming a one-to-many quantum network coding model, which further decreased the resource cost.

8.3.2.1 One-Dimensional Quantum Walk Model on the Line

One-dimensional quantum walk was first proposed by Ambainis in 2001 [17]. One-dimensional quantum walk takes place in a compound Hilbert space comprising position space and coin space. Expressed as

$$H = H_p \otimes H_c, \tag{8.1}$$

where H_p is position space $\{n, n \in Z\}$ and H_c is coin space representing the direction of the coin$\{|0\rangle, |1\rangle\}$.

Every quantum step can be written as

$$W^{(l)} = E^{(l)} \cdot (I \otimes C), \tag{8.2}$$

where $E^{(l)} = S \otimes \ |0\rangle\langle 0| + S^{\dagger} \otimes |1\rangle\langle 1|$, I is identity operator, C is the coin reversal operation in coin space, and shift operators $S = \sum_n \ | \ n + 1\rangle\langle n|$. When the coin operator is $|0\rangle$, coin particles go from $|n\rangle$ to $|n + 1\rangle$, and when the coin operator is $|1\rangle$, coin particles go from $|n\rangle$ to $|n - 1\rangle$.

8.3.2.2 Teleportation of Quantum Walk

Wang [18] and Shang [19] proposed a scheme of teleportation by quantum walk of two-coins. This scheme didn't need to prepare quantum entangled states in advance, but the entanglement of particles was generated in the process of quantum walking through the position space and coin space. Then the quantum teleportation was realized. In this scheme, Alice and Bob only needed to prepare single particle, which would simplify the original initial preparation process and improve the efficiency of quantum teleportation.

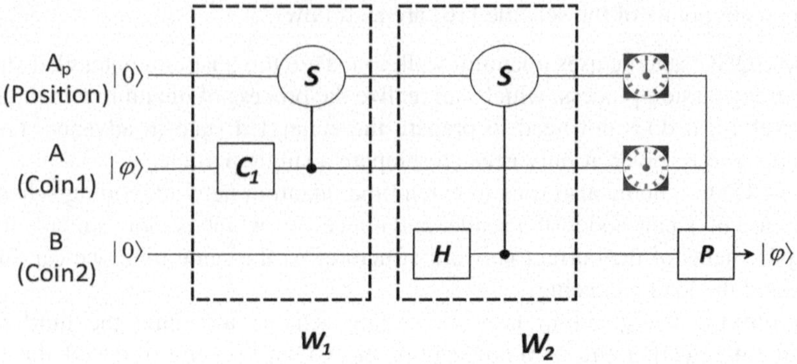

Fig. 8.3 Teleportation with two-coins quantum walks on the line

In this time, we use a one-dimensional quantum linear walking model, is shown in Fig. 8.3. Assume Alice wants to transmit a quantum state $|\varphi\rangle = \alpha \mid 0\rangle + \beta \mid 1\rangle$ (where $|\alpha|^2 + |\beta|^2 = 1$).

Alice needs to prepare two single particles A and A_p to realize the quantum walking scheme, where the particle A_p represents the position space and the particle A represents the coin1 space and carries the information $|\varphi\rangle$ to be transmitted; Bob also needs to prepare a single particle b, which represents the coin2 space, where the initial states of particle A_p and b are $|0\rangle$. Next, quantum state can be transferred in two steps.

Step 1:

$$W^{(1)} = E^{(1)} \cdot \left(A \otimes C_1 \otimes A_p\right), \tag{8.3}$$

where

$$E^{(1)} = S\otimes \mid 0\rangle_1\langle 0|\otimes b + S^{\dagger}\otimes|1\rangle_1\langle 1 \mid \otimes b \tag{8.4}$$

In Eq. 8.3, C_1 represents the operator of particle A, and we choose Pauli operation I as the C_1 operator.

Step 2:

$$W^{(2)} = E^{(2)} \cdot \left(I_p \otimes H \otimes b\right), \tag{8.5}$$

where

$$E^{(2)} = S\otimes \mid 0\rangle_2\langle 0| + S^{\dagger}\otimes|1\rangle_2\langle 1 \mid \tag{8.6}$$

In Eq. 8.5, H represents Hadamard operation on b particle.

Table 8.1 Measurement results and Pauli operations

Measurement results of A and A_p	Pauli operations
1, 1 or −1, −1	I
1, −1 or −1,1	Z
1,0	X
-1,0	ZX

Finally, Alice only needs to measure particle A and particle A_p, respectively and tells Bob the measurement results. Bob can recover the unknown quantum state according to the measurement. The relationship between the measurement results and Pauli operations is shown in Table 8.1.

8.3.2.3 Preparation Phase

In the preparation stage, A1 and A2 are as senders needed to prepare two single particles, respectively, and four single particles need to be generated. Here we label the four single particles as A_{11}, A_{1p} and A_{21}, A_{2p}, where A_{11} and A_{1p} are two single particles of A1 node and A_{21} and A_{2p} are two single particles of A2 node. A_{11} and A_{21} represent the coin space of each node, respectively, and A_{1p} and A_{2p} represent the position space of each node, respectively, and their initial state is $|0\rangle$.

Then A1 and A2, as the senders, need to transmit their states to the controller, which is the third party, and prepare particles for data transmission at the same time. After receiving the message from Alice, the controller informs the receiver Bob, namely, B1 and B2, and store the corresponding relationship between Bob and Alice. That is, if the sender is A1, the receiver is B1, or if the sender is A2, the receiver is B2.

After receiving the message from the controller, B1 and B2, as the receivers, start to prepare the single particle. The particle prepared by B1 is marked as b_1, and the particle prepared by B2 is marked as b_2 (where the initial states of b_1 and b_2 are all $|0\rangle$). Then they inform the third party of being ready. After the preparation, the controller can inform Alice to start transmission.

8.3.2.4 Transmission Phase

In the transmission phase, we will use the teleportation scheme of quantum walk. Suppose that the senders, the Ai node, transmitted the quantum states $|\varphi\rangle_i = \alpha_i \mid 0\rangle + \beta_i \mid 1\rangle$, where $|\alpha_i|^2 + |\beta_i|^2 = 1,\ i \in \{1,2\}$.

In this case, the initial state is

$$\mid \varphi\rangle^{(0)} = (\alpha|000\rangle + \beta|010\rangle)_{A_{pi}A_{1i}b_i} \tag{8.7}$$

First, Step 1 of quantum walk is performed as: $W^{(1)} = E^{(1)} \cdot (A_{1i} \otimes C_1 \otimes A_{pi})$.

We expand the formula to

$$
\begin{gathered}
W^{(1)} = E^{(1)} \cdot \left(A_{1i} \otimes C_1 \otimes A_{pi}\right) = \\
\left(|\, 1\rangle_{pi}\langle 0| \otimes |0\rangle_{1i}\langle 0| + |-1\rangle_{pi}\langle 0| \otimes |1\rangle_{1i}\langle 1|\right) \otimes |0\rangle_{bi} \cdot \\
\left(|0\rangle_{pi} \otimes (\alpha|0\rangle + \beta|1\rangle)_{1i}\right) = \\
(\alpha|100\rangle + \beta|-110\rangle)_{A_{pi}A_{1i}b_i}
\end{gathered} \tag{8.8}
$$

It is not difficult to find that the particles of A_{1i} and A_{pi} are entangled:

$$
|\,\varphi\rangle^{(1)} = (\alpha|100\rangle + \beta|-110\rangle)_{A_{pi}A_{1i}b_i} \tag{8.9}
$$

Second, Step 2 of quantum walk is performed as: $W^{(2)} = E^{(2)} \cdot (I_{pi} \otimes H \otimes b_i)$. We also expand the formula to

$$
\begin{aligned}
W^{(2)} &= E^{(2)} \cdot \left(I_{pi} \otimes H \otimes b_i\right) \\
&= E^{(2)} \cdot \left[(\alpha|10\rangle + \beta|-11\rangle)_{A_{pi}A_{1i}} \otimes (|0\rangle + |1\rangle)/\sqrt{2}\right] \\
&= S \otimes |\,0\rangle_{bi}\langle 0| + S^{\dagger} \otimes |1\rangle_{bi}\langle 1\,| \cdot \\
&\quad (\alpha|100\rangle + \beta|-110\rangle + \alpha|101\rangle + \beta|-111\rangle)_{A_{pi}A_{1i}b_i}/\sqrt{2} \\
&= \left(|\,2\rangle_{pi}\langle 0| \otimes |0\rangle_{bi}\langle 0| + |0\rangle_{pi}\langle 1| \otimes |1\rangle_{bi}\langle 1\,|\right) \cdot \\
&\quad (\alpha|100\rangle + \beta|-110\rangle + \alpha|101\rangle + \beta|-111\rangle)_{A_{pi}A_{1i}b_i}/\sqrt{2} \\
&\quad (\alpha|200\rangle + \alpha|001\rangle + \beta|010\rangle + \beta|-211\rangle)_{A_{pi}A_{1i}b_i}/\sqrt{2}
\end{aligned} \tag{8.10}
$$

In this case, particles A_{pi} and b_i are entangled:

$$
|\,\varphi\rangle^{(2)} = \frac{1}{\sqrt{2}}(\alpha|200\rangle + \alpha|001\rangle + \beta|010\rangle + \beta|-211\rangle)_{A_{pi}A_{1i}b_i} \tag{8.11}
$$

After finishing these two steps, we need to measure the particle A_{1i}, and the measurement result is recorded as r_{1i}. Then particles A_{pi} and b_i will collapse to the corresponding state:

$$
\begin{aligned}
|\,\varphi\rangle^{(3)} = &\left(|\,+\rangle_{A_{1i}}(\alpha|20\rangle + \alpha|01\rangle + \beta|00\rangle + \beta|-21\rangle)_{A_{pi}b_i}\right. \\
&\left.+ (|-\rangle_{A_{1i}}(\alpha|20\rangle + \alpha|01\rangle - \beta|00\rangle - \beta|-21\rangle)\right)_{A_{pi}b_i}
\end{aligned} \tag{8.12}
$$

When we measure particle A_{pi}, the measurement result is recorded as r_{pi}, and particle b_i will collapse to the corresponding state as

$$| \varphi\rangle^{(4)} = | 2'\rangle_{A_{pi}}(\alpha|0\rangle + \beta|1\rangle)_{b_i}/2 + | -2'\rangle_{A_{pi}}(\alpha|0\rangle + \beta|1\rangle)_{b_i}/2 + \\ | 0\rangle_{A_{pi}}(\alpha|1\rangle - \beta|0\rangle)_{b_i}/\sqrt{2} \tag{8.13}$$

After the measurements, we record the results as $X_1 = r_{11}r_{p1}$, $X_2 = r_{12}r_{p2}$. Alice transmits the measured results to C1 node and the controller node, respectively. At C1 node, the measurement results are encoded as $X_1 \oplus X_2$ and then sentt to C2. At C2 node, the encoded results are copied and forwarded to receivers B1 and B2. At controller node, when receiving the measurement results from A1 and A2, the controller stores the measurement results of A1 and A2, respectively, according to the relationship matching in the preparation stage.

8.3.2.5 Recovery Phase

In the recovery phase, if the controller agrees for Bob to obtain the quantum state, the received measurement results are selectively forwarded. For example, *X*1 is sent to B1 and *X*2 is sent to B2. Bob receives the measurement information and performs XOR operation first and then removes the redundant information. Finally, according to Table 8.1, the quantum state $|\varphi\rangle$ transmitted by Alice can be obtained by performing the recovery operation.

8.3.2.6 CQNC Scheme Extended to *N*-Node

In order to make the CQNC scheme more general, we try to extend this CQNC scheme to *n* pairs of transceiver nodes, namely, *N* senders (A1 ~ An) and *N* receivers (B1 ~ Bn), as shown in Fig. 8.4 below.

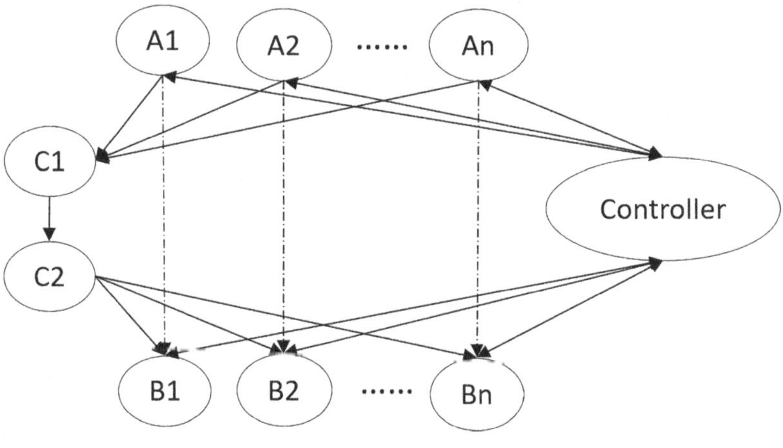

Fig. 8.4 N-CQNC scheme based on *N*-node

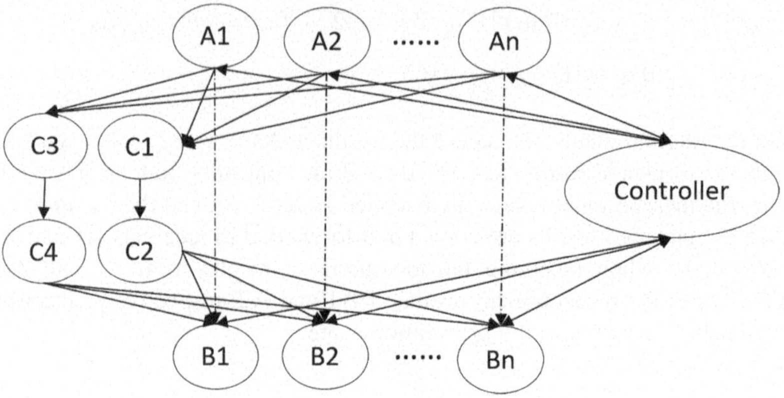

Fig. 8.5 Load balancing scheme

Its basic working principle is similar to that of the above two pairs of transceiver nodes, assuming that the senders, namely: A1, A2…An, want to send a quantum state to Bob, namely: B1, B2…Bn. The workflow is similar to the above. It is worth noting that although it can be extended to *n* pairs of transceiver nodes, each additional pair of nodes will perform an XOR operation at the receiver, which may reduce the communication efficiency. Moreover, due to the limitation of circuit, the throughput of the circuit between C1 and C2 is limited, so there must be an upper limit on the number of transceiver nodes. When this problem occurs, we can use load balancing technology to balance the network load of intermediate nodes. For example, as shown in Fig. 8.5, we add C3 and C4 nodes to split data, which can solve this kind of problem.

8.3.2.7 CQNC Scheme Analysis

This subsection focuses on the performance analysis of the CQNC scheme.

A. Comparative Analysis

1. **Analysis of Rate Region**
   The definition of the rate region of the butterfly network [20] is as follows: If a QNC scheme uses the network *n* times along with other allowed resources and communicates *m1*, *m2* of sizes $n(r_1 - \delta_n)$, $n(r_2 - \delta_n)$ bits/qubits with fidelity at least $1 - \xi_n$ for $\delta_n, \xi_n \rightarrow 0$, then we say that the rate pair (r_1, r_2) is achievable. The achievable rate region is the set of all achievable rate pairs.
   It is assumed that each channel can transmit *1* qubit or *2* classical bits of information at one time. Obviously, the rate region of PE scheme is $\{(r_1, r_2) | r_1, r_2 \leq 1\}$. This means that Hayashi's scheme can transmit two qubits simultaneously by using the network at one time. In CT scheme, because of the increase

of the third party, additional signature information needs to be sent in the network. Therefore, CT scheme needs to use the network *1.5* times to transmit two source qubits, so we can easily obtain$\{(r_1, r_2)\frac{2}{3}\}$. Although the CQNC scheme uses the third party, it will not add extra source qubits, but it needs to measure two more particles in the process of quantum walking, so it needs to transmit classical information once more. So the rate region of our schemes would be $\{(r_1, r_2)\frac{1}{2}\}$.

2. **Analysis of Resource Needed**
   In this CQNC scheme, we try to introduce the quantum walk into the QNC scheme to simplify the initial conditions of the original scheme. We do not need to prepare the quantum entangled state in advance but transform it into a coin space to realize the transmission of the quantum state in the quantum walk. For senders A_i and receivers B_i, they only need to prepare a single particle to realize the QNC on the butterfly network. In quantum communication, the preparation of quantum entangled states consumes many resources, and it is difficult to prepare high purity quantum entangled pairs. Compared with the two-particle entanglement pair and three-particle entanglement pair adopted by PE and CT, the CQNC scheme only needs a single particle to complete communication. The CQNC scheme performs better on resource consumption. That is also one of the innovations.
3. **Analysis of the Third Party**
   About the third party, Shang [14] has discussed its advantages. It cannot only improve the security and reliability of the network but also increase the space for identity authentication and further enhance the security of the scheme. In the CQNC scheme, we still have a third-party control, and we further optimize the role of the third party. In the CT scheme, except for the two particles at the senders, we also need to use the single particle generated by the two third-party controllers to realize two pairs of three-particle entangled states in order to complete the communication. In the CQNC scheme, the third party only needs to process the classical information and verify the identity of the relevant nodes but no longer involves the quantum information processing. Therefore, compared with the two third-party controllers in the CT scheme, the CQNC scheme only needs one control to achieve the same verification effect. And compared with the PE scheme, the third-party authentication will make CQNC scheme more secure.
4. **Scalability Analysis**
   In the PE scheme and CT scheme, the author uses the quantum entanglement pairs in two different ways to achieve the QNC, but it is difficult to expand further. In the CQNC scheme, the quantum walk method not only solves the difficult problem of quantum entanglement manufacturing but can also expand the QNC scheme to *N* pairs of transceiver nodes, which is also the innovation of the scheme.

In conclusion, we compare the comprehensive performance of the three schemes, as shown in Table 8.2.

Table 8.2 Schemes comparison (based on two pairs of transceiver nodes)

Scheme	Transmission rate region	Resources needed	Third-party verification	Scalability
PE scheme	$\{(r_1, r_2) \mid r_1, r_2 \leq 1\}$	Quantum entanglement pair	N	N
CT scheme	$\{(r_1, r_2)\frac{2}{3}\}$	Quantum entanglement pair	Y	N
CQNC scheme	$\{(r_1, r_2)\frac{1}{2}\}$	Single quantum particles	Y	Y

According to the above analysis and Table 8.2, we can see that the CQNC scheme has obvious advantages in terms of resource consumption and scalability. Compared with the PE scheme, the CQNC scheme has stronger security. Compared with the CT scheme, we make the third-party authentication more concise. Although the rate region is reduced a little, on the whole, the CQNC scheme has certain advantages.

B. Security Analysis

In the CQNC scheme, we introduce a third-party control terminal to control the forwarding of key information. Even if the quantum state Alice sends to Bob is intercepted by an eavesdropper, the quantum state cannot be recovered without the third-party measurement information. According to the Heisenberg principle and quantum noncloning theorem, the eavesdropper cannot intercept quantum information without knowing it. According to the CQNC scheme, the receiver Bob will obtain the measurement information from the third party for recovery operation after receiving the quantum state information. Therefore, there is no eavesdropper intercepting quantum information at the same time state information and measurement information. Therefore, on the premise of communication security between the controller and the receiver, the CQNC scheme can effectively resist eavesdropping attacks.

We will further clarify the security of the CQNC scheme from the following two directions:

1. **External Attack**

   As shown in Fig. 8.2, for a classical communication, suppose that an eavesdropper wants to eavesdrop the transmitted information. Whether it is the transmission of A1 (A2)→C1→C2→B1 (B2) or the mutual transmission between A1 (A2), controller, and B1 (B2), the transmitted information is the measurement result. Even if the eavesdropper obtains this part of data, without corresponding quantum state, the real useful information can't be stolen.

   For quantum communication, according to the quantum noncloning theorem, the transmitted quantum state cannot be accurately copied. Moreover, the quantum walk method forms quantum entanglement in the process of transmission and then realizes the transmission of data. If there is an eavesdropper‘s access, it will inevitably lead to the collapse of the whole quantum entanglement system. Even if the eavesdropper obtains the data, it can't recover it, because no one knows where the quantum walk is going. At the same time, there is also no measurement result of the controller, so it is impossible to carry out unitary transformation for recovery operation.

2. **Internal Attack**

   Suppose that one of the intermediate nodes (C1, C2, controller) is dishonest, it will cooperate with the eavesdropper to obtain information. Even if so, the eavesdropper cannot obtain the information to be transmitted. Because for the intermediate node, it transmits the measurement results, which is useless without corresponding quantum information. Assuming that the sender (A1, A2) or the receiver (B1, B2) is dishonest, the third-party controller can find it out in the preparation stage and terminate the communication. Therefore, the CQNC scheme is still secure against internal attacks.

   However, in order to further enhance the security between the third party and the receiving node Bob, we can try to introduce quantum key, identity authentication, quantum secure direct communication, and other technologies to ensure the secure transmission of measurement information; at the same time, we can also add quantum key between Alice and Bob to enhance the security of the receiver.

8.3.2.8 Summary

This section analyzes the mainstream network coding schemes and proposes a CQNC scheme based on quantum walk, which solves the problem that it is difficult to prepare the maximum quantum entangled state in QNC and further extends the butterfly network to *N* pair transceiver nodes. At the same time, the CQNC scheme introduces and optimizes the third-party control terminal, which improves the transmission rate region. Compared with the existing QNC, the CQNC scheme does not need to prepare the quantum entangled state in advance but interact with each other in the process of quantum walking. At present, Lu [21] has implemented the PE scheme in the laboratory, and Chatterjee [22] has also implemented the one-dimensional quantum walk scheme. Therefore, based on the existing technology, it is feasible to implement the CQNC scheme proposed.

8.3.3 *QNC Based on Six-Qubit Entangled State*

Inspired by Li [23] and Shang [24, 25], a QNC scheme based on six-qubit entangled state is proposed and introduced into the butterfly network [26]. In the QNC scheme, we still use the butterfly network model proposed by Hayashi [13] but further optimize the information transmission. We introduce six-qubit entangled state model, which is shown in Fig. 8.6. A1 and A2 are the senders; M1 is the aggregator, which packages and encodes the received messages; M2 is the forwarder, which

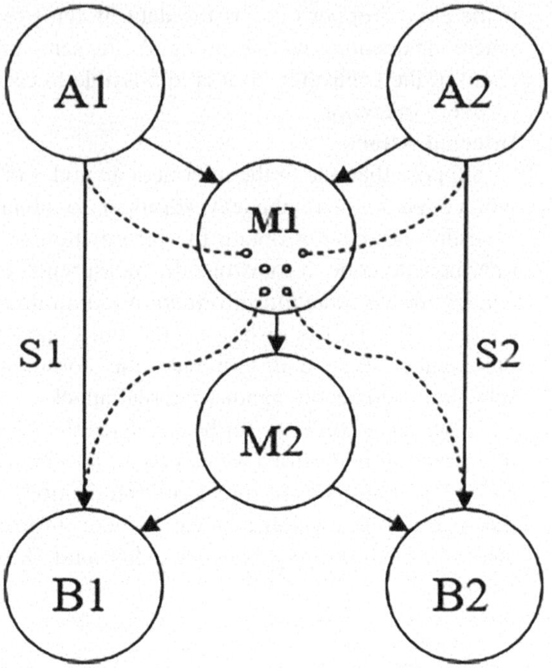

Fig. 8.6 Quantum network coding scheme

performs the replication and forwarding operation; B1 and B2 are the recipients, which integrate the transmitted information and finally obtain the information sent by the sender.

The main points of the QNC scheme are as follows:

1. Using the advantages of quantum coding, the QNC scheme simplifies the information that needs to be transmitted many times into one transmission and further improves the security of network transmission through local verification. Moreover, through analysis, the QNC scheme has advantages compared to other schemes.
2. Based on the anti-pollution attack network coding QNC proposed by Shang [24, 25], the number of single-particle transmission in the network is reduced and the anti-interference ability of information is improved.

8.3.3.1 Quantum Teleportation of Six-Qubit Entangled State

In 2019, Li proposed an effective scheme for the transmission of any two quantum states which were GHZ quantum cluster state and six-particle entanglement state separately [23]. According to the analysis of their paper, the deterministic quantum teleportation of any two-qubit state can be realized in the cavity quantum

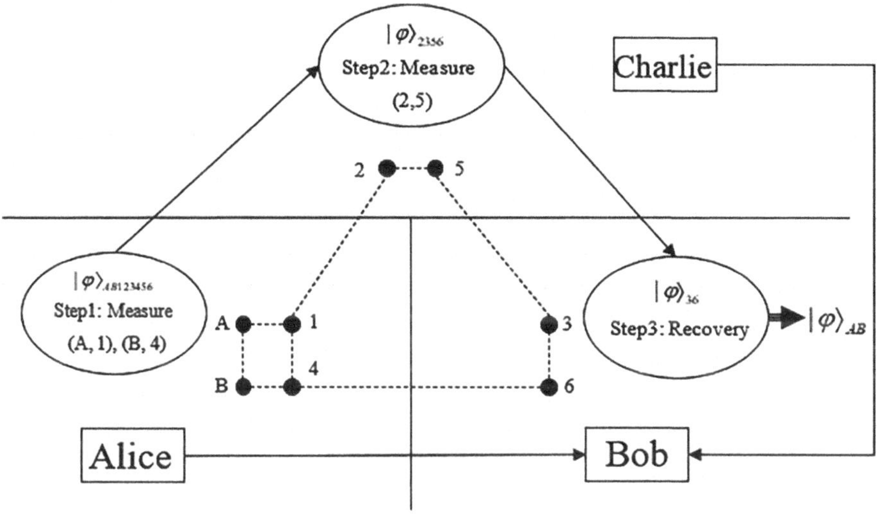

Fig. 8.7 Quantum information splitting for arbitrary two-qubit state

electrodynamics system. Compared with the entangled state, the GHZ quantum cluster state has the largest persistence and is more robust to decoherence; moreover, the scheme is secure against internal and external attacks.

The scheme is as shown in Fig. 8.7. Particles A and B contain the information to be transmitted. One GHZ state is composed of particles 1, 2, and 3, and another GHZ state composed of particles 4, 5, and 6. These particles form a six-particle cluster state. Particles A, B, 1 and 4 belong to Alice, particles 3 and 6 belong to Bob, and particles 2 and 5 belong to Charlie. First, Alice measures the particle pairs (A, 1), (B, 4) and transmits the measurement results to Bob; second, Charlie performs a two-qubit projective measurement on his qubits pairs (2, 5) and transmits the measurement results to Bob; Finally, the initial state $|\varphi\rangle_{AB}$ to be transmitted is restored by the measurement results of Alice and Charlie.

8.3.3.2 Preparation

Step 1: A1 and A2 preparation the particles to be transmitted $|\varphi\rangle_x$, $|\varphi\rangle_y$, and then transmit particles to B1 and B2 through the butterfly network, which is shown in Fig. 8.7.

Step 2: M1 generates two teams of GHZ entangled states:

$$|\varphi\rangle_{123} = \frac{1}{\sqrt{2}}(|000\rangle + |111\rangle) \qquad (3) \tag{8.14}$$

$$|\varphi\rangle_{456}=\frac{1}{\sqrt{2}}(|000\rangle+|111\rangle) \qquad (4) \qquad (8.15)$$

M1 sends its particles *1* and *4* to node A1 and A2, respectively, and sends its particles *2* and *5* to node B1 and B2, respectively, through M2. So far, (A1, A2), M1 and (B1, B2) form a six-particle cluster state:

$$|\varphi\rangle_{123456}=\frac{1}{\sqrt{2}}(|000000\rangle+|000111\rangle+|111000\rangle+|111111\rangle) \qquad (5) \quad (8.16)$$

For the whole system,

$$|\varphi\rangle_{xy123456}=\frac{1}{\sqrt{2}}(|000000\rangle+|000111\rangle+|111000\rangle+|111111\rangle)\otimes|\varphi\rangle_{xy} \qquad (6) \qquad (8.17)$$

8.3.3.3 Transmission

Step 1: In order to complete the quantum information transmission of the intermediate node, A1 and A2 nodes need to measure the particle pairs *(x, 1)*, *(y, 4)*, respectively, and the measurement results recorded as r_1r_x, r_4r_y are transmitted to M1 node. At this time, the quantum state of the whole system will collapse into one of the *16* states, as shown in Table 8.3.

Step 2: After receiving the information transmitted by A1 and A2, M1 measures particles *2* and *5*, and the measurement result is recorded as r_2r_5. At this time, the quantum state of the whole system will collapse into one of the *64* states as shown in *Appendix*. With the measurement results from A1 and A2, M1 encodes r_1r_A, r_4r_B and r_2r_5 into $r_1r_2r_A \oplus r_4r_5r_B$. Then the quantum state of the whole system will collapse to one of the *64* states. See the *Appendix* for more details.

Step 3: Finally, M1 sends the encoded information $r_1r_2r_A \oplus r_4r_5r_B$to M2 and then copies and forwards the data to B1 and B2, respectively.

8.3.3.4 Verify and Recovery

After B1 and B2 receive all the data, but they don't know which set of data is useful, Therefore, they need to ask A1 for verification information.

Then, A1 transmitted the measurement results r_1r_x of the transmission phase to B1 through S1, A2 also transmitted the measurement results r_1r_x to B2 through S2, B1 and B2 By comparing the verification information and coding information, which can get real measurement results. Finally, B1 and B2 could obtain $|\varphi\rangle_x$ and $|\varphi\rangle_y$through unitary transformation.

Table 8.3 The measurement results at A1 and A2, and the corresponding collapse states at M1 and M2

A1, A2 measurement results	State obtained by M1 and M2
$\lvert\Phi^+\rangle_{x1} \mid \Phi^+\rangle_{y4}$	$\lvert\varphi\rangle_{2356}^{1} = \frac{1}{4}[\alpha\lvert 0000\rangle + \beta\lvert 1100\rangle + \gamma\lvert 0011\rangle + \eta\lvert 1111\rangle]$
$\lvert\Phi^+\rangle_{x1} \mid \Phi^-\rangle_{y4}$	$\lvert\varphi\rangle_{2356}^{2} = \frac{1}{4}[\alpha\lvert 0000\rangle + \beta\lvert 1100\rangle - \gamma\lvert 0011\rangle - \eta\lvert 1111\rangle]$
$\lvert\Phi^+\rangle_{x1} \mid \Psi^+\rangle_{y4}$	$\lvert\varphi\rangle_{2356}^{3} = \frac{1}{4}[\alpha\lvert 0011\rangle + \beta\lvert 1111\rangle + \gamma\lvert 0000\rangle + \eta\lvert 1100\rangle]$
$\lvert\Phi^+\rangle_{x1} \mid \Psi^-\rangle_{y4}$	$\lvert\varphi\rangle_{2356}^{4} = \frac{1}{4}[\alpha\lvert 0011\rangle + \beta\lvert 1111\rangle - \gamma\lvert 0000\rangle - \eta\lvert 1100\rangle]$
$\lvert\Phi^-\rangle_{x1} \mid \Phi^+\rangle_{y4}$	$\lvert\varphi\rangle_{2356}^{5} = \frac{1}{4}[\alpha\lvert 0000\rangle - \beta\lvert 1100\rangle + \gamma\lvert 0011\rangle - \eta\lvert 1111\rangle]$
$\lvert\Phi^-\rangle_{x1} \mid \Phi^-\rangle_{y4}$	$\lvert\varphi\rangle_{2356}^{6} = \frac{1}{4}[\alpha\lvert 0000\rangle - \beta\lvert 1100\rangle - \gamma\lvert 0011\rangle + \eta\lvert 1111\rangle]$
$\lvert\Phi^-\rangle_{x1} \mid \Psi^+\rangle_{y4}$	$\lvert\varphi\rangle_{2356}^{7} = \frac{1}{4}[\alpha\lvert 0011\rangle - \beta\lvert 1111\rangle + \gamma\lvert 0000\rangle - \eta\lvert 1100\rangle]$
$\lvert\Phi^-\rangle_{x1} \mid \Psi^-\rangle_{y4}$	$\lvert\varphi\rangle_{2356}^{8} = \frac{1}{4}[\alpha\lvert 0011\rangle - \beta\lvert 1111\rangle - \gamma\lvert 0000\rangle + \eta\lvert 1100\rangle]$
$\lvert\Psi^+\rangle_{x1} \mid \Phi^+\rangle_{y4}$	$\lvert\varphi\rangle_{2356}^{9} = \frac{1}{4}[\alpha\lvert 1100\rangle + \beta\lvert 0000\rangle + \gamma\lvert 1111\rangle + \eta\lvert 0011\rangle]$
$\lvert\Psi^+\rangle_{x1} \mid \Phi^-\rangle_{y4}$	$\lvert\varphi\rangle_{2356}^{10} = \frac{1}{4}[\alpha\lvert 1100\rangle + \beta\lvert 0000\rangle - \gamma\lvert 1111\rangle - \eta\lvert 0011\rangle]$
$\lvert\Psi^+\rangle_{x1} \mid \Psi^+\rangle_{y4}$	$\lvert\varphi\rangle_{2356}^{11} = \frac{1}{4}[\alpha\lvert 1111\rangle + \beta\lvert 0011\rangle + \gamma\lvert 1100\rangle + \eta\lvert 0000\rangle]$
$\lvert\Psi^+\rangle_{x1} \mid \Psi^-\rangle_{y4}$	$\lvert\varphi\rangle_{2356}^{12} = \frac{1}{4}[\alpha\lvert 1111\rangle + \beta\lvert 0011\rangle - \gamma\lvert 1100\rangle - \eta\lvert 0000\rangle]$
$\lvert\Psi^-\rangle_{x1} \mid \Phi^+\rangle_{y4}$	$\lvert\varphi\rangle_{2356}^{13} = \frac{1}{4}[\alpha\lvert 1100\rangle - \beta\lvert 0000\rangle + \gamma\lvert 1111\rangle - \eta\lvert 0011\rangle]$
$\lvert\Psi^-\rangle_{x1} \mid \Phi^-\rangle_{y4}$	$\lvert\varphi\rangle_{2356}^{14} = \frac{1}{4}[\alpha\lvert 1100\rangle - \beta\lvert 0000\rangle - \gamma\lvert 1111\rangle + \eta\lvert 0011\rangle]$
$\lvert\Psi^-\rangle_{x1} \mid \Psi^+\rangle_{y4}$	$\lvert\varphi\rangle_{2356}^{15} = \frac{1}{4}[\alpha\lvert 1111\rangle - \beta\lvert 0011\rangle + \gamma\lvert 1100\rangle - \eta\lvert 0000\rangle]$
$\lvert\Psi^-\rangle_{x1} \mid \Psi^-\rangle_{y4}$	$\lvert\varphi\rangle_{2356}^{16} = \frac{1}{4}[\alpha\lvert 1111\rangle - \beta\lvert 0011\rangle - \gamma\lvert 1100\rangle + \eta\lvert 0000\rangle]$

8.3.3.5 QNC Scheme Analysis

This subsection focuses on the analysis of the transmission rate region, network coding and security of QNC scheme.

A. Transmission Rate Region Analysis

The definition of rate region of butterfly network [20] is as follows: If a scheme uses the network n times along with other allowed resources and communicates *m1*, *m2* of sizes $n(r_1 - \delta_n)$, $n(r_2 - \delta_n)$ bits/qubits with fidelity at least $1 - \xi_n$ for δ_n, $\xi_n \to 0$. Then we say that the rate pair (r1, r2) is achievable. The achievable rate region is the set of all achievable rate pairs.

Suppose that each channel can transmit one qubit or two classical bits of information at a time. Obviously, the transmission rate region of Hayashi's scheme is $(r_1, r_2) = (1, 1)$. Compared to the two-bit quantum information transmitted by Hayashi, the QNC against pollution attacks needs to transmit four additional particles. Due to the capacity of the channel, the QNC scheme needs to use the network to transmit these particles four times, so we can easily get $(r_1, r_2) = \left(\frac{1}{5}, \frac{1}{5}\right)$. The transmission rate region of our scheme is $(r_1, r_2) = \left(\frac{1}{5}, \frac{1}{5}\right)$.

Compared to Shang's scheme, we also use four additional particles, but the difference is that we can realize the information transmission from (A1, A2) to (B1, B2) through four additional particles, while the Shang's scheme can only realize the transmission from (A1, A2) to M2. We want to transfer the information (B1, B2), You need to add four additional particles at the M2 node to copy and forward information.

At the same time, quantum information will weaken the signal strength in the transmission process. In order to reduce the loss in the transmission process, the times of measurement and transmission should be reduced as much as possible so as to improve the anti-interference ability of the QNC scheme. In terms of transmission time, compared to the homomorphic signature scheme proposed by Shang, our QNC scheme reduces the transmission of particles as much as possible. We try to reduce the loss of particles in the transmission process. Except for the construction of quantum entanglement pairs in the preparation stage, our QNC scheme no longer involves the transmission of quantum information. However, in Shang's scheme, the measured particles still need to be transmitted, which will undoubtedly increase and weaken the signal strength.

B. Advantages of Network Coding

1. **Transmission Advantages**
   In the transmission stage, M1 needs to integrate and forward the measurement results $r_1r_2r_A \oplus r_4r_5r_B$. In the traditional network transmission, M1 needs to forward six measurement results one by one. However, in the network coding structure, the information is no longer a single individual, but is regarded as a form of information flow. It is packaged into an information block $r_1r_2r_A \oplus r_4r_5r_B$ at M1 and only needs to be transmitted once.
2. **Security Advantage**
   Our encoded information is not only for higher transmission efficiency, but also to improve the security of transmission. In the verification stage, (B1, B2) receives that the verification information transmitted by M1 is encoded data. At this time, it still needs the measurement results of (A1, A2) to filter the encoded information in order to get really useful information.

C. Security Analysis

Different from the network coding scheme of Hayashi [13] and Shang [14], we use the six-particle entangled state model, which cannot only realize information transmission, but also show the advantages of network coding. In addition, according to the analysis of Li [23], the transmission mode of six-particle entangled state is very reliable and safe in response to external and internal attacks; at the same time, the number of particle transmission and measurement is reduced as much as possible, so as to further improve the anti-interference ability.

We further clarify the security of our QNC scheme from the following two directions,

1. **External Attack**
   As shown in Fig. 8.6, in the transmission process of A1(A2)→M1→M2→B1 (B2), it is assumed that an eavesdropper maliciously tampers with information, which can be found by the verification of (B1, B2).

For example, M1 needs to integrate the measurement results $r_1r_2r_A \oplus r_4r_5r_B$. At this time, the eavesdropper maliciously modifies the data in it, resulting $r_Er_2r_A \oplus r_4r_5r_B$ in the measurement results, which r_E is the data tampered by the eavesdropper. When the tampered data is transmitted to node B1, B1 will verify the data transmitted by M2 according to the verification data transmitted by S1. At this time, the tampered data will be found and the data will be sent again.

Other nodes follow this example. If the eavesdropper maliciously tampers with the data, it will be found in the verification stage.

2. **Internal Attack**

For intermediate node M1 (M2), if there are dishonest nodes, it will cooperate with eavesdroppers to intercept information. Even so, eavesdroppers still cannot get the correct information to transmit.

If M1 is dishonest, according to the information at M1 $r_1r_2r_A \oplus r_4r_5r_B$, the eavesdropper cannot obtain the really useful data. So it must continue to obtain the data r_1r_x transmitted by A1 and carry out the corresponding XOR operation so as to obtain the required unitary transformation operation and finally obtain the useful information. Similarly, if M2 is dishonest, it can only obtain part of the transmitted information and cannot perform the final recovery operation.

Therefore, for internal attacks, our QNC scheme is still secure.

In addition, according to the quantum noncloning theorem, the quantum information transmitted in this scheme is secure. Assuming the existence of eavesdroppers, it will be found in the verification of B1 (B2). Therefore, the proposed QNC scheme is safe and reliable for both quantum information and measurement information.

8.3.3.6 Summary

This section analyzes the current mainstream network coding schemes and proposes a QNC scheme based on six-qubit entangled state. In the transmission process of cross network, six-qubit entangled state are used to replace the original classical information transmission, which improves the security of quantum network coding on butterfly network. And the proposed QNC scheme tries to reduce the occupation of resources as much as possible, such as the measurement and transmission times of single particle are reduced, so as to improve the anti-interference ability of the scheme. According to the analysis and comparison, the QNC scheme is safe for both external and internal attacks, and in some aspects, compared to other schemes, it has significant advantages (Table 8.4).

8.3.3.7 Appendix

Table 8.4 Measurement results at M1, and the corresponding collapsed states at M2

Measurement results on M1	State obtained by M2	Recovery
$\vert+\rangle_2\vert+\rangle_5=\frac{1}{2}(\vert 00\rangle+\beta\vert 01\rangle+\vert 10\rangle+\vert 11\rangle)$	$\vert\varphi\rangle_{36}^{1}=\frac{1}{16}[\alpha\vert 00\rangle+\beta\vert 10\rangle+\gamma\vert 01\rangle+\eta\vert 11\rangle]$	$\mathrm{I}\otimes I$
$\vert+\rangle_2\vert-\rangle_5=\frac{1}{2}(\vert 00\rangle-\beta\vert 01\rangle+\vert 10\rangle-\vert 11\rangle)$	$\vert\varphi\rangle_{36}^{2}=\frac{1}{16}[\alpha\vert 00\rangle+\beta\vert 10\rangle-\gamma\vert 01\rangle-\eta\vert 11\rangle]$	$\mathrm{I}\otimes\sigma_z$
$\vert-\rangle_2\vert+\rangle_5=\frac{1}{2}(\vert 00\rangle+\beta\vert 01\rangle-\vert 10\rangle-\vert 11\rangle)$	$\vert\varphi\rangle_{36}^{3}=\frac{1}{16}[\alpha\vert 00\rangle-\beta\vert 10\rangle+\gamma\vert 01\rangle-\eta\vert 11\rangle]$	$\sigma_z\otimes\mathrm{I}$
$\vert-\rangle_2\vert-\rangle_5=\frac{1}{2}(\vert 00\rangle-\beta\vert 01\rangle-\vert 10\rangle+\vert 11\rangle)$	$\vert\varphi\rangle_{36}^{4}=\frac{1}{16}[\alpha\vert 00\rangle-\beta\vert 10\rangle-\gamma\vert 01\rangle+\eta\vert 11\rangle]$	$\sigma_z\otimes\sigma_z$
$\vert+\rangle_2\vert+\rangle_5=\frac{1}{2}(\vert 00\rangle+\beta\vert 01\rangle+\vert 10\rangle+\vert 11\rangle)$	$\vert\varphi\rangle_{36}^{5}=\frac{1}{16}[\alpha\vert 00\rangle+\beta\vert 10\rangle-\gamma\vert 01\rangle-\eta\vert 11\rangle]$	$\mathrm{I}\otimes\sigma_z$
$\vert+\rangle_2\vert-\rangle_5=\frac{1}{2}(\vert 00\rangle-\beta\vert 01\rangle+\vert 10\rangle-\vert 11\rangle)$	$\vert\varphi\rangle_{36}^{6}=\frac{1}{16}[\alpha\vert 00\rangle+\beta\vert 10\rangle+\gamma\vert 01\rangle+\eta\vert 11\rangle]$	$\mathrm{I}\otimes I$
$\vert-\rangle_2\vert+\rangle_5=\frac{1}{2}(\vert 00\rangle+\beta\vert 01\rangle-\vert 10\rangle-\vert 11\rangle)$	$\vert\varphi\rangle_{36}^{7}=\frac{1}{16}[\alpha\vert 00\rangle-\beta\vert 10\rangle-\gamma\vert 01\rangle+\eta\vert 11\rangle]$	$\sigma_z\otimes\sigma_z$
$\vert-\rangle_2\vert-\rangle_5=\frac{1}{2}(\vert 00\rangle-\beta\vert 01\rangle-\vert 10\rangle+\vert 11\rangle)$	$\vert\varphi\rangle_{36}^{8}=\frac{1}{16}[\alpha\vert 00\rangle-\beta\vert 10\rangle+\gamma\vert 01\rangle-\eta\vert 11\rangle]$	$\sigma_z\otimes\mathrm{I}$
$\vert+\rangle_2\vert+\rangle_5=\frac{1}{2}(\vert 00\rangle+\beta\vert 01\rangle+\vert 10\rangle+\vert 11\rangle)$	$\vert\varphi\rangle_{36}^{9}=\frac{1}{16}[\alpha\vert 01\rangle+\beta\vert 11\rangle+\gamma\vert 00\rangle+\eta\vert 10\rangle]$	$\mathrm{I}\otimes\sigma_z$
$\vert+\rangle_2\vert-\rangle_5=\frac{1}{2}(\vert 00\rangle-\beta\vert 01\rangle+\vert 10\rangle-\vert 11\rangle)$	$\vert\varphi\rangle_{36}^{10}=\frac{1}{16}[-\alpha\vert 01\rangle-\beta\vert 11\rangle+\gamma\vert 00\rangle+\eta\vert 10\rangle]$	$\mathrm{I}\otimes-i\sigma_z$
$\vert-\rangle_2\vert+\rangle_5=\frac{1}{2}(\vert 00\rangle+\beta\vert 01\rangle-\vert 10\rangle-\vert 11\rangle)$	$\vert\varphi\rangle_{36}^{11}=\frac{1}{16}[\alpha\vert 01\rangle-\beta\vert 11\rangle+\gamma\vert 00\rangle-\eta\vert 10\rangle]$	$\sigma_z\otimes\sigma_x$
$\vert-\rangle_2\vert-\rangle_5=\frac{1}{2}(\vert 00\rangle-\beta\vert 01\rangle-\vert 10\rangle+\vert 11\rangle)$	$\vert\varphi\rangle_{36}^{12}=\frac{1}{16}[-\alpha\vert 01\rangle+\beta\vert 11\rangle+\gamma\vert 00\rangle-\eta\vert 10\rangle]$	$\sigma_z\otimes-i\sigma_y$
$\vert+\rangle_2\vert+\rangle_5=\frac{1}{2}(\vert 00\rangle+\beta\vert 01\rangle+\vert 10\rangle+\vert 11\rangle)$	$\vert\varphi\rangle_{36}^{13}=\frac{1}{16}[\alpha\vert 01\rangle+\beta\vert 11\rangle-\gamma\vert 00\rangle-\eta\vert 10\rangle]$	$I\otimes-i\sigma_y$
$\vert+\rangle_2\vert-\rangle_5=\frac{1}{2}(\vert 00\rangle-\beta\vert 01\rangle+\vert 10\rangle-\vert 11\rangle)$	$\vert\varphi\rangle_{36}^{14}=\frac{1}{16}[-\alpha\vert 01\rangle-\beta\vert 11\rangle-\gamma\vert 00\rangle-\eta\vert 10\rangle]$	$-\mathrm{I}\otimes\sigma_x$
$\vert-\rangle_2\vert+\rangle_5=\frac{1}{2}(\vert 00\rangle+\beta\vert 01\rangle-\vert 10\rangle-\vert 11\rangle)$	$\vert\varphi\rangle_{36}^{15}=\frac{1}{16}[\alpha\vert 01\rangle-\beta\vert 11\rangle-\gamma\vert 00\rangle+\eta\vert 10\rangle]$	$\sigma_z\otimes-i\sigma_y$
$\vert-\rangle_2\vert-\rangle_5=\frac{1}{2}(\vert 00\rangle-\beta\vert 01\rangle-\vert 10\rangle+\vert 11\rangle)$	$\vert\varphi\rangle_{36}^{16}=\frac{1}{16}[-\alpha\vert 01\rangle+\beta\vert 11\rangle-\gamma\vert 00\rangle+\eta\vert 10\rangle]$	$-\sigma_z\otimes\sigma_x$
$\vert+\rangle_2\vert+\rangle_5=\frac{1}{2}(\vert 00\rangle+\beta\vert 01\rangle+\vert 10\rangle+\vert 11\rangle)$	$\vert\varphi\rangle_{36}^{17}=\frac{1}{16}[\alpha\vert 00\rangle-\beta\vert 10\rangle+\gamma\vert 01\rangle-\eta\vert 11\rangle]$	$\sigma_z\otimes\mathrm{I}$
$\vert+\rangle_2\vert-\rangle_5=\frac{1}{2}(\vert 00\rangle-\beta\vert 01\rangle+\vert 10\rangle-\vert 11\rangle)$	$\vert\varphi\rangle_{36}^{18}=\frac{1}{16}[\alpha\vert 00\rangle-\beta\vert 10\rangle-\gamma\vert 01\rangle+\eta\vert 11\rangle]$	$\sigma_z\otimes\sigma_z$
$\vert-\rangle_2\vert+\rangle_5=\frac{1}{2}(\vert 00\rangle+\beta\vert 01\rangle-\vert 10\rangle-\vert 11\rangle)$	$\vert\varphi\rangle_{36}^{19}=\frac{1}{16}[\alpha\vert 00\rangle+\beta\vert 10\rangle+\gamma\vert 01\rangle+\eta\vert 11\rangle]$	$\mathrm{I}\otimes I$
$\vert-\rangle_2\vert-\rangle_5=\frac{1}{2}(\vert 00\rangle-\beta\vert 01\rangle-\vert 10\rangle+\vert 11\rangle)$	$\vert\varphi\rangle_{36}^{20}=\frac{1}{16}[\alpha\vert 00\rangle+\beta\vert 10\rangle-\gamma\vert 01\rangle-\eta\vert 11\rangle]$	$\mathrm{I}\otimes\sigma_z$
$\vert+\rangle_2\vert+\rangle_5=\frac{1}{2}(\vert 00\rangle+\beta\vert 01\rangle+\vert 10\rangle+\vert 11\rangle)$	$\vert\varphi\rangle_{36}^{21}=\frac{1}{16}[\alpha\vert 00\rangle-\beta\vert 10\rangle-\gamma\vert 01\rangle+\eta\vert 11\rangle]$	$\sigma_z\otimes\sigma_z$
$\vert+\rangle_2\vert-\rangle_5=\frac{1}{2}(\vert 00\rangle-\beta\vert 01\rangle+\vert 10\rangle-\vert 11\rangle)$	$\vert\varphi\rangle_{36}^{22}=\frac{1}{16}[\alpha\vert 00\rangle-\beta\vert 10\rangle+\gamma\vert 01\rangle+\eta\vert 11\rangle]$	$\sigma_z\otimes\mathrm{I}$
$\vert-\rangle_2\vert+\rangle_5=\frac{1}{2}(\vert 00\rangle+\beta\vert 01\rangle-\vert 10\rangle-\vert 11\rangle)$	$\vert\varphi\rangle_{36}^{23}=\frac{1}{16}[\alpha\vert 00\rangle+\beta\vert 10\rangle-\gamma\vert 01\rangle-\eta\vert 11\rangle]$	$\mathrm{I}\otimes\sigma_z$
$\vert-\rangle_2\vert-\rangle_5=\frac{1}{2}(\vert 00\rangle-\beta\vert 01\rangle-\vert 10\rangle+\vert 11\rangle)$	$\vert\varphi\rangle_{36}^{24}=\frac{1}{16}[\alpha\vert 00\rangle+\beta\vert 10\rangle+\gamma\vert 01\rangle+\eta\vert 11\rangle]$	$\mathrm{I}\otimes I$
$\vert+\rangle_2\vert+\rangle_5=\frac{1}{2}(\vert 00\rangle+\beta\vert 01\rangle+\vert 10\rangle+\vert 11\rangle)$	$\vert\varphi\rangle_{36}^{25}=\frac{1}{16}[\alpha\vert 01\rangle-\beta\vert 11\rangle+\gamma\vert 00\rangle-\eta\vert 10\rangle]$	$\sigma_z\otimes\sigma_x$
$\vert+\rangle_2\vert-\rangle_5=\frac{1}{2}(\vert 00\rangle-\beta\vert 01\rangle+\vert 10\rangle-\vert 11\rangle)$	$\vert\varphi\rangle_{36}^{26}=\frac{1}{16}[-\alpha\vert 01\rangle+\beta\vert 11\rangle+\gamma\vert 00\rangle-\eta\vert 10\rangle]$	$\sigma_z\otimes-i\sigma_y$
$\vert-\rangle_2\vert+\rangle_5=\frac{1}{2}(\vert 00\rangle+\beta\vert 01\rangle-\vert 10\rangle-\vert 11\rangle)$	$\vert\varphi\rangle_{36}^{27}=\frac{1}{16}[\alpha\vert 01\rangle+\beta\vert 11\rangle+\gamma\vert 00\rangle+\eta\vert 10\rangle]$	$\mathrm{I}\otimes\sigma_x$
$\vert-\rangle_2\vert-\rangle_5=\frac{1}{2}(\vert 00\rangle-\beta\vert 01\rangle-\vert 10\rangle+\vert 11\rangle)$	$\vert\varphi\rangle_{36}^{28}=\frac{1}{16}[-\alpha\vert 01\rangle-\beta\vert 11\rangle+\gamma\vert 00\rangle+\eta\vert 10\rangle]$	$I\otimes-i\sigma_y$
$\vert+\rangle_2\vert+\rangle_5=\frac{1}{2}(\vert 00\rangle+\beta\vert 01\rangle+\vert 10\rangle+\vert 11\rangle)$	$\vert\varphi\rangle_{36}^{29}=\frac{1}{16}[\alpha\vert 01\rangle-\beta\vert 11\rangle-\gamma\vert 00\rangle+\eta\vert 10\rangle]$	$\sigma_z\otimes-i\sigma_y$
$\vert+\rangle_2\vert-\rangle_5=\frac{1}{2}(\vert 00\rangle-\beta\vert 01\rangle+\vert 10\rangle-\vert 11\rangle)$	$\vert\varphi\rangle_{36}^{30}=\frac{1}{16}[-\alpha\vert 01\rangle+\beta\vert 11\rangle-\gamma\vert 00\rangle+\eta\vert 10\rangle]$	$-\sigma_z\otimes\sigma_x$
$\vert-\rangle_2\vert+\rangle_5=\frac{1}{2}(\vert 00\rangle+\beta\vert 01\rangle-\vert 10\rangle-\vert 11\rangle)$	$\vert\varphi\rangle_{36}^{31}=\frac{1}{16}[\alpha\vert 01\rangle+\beta\vert 11\rangle-\gamma\vert 00\rangle-\eta\vert 10\rangle]$	$I\otimes-i\sigma_y$
$\vert-\rangle_2\vert-\rangle_5=\frac{1}{2}(\vert 00\rangle-\beta\vert 01\rangle-\vert 10\rangle+\vert 11\rangle)$	$\vert\varphi\rangle_{36}^{32}=\frac{1}{16}[-\alpha\vert 01\rangle-\beta\vert 11\rangle-\gamma\vert 00\rangle-\eta\vert 10\rangle]$	$-I\otimes\sigma_x$
$\vert+\rangle_2\vert+\rangle_5=\frac{1}{2}(\vert 00\rangle+\beta\vert 01\rangle+\vert 10\rangle+\vert 11\rangle)$	$\vert\varphi\rangle_{36}^{33}=\frac{1}{16}[\alpha\vert 10\rangle+\beta\vert 00\rangle+\gamma\vert 11\rangle+\eta\vert 01\rangle]$	$\sigma_x\otimes\mathrm{I}$
$\vert+\rangle_2\vert-\rangle_5=\frac{1}{2}(\vert 00\rangle-\beta\vert 01\rangle+\vert 10\rangle-\vert 11\rangle)$	$\vert\varphi\rangle_{36}^{34}=\frac{1}{16}[\alpha\vert 10\rangle+\beta\vert 00\rangle-\gamma\vert 11\rangle-\eta\vert 01\rangle]$	$\sigma_x\otimes\sigma_z$
$\vert-\rangle_2\vert+\rangle_5=\frac{1}{2}(\vert 00\rangle+\beta\vert 01\rangle-\vert 10\rangle-\vert 11\rangle)$	$\vert\varphi\rangle_{36}^{35}=\frac{1}{16}[-\alpha\vert 10\rangle+\beta\vert 00\rangle-\gamma\vert 11\rangle+\eta\vert 01\rangle]$	$i\sigma_y\otimes\mathrm{I}$
$\vert-\rangle_2\vert-\rangle_5=\frac{1}{2}(\vert 00\rangle-\beta\vert 01\rangle-\vert 10\rangle+\vert 11\rangle)$	$\vert\varphi\rangle_{36}^{36}=\frac{1}{16}[-\alpha\vert 10\rangle+\beta\vert 00\rangle+\gamma\vert 11\rangle-\eta\vert 01\rangle]$	$-i\sigma_y\otimes\sigma_x$
$\vert+\rangle_2\vert+\rangle_5=\frac{1}{2}(\vert 00\rangle+\beta\vert 01\rangle+\vert 10\rangle+\vert 11\rangle)$	$\vert\varphi\rangle_{36}^{37}=\frac{1}{16}[\alpha\vert 10\rangle+\beta\vert 00\rangle-\gamma\vert 11\rangle-\eta\vert 01\rangle]$	$\sigma_x\otimes\sigma_z$
$\vert+\rangle_2\vert-\rangle_5=\frac{1}{2}(\vert 00\rangle-\beta\vert 01\rangle+\vert 10\rangle-\vert 11\rangle)$	$\vert\varphi\rangle_{36}^{38}=\frac{1}{16}[\alpha\vert 10\rangle+\beta\vert 00\rangle+\gamma\vert 11\rangle+\eta\vert 01\rangle]$	$\sigma_x\otimes\mathrm{I}$

(continued)

Table 8.4 (continued)

Measurement results on M1	State obtained by M2	Recovery
$\vert -\rangle_2 \vert +\rangle_5 = \frac{1}{2}(\vert 00\rangle + \beta\vert 01\rangle - \vert 10\rangle - \vert 11\rangle)$	$\vert \varphi\rangle_{36}^{39} = \frac{1}{16}[-\alpha\vert 10\rangle + \beta\vert 00\rangle + \gamma\vert 11\rangle - \eta\vert 01\rangle]$	$-i\sigma_y \otimes \sigma_z$
$\vert -\rangle_2 \vert -\rangle_5 = \frac{1}{2}(\vert 00\rangle - \beta\vert 01\rangle - \vert 10\rangle + \vert 11\rangle)$	$\vert \varphi\rangle_{36}^{40} = \frac{1}{16}[-\alpha\vert 10\rangle + \beta\vert 00\rangle - \gamma\vert 11\rangle + \eta\vert 01\rangle]$	$i\sigma_y \otimes I$
$\vert +\rangle_2 \vert +\rangle_5 = \frac{1}{2}(\vert 00\rangle + \beta\vert 01\rangle + \vert 10\rangle + \vert 11\rangle)$	$\vert \varphi\rangle_{36}^{41} = \frac{1}{16}[\alpha\vert 11\rangle + \beta\vert 01\rangle + \gamma\vert 10\rangle + \eta\vert 00\rangle]$	$\sigma_x \otimes \sigma_x$
$\vert +\rangle_2 \vert -\rangle_5 = \frac{1}{2}(\vert 00\rangle - \beta\vert 01\rangle + \vert 10\rangle - \vert 11\rangle)$	$\vert \varphi\rangle_{36}^{42} = \frac{1}{16}[-\alpha\vert 11\rangle - \beta\vert 01\rangle + \gamma\vert 10\rangle + \eta\vert 00\rangle]$	$\sigma_x \otimes -i\sigma_y$
$\vert -\rangle_2 \vert +\rangle_5 = \frac{1}{2}(\vert 00\rangle + \beta\vert 01\rangle - \vert 10\rangle - \vert 11\rangle)$	$\vert \varphi\rangle_{36}^{43} = \frac{1}{16}[-\alpha\vert 11\rangle + \beta\vert 01\rangle - \gamma\vert 10\rangle + \eta\vert 00\rangle]$	$i\sigma_y \otimes \sigma_x$
$\vert -\rangle_2 \vert -\rangle_5 = \frac{1}{2}(\vert 00\rangle - \beta\vert 01\rangle - \vert 10\rangle + \vert 11\rangle)$	$\vert \varphi\rangle_{36}^{44} = \frac{1}{16}[\alpha\vert 11\rangle - \beta\vert 01\rangle - \gamma\vert 10\rangle + \eta\vert 00\rangle]$	$-i\sigma_y \otimes -i\sigma_y$
$\vert +\rangle_2 \vert +\rangle_5 = \frac{1}{2}(\vert 00\rangle + \beta\vert 01\rangle + \vert 10\rangle + \vert 11\rangle)$	$\vert \varphi\rangle_{36}^{45} = \frac{1}{16}[\alpha\vert 11\rangle + \beta\vert 01\rangle - \gamma\vert 10\rangle - \eta\vert 00\rangle]$	$\sigma_x \otimes -i\sigma_y$
$\vert +\rangle_2 \vert -\rangle_5 = \frac{1}{2}(\vert 00\rangle - \beta\vert 01\rangle + \vert 10\rangle - \vert 11\rangle)$	$\vert \varphi\rangle_{36}^{46} = \frac{1}{16}[-\alpha\vert 11\rangle - \beta\vert 01\rangle - \gamma\vert 10\rangle - \eta\vert 00\rangle]$	$-\sigma_x \otimes \sigma_x$
$\vert -\rangle_2 \vert +\rangle_5 = \frac{1}{2}(\vert 00\rangle + \beta\vert 01\rangle - \vert 10\rangle - \vert 11\rangle)$	$\vert \varphi\rangle_{36}^{47} = \frac{1}{16}[-\alpha\vert 11\rangle + \beta\vert 01\rangle + \gamma\vert 10\rangle - \eta\vert 00\rangle]$	$-i\sigma_y \otimes -i\sigma_y$
$\vert -\rangle_2 \vert -\rangle_5 = \frac{1}{2}(\vert 00\rangle - \beta\vert 01\rangle - \vert 10\rangle + \vert 11\rangle)$	$\vert \varphi\rangle_{36}^{48} = \frac{1}{16}[\alpha\vert 11\rangle - \beta\vert 01\rangle + \gamma\vert 10\rangle - \eta\vert 00\rangle]$	$-i\sigma_y \otimes \sigma_x$
$\vert +\rangle_2 \vert +\rangle_5 = \frac{1}{2}(\vert 00\rangle + \beta\vert 01\rangle + \vert 10\rangle + \vert 11\rangle)$	$\vert \varphi\rangle_{36}^{49} = \frac{1}{16}[\alpha\vert 10\rangle - \beta\vert 00\rangle + \gamma\vert 11\rangle - \eta\vert 01\rangle]$	$-i\sigma_y \otimes I$
$\vert +\rangle_2 \vert -\rangle_5 = \frac{1}{2}(\vert 00\rangle - \beta\vert 01\rangle + \vert 10\rangle - \vert 11\rangle)$	$\vert \varphi\rangle_{36}^{50} = \frac{1}{16}[\alpha\vert 10\rangle - \beta\vert 00\rangle - \gamma\vert 11\rangle + \eta\vert 01\rangle]$	$-i\sigma_y \otimes \sigma_z$
$\vert -\rangle_2 \vert +\rangle_5 = \frac{1}{2}(\vert 00\rangle + \beta\vert 01\rangle - \vert 10\rangle - \vert 11\rangle)$	$\vert \varphi\rangle_{36}^{51} = \frac{1}{16}[-\alpha\vert 10\rangle - \beta\vert 00\rangle - \gamma\vert 11\rangle - \eta\vert 01\rangle]$	$-\sigma_x \otimes I$
$\vert -\rangle_2 \vert -\rangle_5 = \frac{1}{2}(\vert 00\rangle - \beta\vert 01\rangle - \vert 10\rangle + \vert 11\rangle)$	$\vert \varphi\rangle_{36}^{52} = \frac{1}{16}[-\alpha\vert 10\rangle - \beta\vert 00\rangle + \gamma\vert 11\rangle + \eta\vert 01\rangle]$	$\sigma_x \otimes \sigma_z$
$\vert +\rangle_2 \vert +\rangle_5 = \frac{1}{2}(\vert 00\rangle + \beta\vert 01\rangle + \vert 10\rangle + \vert 11\rangle)$	$\vert \varphi\rangle_{36}^{53} = \frac{1}{16}[\alpha\vert 10\rangle - \beta\vert 00\rangle - \gamma\vert 11\rangle + \eta\vert 01\rangle]$	$-i\sigma_y \otimes \sigma_z$
$\vert +\rangle_2 \vert -\rangle_5 = \frac{1}{2}(\vert 00\rangle - \beta\vert 01\rangle + \vert 10\rangle - \vert 11\rangle)$	$\vert \varphi\rangle_{36}^{54} = \frac{1}{16}[\alpha\vert 10\rangle - \beta\vert 00\rangle + \gamma\vert 11\rangle - \eta\vert 01\rangle]$	$-i\sigma_y \otimes I$
$\vert -\rangle_2 \vert +\rangle_5 = \frac{1}{2}(\vert 00\rangle + \beta\vert 01\rangle - \vert 10\rangle - \vert 11\rangle)$	$\vert \varphi\rangle_{36}^{55} = \frac{1}{16}[-\alpha\vert 10\rangle - \beta\vert 00\rangle + \gamma\vert 11\rangle + \eta\vert 01\rangle]$	$\sigma_x \otimes \sigma_z$
$\vert -\rangle_2 \vert -\rangle_5 = \frac{1}{2}(\vert 00\rangle - \beta\vert 01\rangle - \vert 10\rangle + \vert 11\rangle)$	$\vert \varphi\rangle_{36}^{56} = \frac{1}{16}[-\alpha\vert 10\rangle - \beta\vert 00\rangle - \gamma\vert 11\rangle - \eta\vert 01\rangle]$	$-\sigma_x \otimes I$
$\vert +\rangle_2 \vert +\rangle_5 = \frac{1}{2}(\vert 00\rangle + \beta\vert 01\rangle + \vert 10\rangle + \vert 11\rangle)$	$\vert \varphi\rangle_{36}^{57} = \frac{1}{16}[\alpha\vert 11\rangle - \beta\vert 01\rangle + \gamma\vert 10\rangle - \eta\vert 00\rangle]$	$-i\sigma_y \otimes \sigma_x$
$\vert +\rangle_2 \vert -\rangle_5 = \frac{1}{2}(\vert 00\rangle - \beta\vert 01\rangle + \vert 10\rangle - \vert 11\rangle)$	$\vert \varphi\rangle_{36}^{58} = \frac{1}{16}[-\alpha\vert 11\rangle + \beta\vert 01\rangle + \gamma\vert 10\rangle - \eta\vert 00\rangle]$	$-i\sigma_y \otimes -i\sigma_y$
$\vert -\rangle_2 \vert +\rangle_5 = \frac{1}{2}(\vert 00\rangle + \beta\vert 01\rangle - \vert 10\rangle - \vert 11\rangle)$	$\vert \varphi\rangle_{36}^{59} = \frac{1}{16}[-\alpha\vert 11\rangle - \beta\vert 01\rangle - \gamma\vert 10\rangle - \eta\vert 00\rangle]$	$-\sigma_x \otimes \sigma_x$
$\vert -\rangle_2 \vert -\rangle_5 = \frac{1}{2}(\vert 00\rangle - \beta\vert 01\rangle - \vert 10\rangle + \vert 11\rangle)$	$\vert \varphi\rangle_{36}^{60} = \frac{1}{16}[\alpha\vert 11\rangle + \beta\vert 01\rangle - \gamma\vert 10\rangle - \eta\vert 00\rangle]$	$\sigma_x \otimes -i\sigma_y$
$\vert +\rangle_2 \vert +\rangle_5 = \frac{1}{2}(\vert 00\rangle + \beta\vert 01\rangle + \vert 10\rangle + \vert 11\rangle)$	$\vert \varphi\rangle_{36}^{61} = \frac{1}{16}[\alpha\vert 11\rangle - \beta\vert 01\rangle - \gamma\vert 10\rangle + \eta\vert 00\rangle]$	$-i\sigma_y \otimes -i\sigma_y$
$\vert +\rangle_2 \vert -\rangle_5 = \frac{1}{2}(\vert 00\rangle - \beta\vert 01\rangle + \vert 10\rangle - \vert 11\rangle)$	$\vert \varphi\rangle_{36}^{62} = \frac{1}{16}[-\alpha\vert 11\rangle + \beta\vert 01\rangle - \gamma\vert 10\rangle + \eta\vert 00\rangle]$	$i\sigma_y \otimes \sigma_x$
$\vert -\rangle_2 \vert +\rangle_5 = \frac{1}{2}(\vert 00\rangle + \beta\vert 01\rangle - \vert 10\rangle - \vert 11\rangle)$	$\vert \varphi\rangle_{36}^{63} = \frac{1}{16}[-\alpha\vert 11\rangle - \beta\vert 01\rangle + \gamma\vert 10\rangle + \eta\vert 00\rangle]$	$\sigma_x \otimes -i\sigma_y$
$\vert -\rangle_2 \vert -\rangle_5 = \frac{1}{2}(\vert 00\rangle - \beta\vert 01\rangle - \vert 10\rangle + \vert 11\rangle)$	$\vert \varphi\rangle_{36}^{64} = \frac{1}{16}[\alpha\vert 11\rangle + \beta\vert 01\rangle + \gamma\vert 10\rangle + \eta\vert 00\rangle]$	$\sigma_x \otimes \sigma_x$

8.3.4 QNC for All-Optical Quantum Repeater

Currently, all of QNC schemes used EPR pairs between nodes as connections. For the existing technology, it is very difficult to realize the distribution of long-distance high-purity EPR pairs, and the consumption of communication time is also large. To solve this problem, a QNC scheme based on all-optical quantum repeater (AOQR-QNC) is investigated, which not only increase the communication distance but also reduces the consumption of communication time and improves the reliability and security of the channel.

The main contributions of our works can be concluded into following two points:

1. To improve the security of communication, "two-step" deterministic Secure Quantum Communication is introduced into network coding for all-optical quantum repeater.

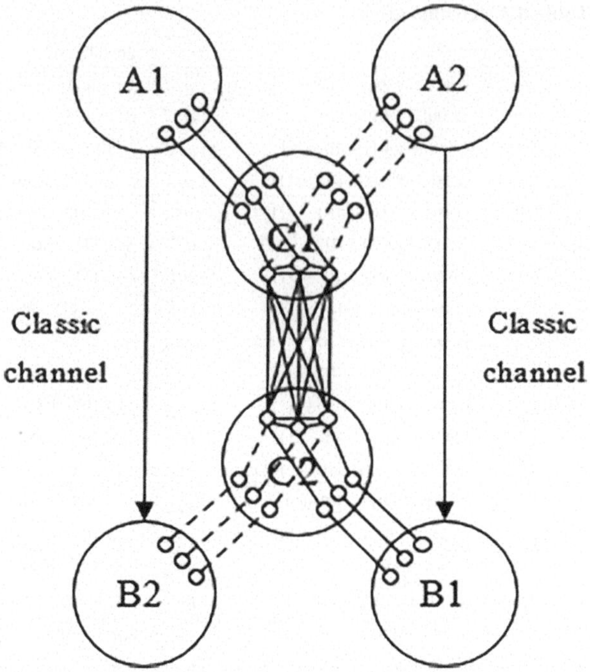

Fig. 8.8 QNC for all-optical quantum repeater

2. All-optical quantum repeater scheme is applied into the butterfly network to establish a more efficient and reliable communication link, which not only reduces the time required for communication but also improves the fault tolerance of communication.

The aim of the presented AOQR-QNC scheme is to combine QNC with quantum repeater, which further improves the transmission efficiency and security of quantum repeater and makes the application of quantum repeater in quantum network more general. As shown in Fig. 8.8, *A1* and *A2* belong to the sender Alice, *B1* and *B2* belong to the receiver Bob, and *C1* and *C2* belong to the intermediate node Charlie. The cross-transmission of information can be explained as S_A: $A1 \rightarrow B1$, S_B: $A2 \rightarrow B2$.

In order to ensure the security of the transmitted information, we refer to the concept of deterministic secure quantum communication proposed by Long [27] and adopt a "two-step" scheme to divide the information to be transmitted into two parts and send it from nodes A1 and A2 to nodes B1 and B2, respectively.

At the same time, we continue to adopt the design of butterfly network by imitating the quantum repeater and QNC scheme proposed by Satoh [11], and in our network, the direct line $A1 \rightarrow B1$ and $A2 \rightarrow B2$ are classical communication channel. $A1$ ($A2$) $\rightarrow C1 \rightarrow C2 \rightarrow B1$ ($B2$) are quantum links used for the establishment of quantum channels and the forwarding of quantum information. At the same time, it also includes a part of classical communication for coding and forwarding. Next, we will implement our AOQR-QNC scheme from three parts: establishing connection; transmitting data; recovering and verifying.

8.3.4.1 All-Optical Quantum Repeater

Long distance quantum communication has always been a hot topic and important research, and quantum repeater (QR) technology makes the long-distance communication possible. From the first-generation scheme to all-optical quantum repeater, quantum repeater has made great progress. The first-generation quantum repeater is shown in Fig. 8.9, the correlation between nodes depends on quantum entanglement pairs, and Alice and Bob establish a remote connection through the measurement of Bell basis. But this method is designed based on quantum entangled pairs between different nodes. It is inevitable to use quantum memory. At the same time, the establishment and distribution of quantum entangled pairs also require a lot of time, and its fault-tolerant ability is very limited. Since then, the second-generation repeater solves this problem and uses quantum error correction codes to reduce quantum purification operations so as to further improve the transmission efficiency.

In 2015, Azuma proposed the scheme of all-optical quantum repeater [28]. They creatively adopted the cluster state $|G_c^m\rangle$ and *2 m*-arm connection mode so that the connection between nodes is no longer based on the measurement of quantum entanglement pairs, thus greatly reducing the time to establish the connection.

In 2019, the research team of China University of science and technology successfully realized this scheme in the laboratory [29]. The experimental results show that the research results of all-optical quantum repeater are progressive. As shown in Fig. 8.10, the *first*-leaf corresponds to the quantum memory at a single repeater node in the traditional repeater. Correspondingly, the *second*-leaf provides entanglement for the adjacent repeater nodes and the *first*-leaf.

For this connection mode, assuming that particles *G* and *J* at *first*-leaf and particles *H* and *I* at *second*-leaf are entangled with each other, respectively, and particles *H* and *I* at *second*-leaf are not entangled with each other, if we want to realize the connection between *G* and *J*, we only need to carry out Bell measurement on particles at *H* and *I* to entangle the particles at *G* and *J*. Thus, the connection from *second*-leaf to *first*-leaf within the relay node is realized. Moreover, due to the use of multiple links, assuming that the Bell measurement of one link fails, only the failed link needs to be removed, which has no impact on other links. So, it greatly improves the fault tolerance of the scheme.

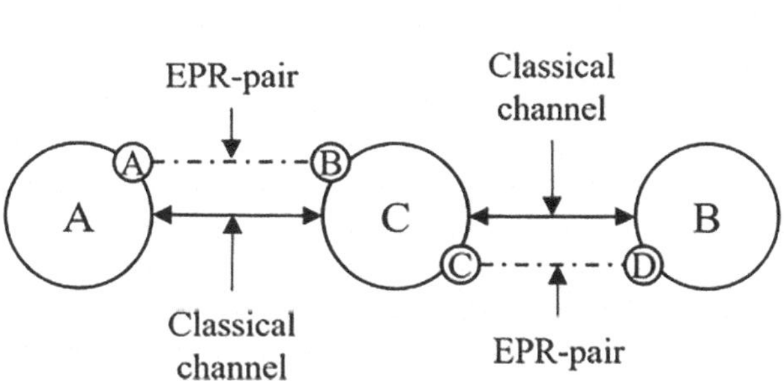

Fig. 8.9 Traditional repeater scheme

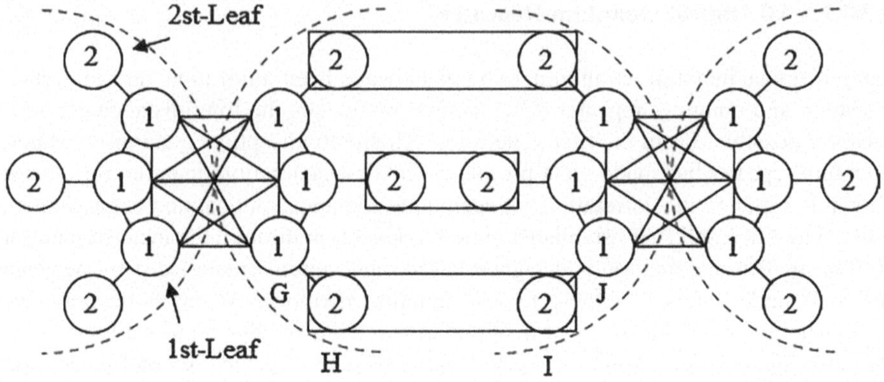

Fig. 8.10 All photonic quantum repeater scheme

Generally, it still takes some time for the *first*-leaf connection between nodes. However, in the all-optical quantum communication scheme, the qubits in the *first*-leaf are composed of single photons, and the *first*-leaf and the *second*-leaf can be sent at the same time, which greatly reduces the time required to establish the connection. However, the *first*-leaf connection between nodes is still a great challenge. Azuma et al. proposed an idea to implement the loss tolerance scheme proposed by Varnava [30, 31]. This connection method sacrificing multiple single photons that guarantees qubits cannot only establish the connection in a very short constant time but also improve the fault tolerance rate of the scheme. Note that Varnava's schemes [30, 31] supports bell basis measurement, which is also consistent with the scheme of all-optical quantum repeater.

8.3.4.2 Establish Connection

In this part, we respectively need to establish remote connections between $A1 \rightarrow B1$ and $A2 \rightarrow B2$ to pave the way for data transmission. Here, we refer to the all-optical quantum communication scheme of Azuma [28].

Step 1: First, Alice (*A1*, *A2*) and Bob (*B1*, *B2*) send a piece of verification information to the corresponding node with the assistance of classical communication. Next, tell the other party to start communication, on the other hand, do verification for subsequent data transmission. Finally, generate three single particles at each node to prepare for the establishment of quantum channel.

Step 2: Between *C1* and *C2* nodes, we need to establish two pairs full connection with the scheme proposed by Varnava [30, 31]. This scheme uses the number of single photons sacrificing polynomials in exchange for the connection of quantum channels. Although it consumes some resources, it establishes a reliable connection in a constant time. (In order to be consistent with the scheme of Azuma [28], three pairs of EPR entangled pairs are chose).

Step3: Using the principle of entanglement swapping, the joint measurement of two particles at (*A1*, *C1*), (*C1*, *C2*), (*B1*, *C2*) and (*A2*, *C1*), (*C1*, *C2*), (*B2*, *C2*) are measured, respectively, and finally a reliable entanglement connection is formed between (*A1*, *B1*), (*A2*, *B2*).

8.3.4.3 Data Transmission

In this subsection, how to carry out a safe and reliable transmission AOQR-QNC scheme is described in detail. Here, we refer to the "two-step" secure direct communication scheme of Long [27].

Step 1: Firstly, for the quantum channel, Alice (*A1*, *A2*) encodes the information to be transmitted into two parts, S_A and S_B, and insert the corresponding verification sequence into S_A according to the verification code transmitted in the previous part. After **Step 1** of transmission completed, node *B1* needs to verify part of the received information, so as to ensure that the transmission of the first section is safe and reliable. If the eavesdropper is detected, the communication terminates immediately. If the transmitted information is not eavesdropped, the second round of transmission continues). Meanwhile, in the previous step, we realized the remote connection between *A1* and *B1*, so S_A can be transmitted from *A1* node to *B1* node by quantum teleportation. The classical information *X1* generated in the transmission process can be transmitted from *A1* node to *C1* and *B2* nodes respectively through the classical channel. Finally, Step1 of the "two-step" scheme has been completed.

Step 2: When ensuring the safety and reliability of the first round of outgoing, the same quantum teleportation scheme is again adopted between nodes *A2* and *B2* to transmit S_B from node *A2* to node *B2*. At the same time, the generated classical information *X2* can be respectively transmitted by node *A2* to nodes *C1* and *B1* through the classical channel.

Step 3: For the classical channel, as shown in Fig. 8.8, when *C1* receives the classical information *X1* and *X2* from Alice*A1 andA2*), respectively, the *C1* node encodes *X1* and *X2* into $X1 \oplus X2$ and then forwards them to the *C2* node. The *C2* node copies and forwards them to Bobs *B1 and B2*, respectively. At this time, the transmission is completed, and the verification and recovery operations can be carried out at the receiving end.

8.3.4.4 Recovery and Verification

In this subsection, the above transmission process and complete data recovery processes are further explained in detail.

After the data transmission of the first two Steps, *B1* has received the information S_A transmitted by Alice A1 the classical information *X2* transmitted by Alice *A2*, and the classical information $X1 \oplus X2$ transmitted by the *C2* node; at the same time, node

B2 possesses the information S_B transmitted by Alice A2, the classical information *X1* transmitted by node *A1*, and the classical information $X1 \oplus X2$ transmitted by node *C2*. For classical information, only the corresponding operation is needed to leave the correct information; thus, the corresponding unitary transformation of quantum information is carried out to complete the final recovery.

8.3.4.5 AOQR-QNC Scheme Analysis

This subsection focuses on the analysis of the transmission and security of AOQR-QNC scheme.

A. Security Analysis

In AOQR-QNC scheme, we adopt the classical "two-step" scheme to forward the information to be transmitted to the receiver in two batches. In the processes of the first transmission, in addition to data transmission, it also includes verifying the effect of the transmission channel, and the second transmission is the quantum information to be transmitted. Because we transmit important information in two steps, eavesdroppers cannot steal two pieces of information at the same time, let alone restore the unitary transformation of quantum information without the assistance of classical information. Hence, according to the quantum noncloning theorem, eavesdroppers cannot steal quantum information without leaving traces. To some extent, our AOQR-QNC scheme is safe.

Next, the security of AOQR-QNC scheme from two aspects: internal attack and external attack, is explained in detail.

1. **External Attack**

   If the eavesdropper eavesdrops on Alice (*A1*, *A2*), it must leave traces. When the first transmission is completed, Bob will detect the existence of the eavesdropper. Accordingly, Bob will inform Alice giving up the communication. Therefore, the eavesdropper cannot obtain any quantum information. Even if the eavesdropper steals the first part of the transmitted information, he cannot transform the quantum information into any valuable information, because the classical information can only be accurately decoded after two-step communication. Furthermore, if the eavesdropper eavesdrops on Bob (*B1*, *B2*) or Charlie (*C1*, *C2*), these results are the same. In summary, the eavesdropper can only steal part of the information at most, but cannot obtain all the information.

   Suppose the eavesdropper maliciously intercepts information at Charlie (*C1*, *C2*), and pretends to be Charlie sending unknown information sequentially. In this case, after the first part of information transmission is completed, Bobs (*B1*, *B2*) will detect and interrupt the communication, because there is a verification sequence in the transmitted information sequence, and Bob will first verify the transmitted information to ensure the security of the information and then continue the second communication. Undoubtedly, if Bob notices the existence of an eavesdropper during the first communication, communication will be interrupted immediately.

2. **Internal Attack**

   Assume that Charlie ($C1$, $C2$) is dishonest, then he will assist the eavesdropper to obtain the transmitted information. Even through this method, it is also can prove that the eavesdropper cannot obtain any valuable information. For Charlie, his main task is to establish a remote connection, and further encode and forward the measurement results without obtaining the transmitted quantum information. Therefore, if the eavesdropper intercepts some information, Bob would detect eavesdropping and interrupt the communication.

   Quantum channel: To some extent, the security of quantum channel itself seems possessing a certain security due to the quantum noncloning theorem. In addition, the two-step scheme is adopted, which further improves the security of presented scheme. Hence, the eavesdropper can't know all the contents of the first transmission information in any cases.

   Classical channel: As shown in the Fig. 8.10, the AOQR-QNC scheme adopted enables the cross-transmission of classical information, that is, $A1 \rightarrow B1$, $A2 \rightarrow B2$. It improves the security of classical information transmission. Because Bobs ($B1$, $B2$) need the information sent by Alice's ($A1$, $A2$) and $C2$ node, respectively, to recover useful information, the eavesdroppers work very hard to steal two parts classical information. Therefore, the eavesdroppers cannot obtain useful quantum information at the quantum channel end, let alone restore.

B. Transmission Analysis

The rate and reliability of data transmission have always been the focus in the communication field. In this subsection, the advantages of our scheme are analyzed in terms of data transmission. In presented AOQR-QNC scheme, the connection between Alice and Bob is built on the principle of all-optical quantum repeater, and use the advantages of QNC to improve the transmission efficiency and security. Compared with the existing schemes, our work has the following three advantages:

1. **Transmission Rate**

   Compared to the traditional quantum repeater scheme, our designed AOQR-QNC scheme fundamentally solves the problem that it is difficult to establish high-purity EPR pairs between nodes. Since the connection inside nodes are established via EPR pairs and full connection method, it significantly reduces the time required to establish EPR pairs among nodes.
2. **Reliability**

   In other schemes, the connection between nodes only depends on a pair of EPR pairs, and it is very complex to distribute high-purity EPR pairs between nodes. At the same time, it is difficult to achieve absolute success in subsequent entanglement swapping, and any error in one step will affect the whole scheme. In contrast, the connection we established adopts the parallel multi-channel AOQR-QNC scheme. As long as a reliable connection is established on one link, Alice and Bob can communicate normally, so the fault tolerance of our scheme is very high.

3. **Channel Utilization**

During the transmission of classical information, we can see that if we need to transmit *1* bit of information to Bob (*B1*, *B2*), respectively. Under normal circumstances, we need two communications to complete the transmission of *2* bit of classical information at the same time. In the presented network coding scheme, only one transmission is needed, which can double the channel utilization.

8.3.4.6 Summary

A novel quantum repeater scheme is presented based on QNC. The whole AOQR-QNC scheme is optimized via introducing an all-optical quantum repeater, which greatly reduces the time of establishing the connection and improves the fault tolerance in the face of special cases. In the meantime, the "two-step" scheme is incorporated, which further improves the security. The analysis shows that the AOQR-QNC scheme is secure against both external attacks and internal attacks, in which eavesdroppers can never obtain the complete information segment. Besides, according to the existing research results, it is easy to know that presented AOQR-QNC scheme can be realized in the laboratory. However, there is still some spaces for further research in terms of resource consumption and assurance.

8.4 Brief Summary

This chapter mainly studies and analyzes the quantum network coding scheme, briefly introduces the classical quantum network coding scheme, summarizes the shortcomings of the existing scheme, and makes some corresponding improvements, three quantum network coding schemes under different scenarios are proposed, which are controlled quantum network coding scheme based on quantum walking, quantum network coding scheme based on six-qubit entanglement, and quantum network coding scheme for all-optical quantum repeater. The analysis shows that the three schemes have high anti-attack ability and practicability and can be applied to secure quantum communication networks.

References

1. R. Ahlswede, N. Cai, S.Y.R. Li, R.W. Yeung, Network information flow. IEEE Trans. Inf. Theory **46**, 1204–1216 (2000)
2. H. Kobayashi, F.L. Gall, H. Nishimura, M. Roetteler, Perfect quantum network communication scheme based on classical network coding, in *2010 IEEE International Symposium on Information Theory*, (IEEE, 2009), pp. 2686–2690

3. H. Kobayasffl, F. Le Gall, H. Nishimura, M. Roetteler, General scheme for perfect quantum network coding with free classical communication, in *Automata, Languages and Programming: 36th International Colloquium, ICALP*, (Springer, Berlin/Heidelberg, 2009), pp. 622–633
4. S.Y. Ma, X.B. Chen, M.X. Luo, et al., Probabilistic quantum network coding of M-qubit states over the butterfly network. Opt. Commun. **283**, 497–501 (2010)
5. J. Li, X.B. Chen, G. Xu, Y.X. Yang, Z.P. Li, Perfect quantum network coding independent of classical network solutions. IEEE Commun. Lett. **19**, 115–118 (2015)
6. J. Li, X. Chen, X. Sun, Z. Li, Y. Yang, Quantum network coding for multi-unicast problem based on 2D and 3D cluster states. Sci. China Inf. Sci. **59**, 1–15 (2016)
7. Z.Z. Li, G. Xu, X.B. Chen, et al., Efficient quantum state transmission via perfect quantum network coding. Sci. China Inf. Sci. **62**, 12501 (2019)
8. Z.Z. Li, G. Xu, X.B. Chen, Y.X. Yang, Secure quantum network coding based on quantum homomorphic message authentication. Quantum Inf. Process **18**, 14 (2019)
9. M. Hayashi, K. Iwama, H. Nishimura, et al., Quantum network coding, in *International Symposium on Theoretical Aspects of Computer Science Marseille*, (2006), pp. 23–25
10. D. Leung, J. Oppenheim, A. Winter, Quantum network communication-the butterfly and beyond. IEEE Trans. Inf. Theor. **56**, 3478–3490 (2010)
11. T. Satoh, F.L. Gall, H. Imai, Quantum network coding for quantum repeaters. Phys. Rev. A **86**, 032331 (2012)
12. T. Shang, G. Du, J.W. Liu, Opportunistic quantum network coding based on quantum teleportation. Quantum Inf. Process **15**, 1–21 (2016)
13. M. Hayashi, Prior entanglement between senders enables perfect quantum network coding with modification. Phys. Rev. A **76**, 040301 (2007)
14. T. Shang, X.J. Zhao, J.W. Liu, Quantum network coding based on controlled teleportation. IEEE Commun. Lett. **18**, 865–868 (2014)
15. T. Shang, K. Li, J.W. Liu, Continuous-variable quantum network coding for coherent states. Quantum Inf. Process **16**, 1–24 (2017)
16. X. Chen, R.G. Zhou, X. Li, X.X. Zhang, Controllable quantum network coding scheme based on quantum walk. Int. J. Theor. Phys. **60**(9), 3363–3374 (2021)
17. A. Ambainis, E. Bach, A. Nayak, A. Vishwanath, J. Watrous, One-dimensional quantum walks, in *Proceedings of the Thirty-Third Annual ACM Symposium on Theory of Computing*, (2001), pp. 37–49
18. Y. Wang, Y. Shang, P. Xue, Generalized teleportation by quantum walks. Quantum Inf. Process **16**, 1–13 (2017)
19. Y. Shang, Y. Wang, M. Li, R.Q. Lu, Quantum communication schemes by quantum walks with two coins. Europhys. Lett. **124**, 60009 (2018)
20. H. Nishimura, Quantum network coding – How can network coding be applied to quantum information? in *International Symposium on Network Coding*, (2013)
21. H. Lu, Z.D. Li, X.F. Yin, R. Zhang, X.X. Fang, L. Li, N.L. Liu, F.H. Xu, Y.A. Chen, J.W. Pan, Experimental quantum network coding. NPJ Quantum Inf. **5**, 89 (2019)
22. Y. Chatterjee, V. Devrari, B.K. Behera, P.K. Panigrahi, Experimental realization of quantum teleportation using coined quantum walks. Quantum Inf. Process **19**, 1–14 (2020)
23. D.F. Li, R.J. Wang, B aagyere, E., Quantum teleportation of an arbitrary two-qubit state by using two three-qubit GHZ states and the six-qubit entangled state. Quantum Inf. Process **18**, 1–15 (2019)
24. T. Shang, X.J. Zhao, C. Wang, J.W. Liu, Quantum homomorphic signature. Quantum Inf. Process **14**, 393–410 (2015)
25. T. Shang, Z. Pei, X.J. Zhao, J.W. Liu, Quantum network coding against pollution attacks. IEEE Commun. Lett. **20**, 1369–1372 (2016)
26. X. Chen, R.G. Zhou, X. Li, Y.C. Li, Quantum network coding based on six-qubit entangled state. Mod. Phys. Lett. B **36**(13), 11 (2022)

27. G.L. Long, X.S. Liu, Theoretically efficient high-capacity quantum-key-distribution scheme. Phys. Rev. A **65**(3), 032302 (2002)
28. K. Azuma, K. Tamaki, H.K. Lo, All-photonic quantum repeaters. Nat. Commun. **6**, 1–7 (2015)
29. Z. Da Li, R. Zhang, X.F. Yin, L.Z. Liu, Y. Hu, Y.Q. Fang, Y.Y. Fei, X. Jiang, J. Zhang, L. Li, N. Le Liu, F. Xu, Y.A. Chen, J.W. Pan, Experimental quantum repeater without quantum memory. Nat. Photonics **13**, 644–648 (2019)
30. M. Varnava, D.E. Browne, T. Rudolph, How good must single photon sources and detectors be for efficient linear optical quantum computation? Phys. Rev. Lett. **100**, 1–4 (2008)
31. M. Varnava, D.E. Browne, T. Rudolph, Loss tolerance in one-way quantum computation via counterfactual error correction. Phys. Rev. Lett. **97**, 120501 (2006)

Index

R.-G. Zhou et al., *Design and Analysis of Secure Quantum Communication Schemes*, https://doi.org/10.1007/978-3-031-78428-6

W

X

Y

Z

The manufacturer's authorised representative in the EU is Springer Nature Customer Service Centre GmbH, Europaplatz 3, 69115 Heidelberg, Germany. If you have any concerns regarding our products, please contact ProductSafety@springernature.com

Printed and bound by CPI Group (UK) Ltd, Croydon, CR0 4YY
30/09/2026
02225569-0010